Molecular Approaches to Learning and Memory

Molecular Approaches to Learning and Memory

Edited by

William L. Byrne

Department of Biochemistry
University of Tennessee Medical Center
Memphis, Tennessee

ACADEMIC PRESS, New York and London 1970

ACADEMIC PRESS, INC.
111 Fifth Avenue, New York, New York 10003

United Kingdom Edition published by
ACADEMIC PRESS, INC. (LONDON) LTD.
Berkeley Square House, London W1X 6BA

Library of Congress Catalog Number: 69-12283

PRINTED IN THE UNITED STATES OF AMERICA

CONTENTS

LIST OF CONTRIBUTORS

Linda B. Adair, *Department of Biochemistry and the Genetics Curriculum, University of North Carolina, Chapel Hill, North Carolina.*

Bernard W. Agranoff, *Mental Health Research Institute, Department of Psychiatry and Biological Chemistry, University of Michigan, Ann Arbor, Michigan.*

Stanley H. Appel, *Departments of Medicine and Biochemistry, Duke University Medical Center, Durham, North Carolina.*

Lucila A. Autilio, *Department of Neurology, Hospital of the University of Pennsylvania, Philadelphia, Pennsylvania.*

Samuel H. Barondes, *Departments of Psychiatry and Molecular Biology, Albert Einstein College of Medicine, Bronx, New York.*

Edward L. Bennett, *Lawrence Radiation Laboratory, University of California, Berkeley, California.*

Robert E. Bowman, *Department of Psychology, University of Wisconsin, Madison, Wisconsin.*

William L. Byrne, *Department of Biochemistry, Duke University, Durham, North Carolina, and Department of Biochemistry, University of Tennessee Medical Center, Memphis, Tennessee.*

Melvin J. Cohen, *Department of Biology, University of Oregon, Eugene, Oregon.*

W. C. Corning, *Animal Behavior Laboratories, Department of Psychology, Fordham University, New York, New York.*

Marian C. Diamond, *Department of Physiology and Anatomy, University of California, Berkeley, California.*

James A. Dyal, *Department of Psychology, University of Waterloo, Waterloo, Ontario.*

Walter B. Essman, *Queens College of the City University of New York, Flushing, New York.*

F. J. Fjerdingstad, *Department of Biochemistry, Duke University, Durham, North Carolina.*

Richard Gay, *Michigan State University, East Lansing, Michigan.*

Edward Glassman, *Department of Biochemistry and the Genetics Curriculum, University of North Carolina, Chapel Hill, North Carolina.*

Arnold M. Golub, *Mental Health Research Institute, University of Michigan, Ann Arbor, Michigan.*

Ward C. Halstead,* *Department of Psychology, The University of Chicago, Chicago, Illinois.*

W. F. Herblin, *Central Research Department, E. I. Du Pont de Nemours & Co., Wilmington, Delaware.*

*Now Deceased.

Murray E. Jarvik, *Department of Pharmacology, Albert Einstein College of Medicine, Bronx, New York.*

E. Roy John, *Brain Research Laboratories, Department of Psychiatry, New York Medical College, New York, New York.*

James V. McConnell, *Mental Health Research Institute, The University of Michigan, Ann Arbor, Michigan.*

D. R. Mobbs, *Department of Biochemistry, Ghana Medical School, Accra, Ghana.*

Thomas Nissen, *Institute of General Zoology, University of Copenhagen, Copenhagen, Denmark.*

Penelope Pettis, *Department of Neurology, Hospital of the University of Pennsylvania, Philadelphia, Pennsylvania.*

Alfred C. Raphelson, *University of Michigan, Flint College, Flint, Michigan.*

S. Reinis, *Department of Physiology, Ghana Medical School, Accra, Ghana.*

D. Riccio, *Animal Behavior Laboratories, Department of Psychology, Fordham University, New York, New York.*

H. H. Roigaard-Petersen, *Institute of General Zoology, University of Copenhagen, Copenhagen, Denmark.*

Frank Rosenblatt, *Section of Neurobiology and Behavior, Cornell University, Ithaca, New York.*

Mark R. Rosenzweig, *Department of Psychology, University of California, Berkeley, California.*

William B. Rucker, *Department of Psychology, The George Washington University, Washington, D.C. and The University of Chicago, Chicago, Illinois.*

Harold Salive, *Mental Health Research Institute, The University of Michigan, Ann Arbor, Michigan.*

Tsuyoshi Shigehisa, *Mental Health Research Institute, The University of Michigan, Ann Arbor, Michigan.*

G. Ungar, *Department of Anesthesiology, Baylor University College of Medicine, Houston, Texas.*

Kenneth P. Weiss, *Department of Psychology, Kings College, Wilkes-Barre, Pennsylvania.*

John Eric Wilson, *Department of Biochemistry and the Genetics Curriculum, University of North Carolina, Chapel Hill, North Carolina.*

O. L. Wolthuis, *Medical Biological Laboratory of the National Defense Research Organization TNO, Rijswijk Z. H., The Netherlands.*

John W. Zemp, *Department of Biochemistry and the Genetics Curriculum, University of North Carolina, Chapel Hill, North Carolina.*

INTRODUCTION

William L. Byrne

Department of Biochemistry
University of Tennessee Medical Units

In August of 1966 a brief article was published in *Science* (p. 658) which indicated that a number of laboratories had tried to replicate a series of memory transfer experiments which were carried out in the laboratory of Allan Jacobson at the University of California in Los Angeles. The first publication resulting from this series of experiments was titled "Transfer of a Response to Naive Rats by Injection of Ribonucleic Acid Extracted from Trained Rats." Three additional articles were published in the next few months which extended their results to task specific transfer and even a cross-species transfer. The "failure to replicate" article was cautiously phrased, as illustrated by the concluding paragraph:

> Our consistently negative findings do not, of course, bear directly on the possibility that RNA may be involved in the mechanism of memory. They indicate only that results obtained with one method of evaluating this possibility are not uniformly positive. Furthermore, we feel that it would be unfortunate if these negative findings were to be taken as a signal for abandoning the pursuit of a result of enormous potential significance. This is especially so in the light of several other related but not identical experiments that support the possibility of transfer of learning by injection of brain-extract from trained donors. Failure to reproduce results is not, after all, unusual in the early phase of research when all relevant variables are as yet unspecified.

This article was signed by 23 authors representing 7 laboratories, and the detailed data are available.* Irrespective of the cautious phraseology, as indicated by the paragraph quoted above, it was not surprising that a failure to replicate article by 23 authors within 12 months of the initial publication would cast serious doubt on the initial observations, and it was also predictable that this challenge would carry over to the other original and independent reports of memory transfer in rodents. Three earlier publications had also indicated the inability of additional laboratories to observe the memory transfer phenomenon.

From the massive attempt to replicate a specific series of experiments, it is apparent that the memory transfer phenomenon had attracted the interest of the scientific community. A series of experimental observations, which were by their very nature controversial, had not only attracted interest, but the initial failures to achieve replication had also enhanced the skepticism of the general scientific community, in particular a peer group in the neurosciences. By the summer of 1967, it was clear that the status of the memory transfer phenomenon was a controversial subject which could not be properly presented by one individual or one laboratory. Therefore, in an attempt to provide a format for the presentation and evaluation of the memory transfer phenomenon, a symposium was organized with a title of "Molecular Approaches to Learning and Memory." This symposium was held in New York City in December of 1967 as a component of the annual meeting of the American Association for the Advancement of Science. The multiple objectives of the symposium were presented in the presymposium publicity in *Science* (Byrne 1967):

> A molecular approach to learning and memory does not assume that a detailed mechanism will be outlined and documented in the immediate future, but it does mean that experimental efforts can be placed in a framework which is consistent with current concepts of molecular biology. This framework may then be used to design experiments. For those who feel that molecular approaches are premature, it should be pointed out that Ward C. Halstead, in 1947, proposed 'molecular models' based on protein organization and suggested specific experimental designs which were subsequently pursued in a number of laboratories (J. J. Katz and W. C. Halstead, *Comp. Psychol. Monogr.* **20**, 1–38, 1950). With the support of historical perspective and the confidence engendered by recent advances in our understanding of living processes at the molecular level, a symposium was organized in order to present and evaluate current experimental procedures which may be considered as molecular approaches to learning and memory.

*Document 8990, ADI Auxiliary Publication Project, Library of Congress, Washington, D.C. 20025.

Emphasis has been placed on procedures which have made it possible to detect multiple stages in the process of memory storage. These procedures also suggest the involvement of macromolecules. The multiple stages (short-term memory and two or more stages of long-term memory) serve as a working model for attempts to correlate specific chemical and anatomical changes with the storage process. This emphasis, together with the emphasis on 'memory transfer,' meant that many basic types of information which are fundamental to current working models would not be properly represented. One of these types, however, the body of information which strongly indicates the specificity of the developmental process is represented by nerve regeneration experiments carried out with the transplanted metathoracic ganglion of the cockroach. Macromolecular synthesis in synaptosome preparations must also serve as a representative contribution for past and present fundamental studies on the role of the synapase in the integrated nervous system.

Particular emphasis has been placed on the new and controversial field of behavioral modification by injection of brain-derived materials, so called 'memory transfer.' If these observations are related to the molecular basis of long-term memory, they represent a direct experimental approach at a time when other approaches must depend on correlations. It is clear that these experiments are reproducible within individual laboratories over a period of one to two years. Additional laboratories are finding behavioral modifications which are consistent with 'memory transfer.' However, even though these positive reports include a great variation in the type of material injected, the route of injection and behavioral methodology, the general acceptance, and usefulness of these observations require procedures which are reproducible within a single laboratory and reproducible in other laboratories. Several laboratories have organized a cooperative attempt to develop and test specific experimental procedures. The symposium will include a progress report on this cooperative effort.

The afternoon program on 29 December includes representatives of three of the first five laboratories which independently published data indicating that the injection of brain-derived materials resulted in behavioral modification which were consistent with 'memory transfer.' These representatives will review and extend their published observations. The published results of the two laboratories which are not represented will be reviewed as part of the presentation on the current status of 'memory transfer.' The first presentation in the afternoon will review and summarize the information available on planarians as experimental animals for experiments on learning and memory.

The participants in the symposium were invited to submit manuscripts and these manuscripts formed the basis for this book. Two additional contributions were solicited from Ward C. Halstead and E. Roy John. These contributions have been included as first and last chapter, respectively, "The Molecular Biology of Memory" by Ward C. Halstead and William B. Rucker, and "Summary, . . . " by E. Roy John.

From a historical point of view, it seemed apparent that the memory transfer experiments with planarians were responsible for the memory transfer experiments in rodents. This historical responsibility (or credit) was documented by listing the references cited by the five original and independent descriptions of memory transfer in rodents; see Table I. The references cited in this original series included articles by a number of investigators and laboratories, but all five original descriptions cited publications dealing with planarians.

TABLE I

HISTORICAL ORIGINS OF MEMORY TRANSFER
LITERATURE CITATIONS USED IN THE ORIGINAL PUBLICATIONS[a]

	Year[b]	Copenhagen	Pilsen	Houston	Los Angeles	Montreal
Hyden *et al.*	(62, 63, 64)	Cited	–	Cited	Cited	–
McConnell *et al.*	(59, 62)	Cited	Cited	–	–	–
Zelman *et al.*	(63)	Cited	–	–	Cited	Cited
Smith	(63)	Cited	–	–	–	–
Westerman	(63)	Cited	–	Cited	–	–
Ziegler	(63)	Cited	–	–	–	–
Kabat	(64)	Cited	–	–	–	–
John	(61)	Cited	–	–	–	–
Dingman and Sporn	(61)	Cited	–	–	Cited	–
Cameron *et al.*	(61)	Cited	–	–	–	–
Morrell	(62)	Cited	–	–	–	–
Cook *et al.*	(63)	Cited	–	–	–	–
Ghosh *et al.*	(64)	Cited	–	–	–	–
Ungar	(65)	–	–	Cited	–	–
Fried and Horowitz	(64)	–	–	–	Cited	–

[a]Original publications of memory transfer in rodents:

Copenhagen–Fjerdingstad, E. J., Nissen, Th., and Røigaard-Peterson, H. H. (1965). Effect of ribonucleic acid extracted from the brain of trained animals on learning in rats. *Scand. J. Psychol.* **6**, 1–6.

Pilsen –Reiniš, S. (1965). The formation of conditioned reflexes in rats after the parenteral administration of brain homogenate. *Activitas Nervosa Super.* **7**, 167–168.

Houston –Ungar, G. and Oceguera-Navarro, C. (1965). Transfer of habituation by material extracted from brain. *Nature* **207**, 301–302.

Los Angeles–Babich, F. R., Jacobson, A. L., S. and Jacobson, A. (1965). Transfer of a response to naive rats by injection of ribonucleic acid extracted from trained rats. *Science* **149**, 656–657.

Montreal –Albert, D. J. (1966). Memory in mammals: Evidence for a system involving nuclear ribonucleic acid. *Neuropysychologia* **4**, 79–92.

[b]List of references cited (limited to references indicating historical origin):

Cameron, D. E., Solyom, L., and Beach, L. (1961). Further studies on the effects of the administration of ribonucleic acid in aging patients suffering from memory (retention) failure. *Neuropsychopharmac.* **2**, 351–355.

Cook, L., Davidson, A. B., Davis, D. J., Green, H., and Fellows, E. J. (1963). Ribonucleic acid: Effect on conditioned behavior in rats. *Science* **141**, 3577, 268–269.

Dingman, W. and Sporn, M. B. (1961). The incorporation of 8-azaguanine into rat brain RNA and its effect on maze learning by the rat: An inquiring into the chemical basis of memory. *J. Psychiat. Res.* **I**, I–II.

Fried, C. and Horowitz, S. (1964). Contraction–a leaRNAble response. *Worm Runner's Digest* **6**, 3–6.

Ghosh, J. J., Datta, R. K., Ghosh, S. Sikpar, K., Roy, S., and Bhattacharyya, D. (1964). Action of phenothiazine group of tranquillizers on cytoplasmic ribonucleoprotein constituents of nerve cells. *Proc. 2nd Intern. Congr. Histo- and Cytochem.* Frankfurt am Main, 230.

Hyden, H. (1962). A molecular basis of neuron-glia interaction. *In* F. O. Schmitt, (Ed.) "Macromolecular Specificity and Biological Memory," pp. 55–69. M.I.T. Press, Cambridge, Mass.

Hyden, H. (1963). Biochemical and functional interplay between neuron and glia. Proc. 18th Ann. Convention and Sci. Program Soc. Biol. Psychiatry, Atlantic City, N. J., June 7–9, 1963, *Recent Adv. Biol. Psychiat.* **4**, 31–54.

Hyden, H. and Egyhazi, E. (1962). Nuclear RNA changes of nerve cells during a learning experiment in rats. *Proc. Natl. Sci. U.S.A.* **48**, 1366–1377.

Hyden, H. and Egyhazi, E. (1962). Glial RNA changes during a learning experiment in rats. *Proc. Natl. Acad. Sci. U.S.A.* **49**, 618-624.

Hyden, H. and Egyhazi, E. (1964). Changes in RNA content and base composition in cortical neurons of rats in a learning experiment involving transfer of handedness. *Proc. Natl. Acad. Sci. U.S.A.* **52**, 1030–1035.

John, E. R. (1961). High nervous functions: Brain functions and learning. *Ann. Rev. Physiol.* **23**, 451.

Kabat, L. (1964). Transfer of training through ingestion of conditioned planarians by unconditioned planarians. *Worm Runners Digest* **6**, 23–27.

McConnell, J. V. (1962). Memory transfer through cannibalism in planarians. *J. Neuropsychiat.* **3**, 42–48.

McConnell, J. V., Jacobson, A. L. and Kimble, D. P. (1959). The effects of regeneration upon retention of a conditioned response in the planarian. *J. Comp. Physiol. Psychol.* **52**, **I**, 1–5.

Morrell, F. (1962). Electrochemical mechanisms and information storage in nerve cells. *In* F. O. Schmitt, (Ed.) "Macromolecular Specificity and Biological Memory," pp. 73–79. M.I.T. Press, Cambridge, Mass.

Smith, S. J. (1963). An attempt to replicate the cannibalism studies. *Worm Runners Digest* **5**, 49–54.

Ungar, G., and Oceguera-Navarro, C. (1965). Transfer of habituation by material extracted from brain. *Nature* **207**, 301–302.

Westerman, R. A. (1963). Somatic inheritance of habituation of response to light in planarians. *Science* **140**, 676–677.

Zelman, A. Kabat, L., Jacobson, R., and McConnell, J. V. (1963). Transfer of training through injection of conditioned RNA into untrained planarians. *Worm Runners Digest* **5**, 14–21.

Ziegler, R. A. (1963). Is knowledge edible? *Worm Runners Digest* **5**, 55–57.

Parenthetically, it also seems appropriate to attribute responsibility (or credit) for the interest in the molecular approach to a book entitled "Macromolecular Specificity and Biological Memory" edited by F. O. Schmitt (1962). From a historical point of view, it seems appropriate to point out that, within two months of the publication of the failure to replicate article in *Science,* two of the authors presented "positive" results which were consistent with memory transfer. These "positive" results, *using modified procedures*, were presented before a regional meeting of the National Academy of Science through the sponsorship of Melvin Calvin (Byrne and Samuel, 1966). These results in themselves did not clarify the nature of the transfer phenomenon in terms of a task specific transfer as opposed to a "hormone-like" effect, but they were from one of the laboratories which had failed to replicate the series of studies from Allan Jacobson's laboratory, and they were carried out with a rigid set of criteria. In addition, these studies were a stimulus which resulted in the formulation of a molecular mechanism (Byrne and Hughes, 1967) which might explain the memory transfer phenomenon. A specific molecular mechanism is in many ways premature; it would be a mechanism to explain a phenomenon which is still in the process of being defined in terms of its nature and replicability, but it is also clear that a proper evaluation of the transfer phenomenon has been handicapped by the lack of a hypothesis. A hypothesis is needed which would predict that molecules have information in a form which would play a role in long term memory and that this information would be such that the molecules would also effectively modify the behavior of an injected animal. With these objectives in mind, a hypothesis was developed which predicts that the information containing molecules modify the behavior of the injected animal by mimicking one or more stages of the process which normally occur in the process of consolidation in the donor animal.

The hypothesis requires three assumptions. (*a*) Long term memory results from synaptic modification. (*b*) The synaptic modification associated with long term memory is the consequence of the synthesis of new molecules. (*c*) The nervous system is "prewired" by the developmental process. R. W. Sperry suggested the term "prewired" and further suggested that the developmental specificity would result in a chemical specificity of the ultimate system. This chemical specificity or individuality of specific anatomical regions, according to Sperry, could even be thought of as a region with an identification tag or a serial number. Further assumptions are not necessary to provide a minimal hypothesis for the role of molecules in long term memory, but the final hypothesis was formulated, and the necessary assumptions were selected with the additional implied understanding that there are multiple stages in the process of learning. See Fig. 1.

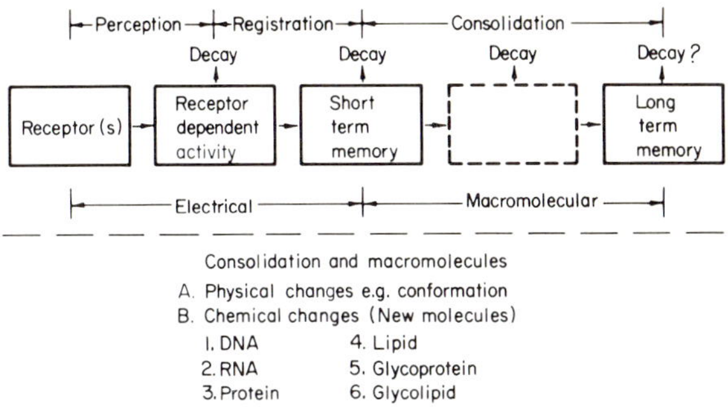

FIGURE 1 *Multiple Stages in Learning.* Learning may be considered as a multiple stage process. It would appear that the initial learning (short term memory) does not require the synthesis of new molecules; this is indicated as an electrical (electrochemical) process. Consolidation may involve conformational changes in macromolecules (postulate A) or it may involve the synthesis of new molecules (postulate B). It should be noted that the arrow indicating decay of long term memory is given with a question mark. Direct experimental attempts to induce permanent amnesia following consolidation have failed. It seems possible that living systems have paid a high price for long term memory and there is no mechanism of erasure. If long term memory has a molecular basis involving specific anatomical sites, this suggests that there is a finite limit to long term memory capacity and this finite capacity could conceivably become a significant (limiting) variable in behavior.

This particular conceptualization was used to introduce the symposium and is presented here as a means of outlining a role for macromolecules in consolidation. Specifically, the second postulate states that consolidation requires the synthesis of new molecules, as opposed to a conformational change, and for the present hypothesis it is proposed that the new molecules are mRNA and protein. Short term memory would not require the synthesis of new macromolecules and long term memory would depend on the synthesis of proteins, which would self-assemble into the membranes of the synapase. This particular scheme includes one stage between short term and long term memory which might be mRNA. The mRNA in itself would not be associated with a synaptic modification and therefore would not be associated with a behavioral counterpart; therefore behavioral detection would be dependent on conversion to the next stage–long term memory. Figure 2 was also presented in the symposium and is included here to illustrate a specific suggestion for information in macromolecules.

Basically, the "central dogma" of molecular biology states that the information required to form a molecular complex, such as a regulated enzyme, resides in DNA in the form of a genetic code and this code is transcribed to mRNA, translated to a polypeptide, and the polypeptide has enough

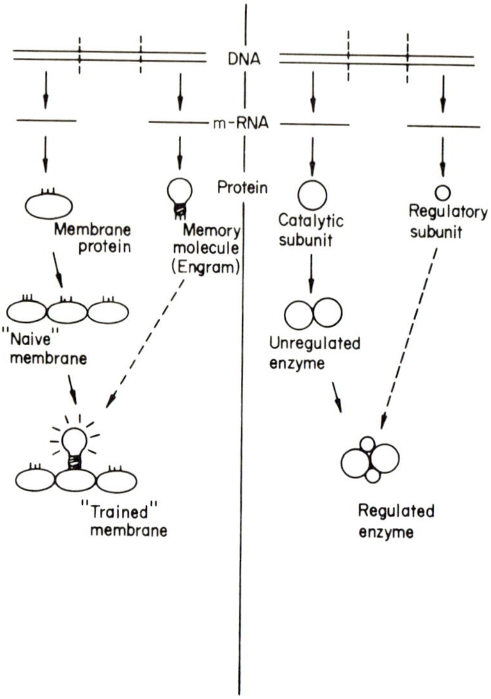

FIGURE 2 *Information in Macromolecules.* Genetic information can be used to synthesize molecular complexes. This can be illustrated with the synthesis of a regulated enzyme (regulatory or allosteric enzyme) and the final complex represents the interaction of nonidentical subunits. An analogous process is proposed for synaptic modification as the basis for inherited behavior and long term memory. The developmental process would provide the "prewired" nervous system with the "naive synapse," and development (inherited behavior) or environment (acquired behavior) would induce the synthesis of the memory molecule as the first step in consolidation.

information in its sequence of amino acids to form a specific three dimensional structure. The three dimensional structure formed from a single polypeptide, a subunit, can also have the information needed to specifically interact with other identical or nonidentical subunits of complimentary specificity. The synthesis of a regulated enzyme beginning with genetic information in DNA in Fig. 2 is an example which ultimately depends upon the interaction of nonidentical subunits. Consolidation is illustrated in Fig. 2 as a process which is analogous to the synthesis of a regulated enzyme including the interaction of nonidentical subunits. The membrane structure of the synapase would be genetically determined through the assembly of protein subunits, and consolidation would lead to the synthesis of a new, nonidentical subunit, which would self-assemble into a synaptic membrane converting it from a "naive" to a "trained" membrane. In Fig. 2 this new subunit is designated as a memory molecule (engram). The cistron corresponding to the new protein subunit would be "turned on" as the initial step in consolidation. The model for consolidation in Fig. 2 can also serve as a model for inherited behavior if one assumes that the cistron is "turned on" by the developmental process. Thus, acquired and inherited behavior, at the molecular level, would be fundamentally the same process.

This particular hypothesis, a molecular basis for long term memory, would predict that the memory transfer phenomenon is a consequence of information containing molecules which can mimic the normal consolidation process. It does not require that all animals utilize a unique combination of pathways in learning a given task, but it does require that the combination of pathways utilized by the donor would also be an effective combination in the recipient. For the transfer phenomenon, it is predicted that the behavioral modification would be a direct consequence of the injection of the nonidentical protein subunits. However, if the transfer mimics the consolidation process, it is also predicted that mRNA could be effective if it led to the synthesis of the nonidentical subunits, and, finally, the hypothesis (see Fig. 2) states that a given region of the genome is activated as the first step in consolidation. The molecular mechanism for the initiation of gene expression in animals is unknown, but this process could also involve information containing (specific) molecules which could lead to the transfer phenomenon.

A hypothesis of this type can be pursued in many ways, e.g., temporary or permanent behavioral changes following injection, but the hypothesis is dependent on the specificity of membrane structure and the potential of membrane subunits to interact, self-assemble, into preexisting structures. A series of experiments has been carried out in our laboratory to test this component of the hypothesis.

Preliminary experiments by R. J. Porter indicated that homogenates of isotopically labeled liver or kidney, following intraperitoneal injections, would preferentially localize in the corresponding organ. This preferential localization confirmed the earlier studies of Ebert (1954) which demonstrated a selective transfer of radioactivity from adult chicken spleen and kidney grafts, using ^{35}S– methionine, into the corresponding tissues of host chick embryos. Also, Walter *et al.* (1956) showed a preferential localization of injected adult liver and heart homogenates into the homologous tissues of embryonic chicks.

A more detailed study of preferential localization of injected homogenates has been carried out by E. F. Stilwell in our laboratory, and will be published in the proceedings of the "Symposium on Biology of Memory" which was sponsored by the Biological Section of the Hungarian Academy of Sciences and organized by G. Adam and J. Salanki. This symposium was held in Tihany, Hungary, on September 1–4, 1969, and will be published by Akademiai Kiado, Budapest. In this study, intraperitoneal injection of ^{3}H– leucine-labeled homogenates prepared from individual organs resulted in a preferential localization of the labeled material(s) in the homologous organ, and the specificity was heat labile. This specificity was demonstrated for kidney, lung, olfactory lobe and brain (without the olfactory lobe) in the young adult rat. This preferential localization of injected material was predicted by the hypothesis, but it is not proof of the hypothesis. These experiments should be confirmed and extended before they can be confidently interpreted in terms of a specific hypothesis, but the hypothesis and the preferential localization hopefully will be of at least transient value in the development of our understanding of learning and memory.

As indicated above, the symposium which led to this volume was organized in an attempt to provide a format for the presentation and evaluation of the memory transfer phenomenon and was broadened to provide a representative view of molecular approaches. The value of the volume may ultimately reside in the chapters which do not include memory transfer experiments, but the emphasis on memory transfer led to the decision to include a guide to the memory transfer literature. This is included as a table by Robert W. Lonon which summarizes the experimental evidence as of the time of the symposium. In addition, a list of current publications is included. These publications and this volume should serve the purpose of documenting the transfer phenomenon as of October 1969.

In concluding this introduction, it seems appropriate to comment that there are new developments which cannot be included because there is no means of overcoming the time lag between experimental data, oral presentation, and publication. However, this is being written following two symposia which emphasized molecular approaches to learning and memory including memory transfer, and with this background one can evaluate the contents of this volume, and attempt to capture the essence of these symposia. The proceedings of the Symposium on the Biology of Memory, held in Hungary, will be published. The second symposium, "The Chemistry of Learning and Memory," was held in Memphis and was sponsored by IDEA (Institute for Development of Education Activities, Melbourne, Florida). The results of this latter meeting will not be published in detail, but five participants—J. V. McConnell, E. J. Fjerdingstad, G. Ungar, O. L. Wolthuis, and F. Rosenblatt—spoke on memory transfer. These oral presentations plus the publications from new groups of investigators (see list of current publications) were convincing in terms of substantial progress and future potential. In spite of this progress, however, there was no breakthrough which outdated the contents of this volume. At the symposium in Memphis, the discussants emphasized the need for more experiments, better procedures, further emphasis on replicability and chemical characterization, but the basic phenomenon went unquestioned. One discussant even implied that the role of molecules in memory had reached a definitive stage of development, and thereupon called attention to the broader view of learning and memory.

As indicated above, it seems very clear that the transfer experiments with planarians led to the transfer studies in rodents, and it now seems equally clear that these experiments will eventually be accepted as the first direct evidence for behavioral information in molecules. In this same vein, one might also predict that the planarian will once again be accepted as a member of the animal kingdom which is suitable for serious scientific endeavors.

ACKNOWLEDGEMENT

The publication of this volume would not have been possible without the direct and indirect support of many institutions and individuals. The symposium was organized with the encouragement and support of Walter Berl (American Association of Advancement of Science) and NIMH. In particular, I would like to express my sincere appreciation to L. E., T. A., S. N., V. A., M. H., R. W., A. G., S. D., E. G., D. K., M. R., E. B., A. H., M. C., E. C., D. S., L. B., and F. B.

At the time the symposium was organized, Ward C. Halstead was seriously ill and could not participate, but two chapters bear his name and his enthusiastic support is gratefully acknowledged. Thanks are also due to R. A. McCleary, S. L. Chorover, and D. Samuel for their participation as session chairmen in connection with the symposium.

REFERENCES

Byrne, W. L. (1967). *Science* **158**, 1082.

Byrne, W. L. and Samuel, D. (1966). *Science* **154**, 418.

Byrne, W. L. and Hughes, A. (1967). *Fed. Proc.* **26**, 676.

Byrne, W. L., Samuel, D., Bennett, E. L., Rosenzweig, M. R., Wasserman, E., Wagner, A. R., Gardner, F., Galambos, R., Berger, B. D., Margules, D. L., Fenichel, R. L., Stein, L., Corson, J. A., Enesco, H. E., Chorover, S. L., Holt, C. E. III, Schiller, P. H., Chiappetta, L., Jarvik, M. E., Leaf, R. C., Dutcher, J. D., Horovitz, Z. P., and Carlson, P. L. (1966). *Science* **153**, 658.

Ebert, J. D. (1954). *Proc. Natl. Acad. Sci., U.S.A.* **40**, 337.

Schmitt, F. O. (Ed.) (1962). "Macromolecular Specificity and Biological Memory." M. I. T. Press, Cambridge, Mass.

Walter, H., Allmen, D. W. and Mahler, H. R. (1956). *Science* **154**, 1251.

BIBLIOGRAPHY

Adam, G., and Faiszt, J. (1967). Conditions for successful transfer effects. *Nature* **216**, 198–200.

Bozhko, G. Kh. (1968). Rate of conditioning and turnover of brain RNA after administration of homologous RNA from trained rats. *Zh. Vysshei Nerv. Deyat. Pavlov* **18**, 1085–1087.

Chapouthier, G. (1968). Le transfert de molécules de mémoire chez les vertébres. *Ann. Biol. Clin.* **7**, 275–285.

Chapouthier, G. and Ungerer, A. (1968). Effect de l'injection d'extraits de cerveau conditionné sur l'apprentissage *C. R. Acad. Sci.* [D] (Paris) **267**, 769–771.

Corson, J. A. and Enesco, H. E. (1968). Attempts to obtain "information transfer" in rats with brain extracts. *J. Biol. Psychol.* **10**, 10–23.

Daliers, J. and Rigaux-Motquin, M. L. (1968). Transfer of learned behavior by brain extracts I. Audiogenic seizure. *Arch. Intern. Pharmacodyn.* **176**, 461–463.

Faiszt, J. and Adam, G. (1968). Role of different RNA fractions from the brain in transfer effect. *Nature* **220**, 367–368.

Fjerdingstad, E. J. (1969). Chemical transfer of two positive reinforcements training schedules with control for the effects of deprivation and exposure to the experimental environment. *Fed. Proc.* **28**, 641.

Fjerdingstad, E. J. (1969). Chemical transfer of learned preference. *Nature* **222**, 1079–1080.

Fjerdingstad, E. J. (in press). Chemical transfer of alternation training in the Skinner box. *Scand. J. Psychol.*

Gibby, R. G., Crough, D. G., and Thios, S. J. (1968). RNA-enhancement of a single discrimination task. *Psychonom. Sci.* **12**, 295–296.

Golub, A. M. (1968). Negative results in memory transfer studies: a response to Drs. Carson and Enesco. *J. Biol. Psychol.* **10**, 24–25.

Golub, A. M. and McConnell, J. V. (1968). Transfer of response bias by injection of brain homogenates: A replication. *Psychonom. Sci.* **11**, 1–2.

Kleban, M. H., Altschuler, H. Lawton, M. P., Parris, J. L., and Lords, C. A. (1968). Influence of donor-recipient brain transfer on avoidance learning. *Psychol. Rep.* **23**, 51–56.

Lagerspetz, K. M. J., Raitis, P., Tirri, R., and Lagerspetz, K. Y. H. (1968). Experiments concerning the transfer of a learned response and activity level by brain homogenates. *Scand. J. Psychol.* **9**, 225–229.

LeVan, H., Hebron, D. L., Moos, W. S., and Mason, H. C. (1969). Transferability of post-irradiation conditioned avoidance behavior by intraperitoneal injection of brain tissues. *Radiation Research* **39**, 551.

LeVan, H., Moos, W. S., and Mason, H. C. (in press). Induction of post-irradiation conditioned avoidance behavior by intraperitoneal injection of brain tissues. *Acta Radiologie.*

McConnell, J. V., Shigehisha, T., and Salive, H. (1968). Attempts to transfer approach and avoidance responses by RNA injections in rats. *J. Biol. Psychol.* **10**, 32–50.

Moos, W. S., LeVan, H., and Mason, H. C. (in press). Radiation induced avoidance behavior transfer by brain extracts of mice. *Experientia.*

Reiniš, S. (1968). Block of "memory transfer" by actinomycin D. *Nature* **220**, 117–178.

Reiniš, S. and Kolousek, J. (1968). Effects of methionine sulphoximide on memory transfer. *Nature* **217**, 680–681.

Røigaard-Petersen, H. H., Fjerdingstad, E. J., and Nissen, Th. (1968). The effects of ribonucleic acid (RNA) extracted from the brain of trained animals on learning in rats III. Results obtained with an improved procedure. *Scand. J. Psychol.* **9**, 1–16.

Rosenthal, E., and Sparber, S. B. (1968). Transfer of a learned response by chick brain homogenate fed to naive donors. *The Pharmacologist* **10**, 168.

Ungar, G. (1968). Learned fear transferred by a brain peptide. *Fed. Proc.* **27**, 223.

Ungar, G. (1968). Molecular mechanisms in learning. *Perspect. Biol. Med.* **11**, 217–223.

Ungar, G. (1969). Chemical transfer of passive avoidance. *Fed. Proc.* **28**, 647.

Ungar, G. and Galvan, L. (1969). Conditions of transfer of morphine tolerance by brain extracts. *Proc. Soc. Exp. Biol. Med.* **130**, 287–291.

Ungar, G. Galvan, L., and Clark, R. H. (1968). Chemical transfer of learned fear. *Nature* **217**, 1259–1261.

Ungar, G. and Fjerdingstad, E. J. (1969). RNA-bound peptides in chemical transfer of learned behavior. *Molecular Neurobiology Bulletin* **2**, 9.

Wolthuis, O. L. (in press). Interanimal information transfer by brain extracts. *Arch. Intern. Pharmacodyn.*

Zippel, H. P. and Domagk, G. F. (in press). Versuche zur chemischen Gedächtnisübertragung von farbdressierten Goldfischen auf undressierte Tiere. *Experientia.*

TABULAR SUMMARY

Robert W. Lonon

LIST OF ABBREVIATIONS USED IN TABLE

IP = intraperitoneal
IC = intracisternally
ICbr = intracerebral
IV = intravenous

W = Wilcoxon
M-W = Mann-Whitney
K-W = Kruskal Wallis

No.	Experimenter	Donor	Recipient	Extraction Procedure
1	Ádám Faiszt '67	Rat (♂, Wistar) trained over varying time periods to jump to shelf in response to bell paired with electric shock (5 sec delay for shock)	Rat (like donor) preconditioned in various ways, injected, then tested in the training situation	RNA extract as in 3
2	Albert '66	Rat (♂, hooded, 225–275 g) trained in shock avoidance task with one hemisphere under spreading depression (escape to black or white)	Same rat with medial cortex of trained hemisphere removed, trained in the same task and the savings in learning measured	Homogenize brain tissue, fractionate into subcellular components, medial cortex removed under anesthesia 1 to 2 days after training[a]
3	Babich Jacobson Bubash Jacobson '65	Rat (♂, Sprague-Dawley, 250 g) trained to approach food cup at sound of food dispenser with food reinforcement	Rat (like donors) with food deprivation (28 hr) approaches food cup at sound of food dispenser but is not reinforced	RNA extract: aqueous layer of saline-phenol extract precipitated with ethanol; animals killed by ether[b]
4	Jacobson Babich Bubash Jacobson '65	As in No. 3 but with one group trained to respond to light	Rat (like donors) with food deprivation (24 hr) approaches food cup in response to stimulus but is not reinforced	RNA extract as in 3
5	Babich Jacobson Bubash '65	Hamster (♂, 100 g) trained as in 3 but with independent control of food reinforcement	Rat (♂, Sprague-Dawley, 220 g) with food deprivation (24 hr) approaches food cup at sound of food dispenser but is not reinforced	RNA extract as in 3
6	Jacobson Babich Bubash Goren '66	Rat (♀, Sprague-Dawley, 250 g) trained for light-dark discrimination in 2-alley maze with food reinforcement	Rat (like donor) chooses light or dark alley but response is not reinforced	RNA extract as in 3
7	Dyal Golub Marrone '67	Rat (♂, Sprague-Dawley) trained to bar press for food reinforcement; CRF, yoked, and CRF with extinction and reacquisition groups	Rat (like donor) is tested 24 hr post injection with food deprivation (48 hr); score magazine entries (a) unreinforced in response to click and (b) on CRF schedule	Whole brain: homogenize brain in physiologic saline (1.0 ml/brain)
8	Fjerdingstad Nissen Røigaard-Petersen '65	Rats (♂, Wistar, 200 g) trained in 2 alley runway to choose lighted runway with water reinforcement	Rats (like donor) trained in same task and errors per hundred reinforcements used as index of learning	RNA extract: method of Laskov *et al.* modified by precipitation of RNA with isopropanol; brains removed under anesthesia and frozen in dry ice

Injection	Control Injection	Nature of Transfer	Statistical Test	P value	Group Size; t = test, c = control, r = recipients
IP	comparable extract from untrained donors	Transfer less significant with poorly or overtrained animals. State of habituation before injection affects significance of transfer	W U-test		191 t 171 c
IP	no injection or posterior cortex	Interanimal transfer did not work. Consolidation complete at 7 hrs		<.01 <.05 <.01	15 t 15 t 16 t
IP	comparable extract from untrained donors		M-W U-test one-tailed	<.002	8 t
IP	compare click vs light trained for control	Transfer is stimulus specific	M-W U-test one-tailed	<.001 for light conditioned. <.002 for click conditioned	8 t 8 t
IP	comparable extract from yoked donors	Interspecific transfer can take place	M-W U-test	<.001	8 t
IP	comparable extract from untrained donors	Transfer is specific; applies to at least several training tasks	X^2 one-tailed	.02	t + c = 42
IP 3.2 ml	extract from yoke rats	Extinguished and retrained animals showed greatest transfer	K-W one-way analysis of variance	<.01	6/group 3 groups
IC, RNA from 1 brain	comparable extract from untrained donors	Animals injected with extract from trained donors showed fewer errors/100 reinforcements	M-W U-test	.027 (.014)	4 t 4 c 5 noninjected

(continued)

TABULAR SUMMARY *(continued)*

No.	Experimenter	Donor	Recipient	Extraction Procedure
9	Nissen Røigaard-Petersen Fjerdingstad '65	Rats trained as in 8 but different groups choosing lighted or dark runways for water reinforcement	Rats (like donor) habituated to runway prior to injection then tested by training either with or against the donors training, then training reversed	RNA extract as in 8[c]
10	Røigaard-Petersen Nissen Fjerdingstad '66	Rats as in 8	Rats as in 8	RNA extract as in 8
11	Røigaard-Petersen Nissen Fjerdingstad '68	Rats as in 8	Rats as in 8	RNA extract: precipitated from aqueous phase of phenol type extract with isopropanol, redissolved, dialyzed, reprecipitated with isopropanol, and dried[d]
12	Gay Raphelson '67	Rat (♂ and ♀, hooded) shocked in black box and allowed to escape	Rat (like donor); measure time spent in black box before and after injection	RNA extract: aqueous phase of saline phenol extract precipitated with ethanol
13	Reiniš '66	Rat (Wistar strain) is conditioned to tone or light with food reinforcement	Rat (Wistar); correct responses and activity recorded under food deprivation (48 hr)	Whole brain: saline homogenate rats killed by decapitation
14	Rosenblatt Farrow Herblin '66	Rat (♀, albino, 200 g) trained to approach food cup in response to click of food dispenser with food reinforcement	Rat (as donor); unreinforced approaches to food cup in response to food dispenser are measured	RNA extract: extraction is according to Babich with modifications; the brain grossly divided before extraction[e]
15	Rosenblatt Farrow Rhine '66	Rats (♀, Holtzman albino, 200 g) trained in various reward, avoidance, spatial and discrimination tasks	Rats (like donor); unreinforced correct responses measured in the training situation	RNA extract: as in 14, rat sacrificed by decapitation[f]
16	Rosenblatt '66	Rats (♀, Holtzman albino, 200 g) trained in various discrimination and activity tasks	Rats (like donor); correct responses measured in the training situation	Varying extracts: homogenization in Tris-saline followed by various means of extraction and fractionation[g]
17	Ungar Oceguera-Navarro '65	Rats (♂, Sprague-Dawley, 200 g) habituated to sound as measured by the absence of the startle reflex with sudden loud sounds	Mice (♂, Swiss albino, 20-30 g); trials to reach 50% habituation measured	Protein extract: brains homogenized, dialyzed, then precipitated with acetone (and in some cases, partitioned with phenol)
18	Ungar '67	Rats (as in 17) trained in various tasks including sound habituation, avoidance, and discrimination	Mice (as in 17) tested according to the particular task and in general not reinforced	Protein extract as in 17, animals sacrificed by decapitation[h]
19	Ungar Irwin '67	Rats (as in 17) trained to escape electric shock in a Y-maze by going to lit, unlit, right, or left arms	Mice (as in 17) pretested to establish their natural bias, then tested unreinforced to measure transfer effect	Supernate of saline homogenate is used for injection
20	Ungar '68	Rats (as in 17) trained similarly to 12	Mice (as in 17) tested as in 12, after screening for dark preference	Dialysate of homogenate; in some cases acetone precipitated, phenol partitioned, and partially purified by gel filtration

(*a*) Transfer activity is the insoluble portion of the nuclear fraction. Heat and RNAse kills the activity; trypsin has no effect.

(*b*) There is no detectable DNA present by diphenylamine or protein by biuret.

(*c*) DNA is less than 15 percent.

(*d*) There is no detectable DNA by the indole test.

(*e*) RNAse does not affect transfer. The folin-ciocaltean test shows a protein level of 1.1 mg/ml.

Injection	Control Injection	Nature of Transfer	Statistical Test	P value	Group Size; t = test, c = control, r = recipients
IC, RNA from 1 brain	trained groups used as mutual controls	Specific reversal of the transfer effect	M-W U-test	.055 .013	22 animals in 4 groups
IC	comparable extract from untrained donors	Transfer effect decreases with increasing time that RNA extract is kept redissolved, prior to injection	M-W U-test	<.001	
IC, 1.0 mg RNA	comparable extract from untrained donors	Initial reverse effect of transfer changes to expected result after several days	M-W U-test	.014	28 t 34 c
IP	comparable extract from unshocked donors	Sex does not affect transfer; distinct difference in time spent in black box by test and control animals			8 ♂ and 8 ♀ r
IP	comparable extract from untrained donors	Change in reactivity of recipients of extract from trained donors	Student's t-test	<.01	175 animals in 8 groups
IP (varying dose)	comparable extract from untrained donors	Transfer effected by various portions of brain	Fisher exact-probability test	.004	80 r
IP >1 brain	comparable extract from untrained donors	All tasks showed transfer; most were statistically significant. Transfer greater from cerebrum but present from cerebellum.	M-W U-test	.0006 overall	487 r
IP, IV, ICbr & intracerebral varying doses	comparable extract from untrained donors	Dose response relationship with inversion of transfer effect at low doses; transfer of discrimination seems to be better measure than transfer of activity	M-W U-test	.02 for unreinforced test (overall)	~100 t r
IP 140–340 mg of brain	comparable extract from normal donor	Definite savings in trials to criterion in mice receiving injections from habituated donors	t-test	<.001	45 r
IP >750 mg brain for maximum transfer	comparable extract from untrained donor		X^2 U-test	.001 overall	374 r of trained extract
IP varying dose	comparable extract from untrained donor	Optimum dose for transfer. Optimum duration of donor training. Strains of slower learning donors give best results, recipient bias is critical factor.	X^2 M-W U-test	<.005 (Right-Left) <.05 (dark-light)	~180 animals ~50 animals
IP, varying dose	as in 17, also yoke shocked in light	Confirmation of 12, transfer quantitatively related to dose, optimum duration of training by donors	t-test	<.001	~500 r

(*f*) RNAse does not affect transfer. The lower molecular weight fraction from sephadex contains most of the activity incubation.

(*g*) Trypsin incubation seems to have an inverting effect on the transfer.

(*h*) The "transfer molecule" is soluble in H_2O but not in acetone or 95 percent ethanol. It is dializable and partitions to the phenol phase. Chymotrypsin destroys activity but RNAse does not.

BIBLIOGRAPHY

Adam, Gyorgy and Faiszt, Jazsef (1967). Condtions for successful transfer effect. *Nature* **216**, 198–200.

Albert, D. J. (1966). Memory in mammals: Evidence for a system involving nuclear ribonucleic acid. *Neuropsychologia* **4**, 79–92.

Babich, F. R., Jacobson, A. L., and Bubash, S. (1965). Cross-species transfer of learning: Effect of ribonucleic acid from hamsters on rat behavior. *Proc. Natl. Acad. Sci.* **54**, 1299–1302.

Babich, F. R., Jacobson, A. L., Bubash, S., and Jacobson, A. (1965). Transfer of a response to naive rats by injection of ribonucleic acid extracted from trained rats. *Science* **149**, 656–657.

Byrne, W. L. and Samuel D. (1966). Behavioral modification by injection of brain extract prepared from a trained donor. *Science* **154**, 418.

Byrne, W. L. and Hughes, A. (1967). Behavioral modification by injection of brain extract prepared from trained donors. *Fed. Proc.* **26**, 676.

Dyal, J. A., Arnold, M. G., and Marrone, R. L. (1967). Transfer effects of intraperitoneal injection of brain homogenates. *Nature* **214**, 720–721.

Essman, W. B. and Lehrer, G. M. (1966). Is there a chemical transfer of training? *Fed. Proc.* **25**, 208.

Essman, W. B. and Lehrer, G. M. (1967). Facilitation of maze performance by "RNA extracts" from maze trained mice. *Fed. Proc.* **26**, 263.

Fjerdingstad, E. J., Nissen, Th., and Roigaard-Petersen, H. H. (1965). Effect of ribonucleic acid extracted from the brain of trained animals on learning in rats. *Scand. J. Psychol.* **6**, 1–6.

Gay, R. and Raphelson, A. (1967). "Transfer of learning" by injection of brain RNA: A replication. *Psychonom. Sci.* **8**, 369–370.

Jacobson, A. L., Babich, F. R., Bubash, S., and Jacobson, A. (1965). Differential approach tendencies produced by injection of ribonucleic acid from trained rats. *Psychonom. Sci.* **4**, 3–4.

Jacobson, A. L., Babich, F. R., Bubash, S., and Goren, C. (1966). Maze preferences in naive rats produced by injection of ribonucleic acid from trained rats. *Psychonom. Sci.* **4**, 3–4.

Nissen, Th., Røigaard-Petersen, H. H., and Fjerdingstad, E. J. (1965). Effect of ribonucleic acid extracted from the brain of trained animals on learning in rats. II Dependence of the RNA effect on training condition prior to RNA extraction. *Scand. J. Psychol.* **6**, 265–270.

Reiniš, S. (1965). Formation of conditioned reflexes in rats after the parenteral administration of brain homogenate. *Activitas Nervosa Super.* **7**, 167–168.

Reiniš, S. (1966). Influence of brain homogenate injection on the speed of the formation of alimentary conditioned reflex in rats. *Worm Runner's Digest* **8**, 7–24.

Røigaard-Petersen, H. H., Nissen, Th., and Fjerdingstad, E. J. (1966). Experimental investigations on the role of RNA in learning. International Congress Series No. 129, p. 176–184. *Excerpta Med.*

Røigaard-Petersen, H. H., Nissen, Th., and Fjerdingstad, E. J. (1967). Effect of ribonucleic acid (RNA) extracted from the brain of trained animals on learning in rats. III Results obtained with an improved procedure. *Scand. J. Psychol.* **8**.

Rosenblatt, F. and Miller, R. G. (1966). Behavioral assay procedures for transfer of learned behavior by brain extracts, I and II. *Proc. Natl. Acad. Sci.* **56**, 1423–1430, 1683–1688.

Rosenblatt, F., Farrow, J. T., and Herblin, W. F. (1966). Transfer of conditioned responses from trained to utrained rats by means of a brain extract. *Nature* **209**, 46–48.

Rosenblatt, F., Farrow, J. T., and Rhine, S. (1966). The transfer of learned behavior from trained to untrained rats by means of brain extracts, I and II. *Proc. Natl. Acad. Sci.* **55**, 548–555, 787–792.

Ungar, G. (1965). Configurational and hydrolytic changes of proteins on excitation: their role in information processing. In "Protides of Biological Fluids" (H. Peeters, Ed.) Elsevier, Amsterdam.

Ungar, G. (1966). Chemical transfer of learning; its stimulus specificity. *Fed. Proc.* **25**, 207.

Ungar, G. (1967). Transfer of learned information by brain extracts. *Fed. Proc.* **26**, 263.

Ungar, G. (1967). Transfer of learned behavior by brain extracts. *J. Biol. Psychol.* **9** (1), 12–27.

Ungar, G. (1967). Chemical transfer of learned information. CINP Symposium, Washington, 1966, *Excerpta Med.*, Amsterdam, pp. 169–175.

Ungar, G. (1968). Molecular mechanisms in learning. *Persp. Biol. Med.*

Ungar, G. (1968). Learned fear transferred by a brain peptide. *Fed. Proc.*

Ungar, G. and Irvin, L. N. (1967). Transfer of acquired information by brain extract. *Nature* **214**, 453–455.

Ungar, G. and Oceguera-Navarro, C. (1965). Transfer of habituation by material extracted from brain. *Nature* **207**, 301–302.

THE MOLECULAR BIOLOGY OF MEMORY

Ward C. Halstead*

Department of Psychology
The University of Chicago
Chicago, Illinois

William B. Rucker

Department of Psychology
The George Washington University
Washington, D.C.

The success of a symposium is measured by its catalysis of scientific strategies. The time required for incubation may vary from one to fifty years or more. To the substrate of this symposium, *Molecular Approaches to Learning and Memory*, Dr. Byrne, as chairman and chief chemist, has added optimal amounts of such cofactors as preplanning and expedition. From the heat generated during both formal and informal meetings of the symposium in New York, it would appear that at least intermediary reaction products are being formed.

One such intermediary product can be seen in the numerous reports in this volume dealing with the behavioral results obtained by injecting one animal with an extract prepared from the brain (or perhaps the liver) of a second animal which has been trained to perform some specific task. The transition product appears as a shift in emphasis, exemplified by a new term to describe this kind of experiment—"behavioral modification" beginning

*After a long illness, Ward Campbell Halstead died on March 25, 1969.

implied that performance of cognitive tasks depended on operations conducted upon the memory store, and that focal brain trauma could selectively disrupt and thereby identify these operations. What seemed to be missing was some clear idea as to how the individual neurons could retain their gating instructions to carry out these operations.

Joe Katz and Halstead began a dialog on this problem in 1947. Following their chemical intuitions, they implicated the nucleoprotein ultrastructure as the permanent component of the neuronal contribution to the trace. At the Hixon Symposium in 1948 (Halstead, 1951), George Beadle and Linus Pauling encouraged the development of these ideas and their subsequent publication (Katz and Halstead, 1950). The following outline may suggest some of the flavor of these primitive formulations.

> We wish to suggest that an essential feature in the genesis of the memory trace is the formation, as a result of individual experience, of geometrically ordered protein molecules in the neurons of the cerebrum. These new ordered molecules then function as template molecules and thus now have the same relationship to the cell as do the original genes. The memory trace itself we conceive to be a protein lattice. . . . The various memory traces are characterized and differentiated by the chemical composition and configuration of the repeating units of the particular protein lattice [Katz and Halstead, 1950, 4–5].

The postulate that macromolecules controlled gating operations at the synapse was based on these assumptions: (1) A neuron does not become a member of a trace network simply because of a shared synapse; (2) the neuron becomes a functioning member of the long-term trace only as a result of primary (or perhaps secondary) changes in its chemical ultrastructure; and (3) these changes involve completion of the neuronal membranes on both sides of the synapse in order to provide a certain degree of matching of their respective positions in the network. This rather daring return to the Doctrine of Specific Nerve Energies seemed appropriate, since bidirectional electrical conduction was still the favored alternative to one-way chemical transmission even at the vertebrate synapse (Eccles, 1946).

When formulated, the model reflected the current state of molecular genetics (*e.g.*, Pauling, 1940) in that sometimes the *de novo* template was identified as protein and sometimes as nucleoprotein:

> The mechanisms of racial and individual memory have in common, then, the formation of ordered protein molecules from a template; they differ in that the template in the genetic apparatus is a natural endowment, whereas in individual memory the template molecules arise from external stimuli [Katz and Halstead, 1950, 5].

For all of these handicaps, the model succeeded in identifying the synapse as the basic unit of neural information processing, and in directing attention towards the nucleoproteins as participating in a self-regenerating feedback loop which must be assumed to maintain the single cell's contribution to the engram. Nostradamically, the Katz and Halstead paper hinted at: environmentally initiated consolidation (Halstead, 1940; Davis and Agranoff, 1966; Zinkin and Miller, 1967); redundant expansion of the trace with the passage of time (Albert, 1966; Flexner *et al*, 1967); and a three-phase model of memory (Fig. 1; Halstead *et al.*, 1967; Halstead and Rucker, 1968).

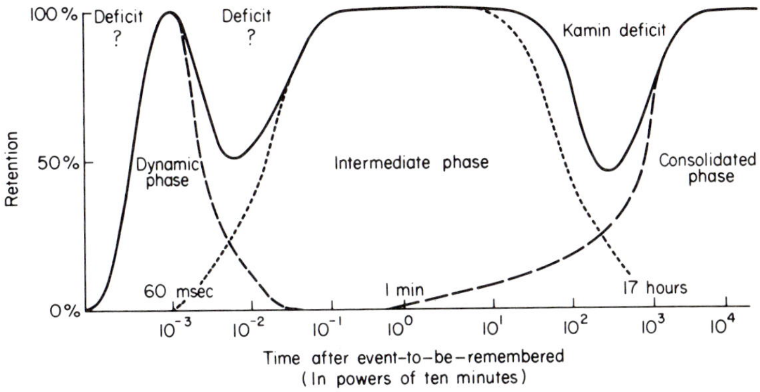

Figure 1 Three possible phases of memory retention. The model attributes the Kamin (1963) deficit to a rate of consolidation too slow to compensate for the decline of the intermediate phase. Other similar deficits may occur for similar reasons (Halstead and Rucker, 1968).

Our current conception of the molecular events which establish long-term memory came about through an exchange of ideas with other workers (among others: Davis and Agranoff, 1966; Flexner *et al.*, 1967). Fig. 2 represents a prototype of the macromolecular system which we think initiates and maintains the neuron's contribution to the consolidated phase of an engram. A

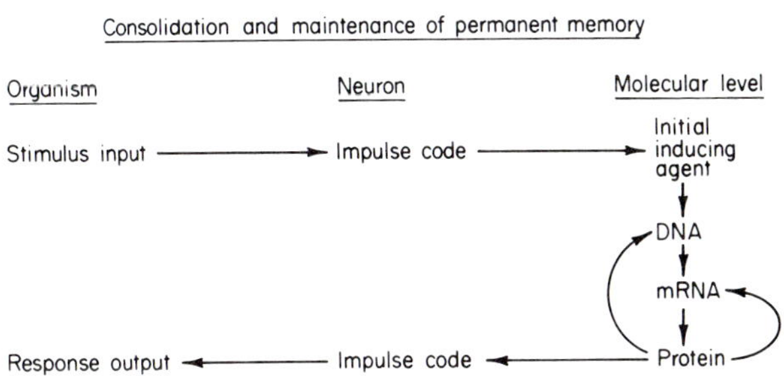

Figure 2 Consolidation and maintenance of permanent memory. Two feedback loops may have evolved to protect permanent memory at the molecular level (Halstead and Rucker, 1968).

novel characteristic of this model is the proposition that a second macromolecular feedback loop may have evolved to maintain and to regulate an adequate level of the proteins essential for trace retrieval, and to protect the chain at its weakest link. Since relatively little RNA need be produced to maintain an adequate level of the essential protein (once consolidation has passed the rapid protein synthesis stage), and since RNA typically has the shortest life of any of the other macromolecules, its loss is more probable than that of the other links in the chain.

For these reasons, actinomycin D, which greatly reduces the rate of DNA-dependent RNA synthesis in brain (Barondes and Jarvik, 1964), and magnesium pemoline (Cylert), which *may* enhance that same synthesis (Glasky and Simon, 1966; Morris *et al.*, 1967) would both be expected to have profound effects on the consolidation phase of memory. But neither drug has been found to influence memory reliably (*e.g.*, Batkin *et al.*, 1966; Beach and Kimble, 1967; Goldsmith, 1967; Smith, 1967). On the other hand, 1,1,3-tricyano-2-amino-1-propene (U9189 or TCAP), which appears to produce an increase in brain RNA *without* facilitating the action of DNA-dependent RNA-polymerase (Barondes, 1964), has been found, rather reliably, to improve the consolidation phase of memory (Essman, 1966; Daniels, 1967). For these reasons, we have proposed a second macromolecular loop which may depend either on an RNA-dependent RNA-polymerase or on the formation of protected, double strands of mRNA (Halstead and Rucker, 1968; Lodish and Zinder, 1966).

The belief that single cells within the synaptic matrix of the trace perform the basic, unitary steps in information processing has been strengthened by two lines of evidence: observations of instrumental conditioning of the rate of firing in single neurons (Olds, 1965); and experiments finding that individual one-celled animals may not only be taught simple tricks, but that they may also retain their training in a manner grossly similar to that of mammals. (A Kamin-like deficit in retention, demonstrated by Gelber, 1965, and other examples of protozoan learning have been ably reviewed by McConnell, 1966.)

BEHAVIORAL MODIFICATION, NÉE TRANSFER OF TRAINING

At the neuronal level, consolidation of memory presumably results in changes in the operating characteristics of synapses in the set of neurons which has recorded the memory. Although it is possible—and even likely—that the same information may have been recorded in more than one set of neurons, the synaptic links in any given set must contribute components of the information held by the entire engram. That is, the information of the entire system is made up of information held by particular synapses operating between particular neurons. The information held by a given synapse is determined by the identity of the neurons connected, by the identity of the synapse, and by the new state of the operating characteristic of that synapse. If the new state of the operating characteristic of each synapse is maintained by a macromolecular feedback loop, then the memory trace will be characterized by the particular macromolecules produced in the set of neurons holding the trace. It then becomes reasonable to ask if different loops are activated by different neurons and also if the loops activated for different synapses on a single neuron are themselves different. In such a frame of reference, the pattern of macromolecular synthesis within any set of neurons maintaining an engram depends on the way in which synapses are chemically identified or named. There are four possible ways to identify a synapse chemically; either by the neurons to which they belong, by the position they occupy on these neurons, by both of these methods, or possibly by neither method. The amount of information chemically extractable from donor brains for transfer experiments will be restricted accordingly.

For this model we can specify conditions for a "perfect" transfer of training experiment: The brains of both donors and recipients, regardless of species, must by effectively "pre-wired" (Gaze, 1967) and the operating characteristics of each synapse set at zero. The macromolecular code for protein production must be the same for both donors and recipients, regardless of species (Marshall *et al.*, 1967). The donors will be divided into two groups, those who do and those who do not receive training. Training will establish a unique change in the matrix of the operating characteristics of donor synapses, and this change will be reflected in a concomitant pattern of protein production. The degree of equivalence between donor training and the specific proteins so produced will depend on one of the four ways by which synapses may be named.

All of the protein should be extracted from the brains of donors, and each molecule, after recipients have been injected, should be equally likely to be taken up and utilized by any recipient neuron. Under these conditions, if enough recipients are tested on the same task to which donors were trained, the a priori odds that a transfer effect will be observed are very high (but not perfect).

If we wish to discover by which of the four ways synapses are actually named, we may do so by counting the number of different kinds of donor training that can be distinguished by the differential performance of recipients tested on the same task to which donors were trained. If the kind of donor or recipient training is found to be irrelevant, then only one type of synapse could be counted and all of the protein specifically produced during consolidation would be equivalent and would have but one molecular structure. To conclude that these proteins are not equivalent, we need find only two kinds of donor training that produce different responses in recipients (*e.g.*, Nissen *et al.*, 1965; Rucker and Halstead, 1968). A number of experiments may be necessary, in that event, to determine which of the remaining three ways of synapse-naming is actually operating.

> 10. The Dirac Method. We observe that wild lions are, *ipso facto*, not observable in the Sahara Desert. Consequently, if there are any lions in the Sahara, they are tame. The capture of a tame lion may be left as an exercise for the reader [Petard, 1938].

SIMULTANEOUS EQUATIONS FOR MULTIPLE UNKNOWNS

In the example of a "perfect" transfer experiment given above, a very simple model of the competence of macromolecules to participate in memory is imbedded in an incredibly naive performance model. It would appear from this line of thinking that a number of factors may operate against the detection of any transfer effects; unfortunately, the model does not tell us which experiments should succeed!

A salient feature of the performance model is that the recipient's response to the extract need not be limited to the extract's original function in the donor. If, for example, the transfer agents are specific only to the type of neuron from which they were derived, and not to the particular synapse, they would be competent to change the operating characteristics of a number of synapses which might either be unrelated to the recipient's performance of the donor's task, or which might in fact support a new and competing response. Not only would this add noise to the experimental test, but it would also imply that no result of a transfer experiment may simply be interpreted in terms of how the original donor has utilized the transfer agents. The recipient's response may be expected to be much more complicated than the training of the donor.

Consequently, even an incredibly naive performance model allows us to be somewhat happier with the three kinds of "behavioral modification" results which do not correspond to a simple and direct "transfer of training": (1) failure to demonstrate any net transfer effect; (2) successful demonstration of a transfer effect in which the recipient's response on some other, perhaps related and opposed task, may be equally affected (nonspecific transfer); and (3) antagonistic interaction between transfer agents. But the analysis is still no closer to specifying the conditions under which these various effects will occur.

> So Eddington argues "we need a super-mathematics in which the operations are as unknown as the quantities they operate on, and a super-mathematician, who does not know what he is doing when he performs these operations. Such a super-mathematics is the Theory of Groups" [Singh, 1959, 143].

WHAT KIND OF UNKNOWN OPERATIONS?

Early warning and efficient search have been so favored by natural selection that higher animals have necessarily evolved extraordinary capacities to filter the resultant avalanche of information (Ashby, 1966). Furthermore, the human race is competent, chromosomally, to store information at instantaneous rates almost matching the tremendous channel capacity of the sensory systems. Yet eidetic memories are scarce. It would appear that the remarkably detailed record held by the eideticer makes serious demands on other aspects of his ability to process information (Klüver, 1932). Perhaps the ability of the race to process information quickly and accurately has been limited over the course of evolution not so much by the channel capacity of sensory systems and the ultimate potential of the race to produce eidetic brains, as by other rate-limiting functions such as the four categories of biological intelligence discussed above.

There is, to our knowledge, no factor-analytic study of the Biological Intelligence of any animal other than man. But the information processing ability of other animals is even more limited than that of man, and these limitations presumably may be attributed to inadequacies in the systems analogous to those of man: systems involving operations performed on input, operations performed during storage, operations performed during retrieval, and operations performed on operations. That retrieval rules, for example, can be manipulated in animal experiments, independent of the content information to be retrieved, is amply illustrated in studies of behavioral fixations (Feldman and Green, 1967), of state-dependent learning (John, 1967), and of at least some of the effects of spreading cortical depression (Schneider, 1967). These phenomena may provide an excellent substrate for the elucidation of the macromolecular basis of information processing, particularly by the method of transfer of training experiments.

It is interesting to speculate as to whether some transfer of training experiments may have already separated content information and retrieval instructions. Jacobson *et al.* (1966), for example, tested the effects of RNA extracts prepared from planaria which had been either trained, trained and extinguished, pseudo-conditioned, or given no special training. The worms receiving extracts from either trained *or* trained-extinguished donors conditioned faster than controls. Was this due to a failure to transfer content information, a failure to transfer retrieval instructions, or a failure to test the recipients in a manner which would invoke the retrieval instructions? Ungar and Irwin (1966), who used rodents, reported that if donors had been trained to turn in the direction opposite to their natural bias, then the behavior of recipients could be described as a simultaneous transfer of both content information and a retrieval instruction: *i.e.*, "turn in a particular direction (left or right)" plus "turn against your natural preference," respectively. When the recipient's bias corresponded to the direction of donor training, there was no net transfer effect.

There have been several reports of negative transfer, defined as behavior of recipients in the direction opposite to the training of their respective donors (Halstead and Rucker, 1968; Nissen *et al.*, 1965; Rosenblatt and Miller, 1966; Ungar and Irwin, 1967). We suggested above that, as one possible mechanism for transfer effects, the macromolecules supporting a change in the operating characteristic of a particular synapse on a particular neuron of the donor may later act in the recipient to potentiate any synapse on a particular neuron. Negative transfer might arise in this case if the transfer agent was more likely to affect those synapses which monitor and regulate neural activity (for example, axo-axonal synapses which produce presynaptic inhibition), than to affect those synapses which produce the activity so monitored.

In earlier studies, Fjerdingstad and his co-workers had shown that recipients rewarded for responding to the same light or dark stimulus as had been used in the training of their respective donors, learned faster than control recipients injected either with buffer or with an extract prepared from untrained donors. They also reported negative transfer in an experiment in which some recipients were trained to the same discrimination as their donors while other recipients were trained to the opposite or reversed discrimination. The curious result was that recipients trained in the direction opposite to that of their respective donors learned even faster than the recipients trained in the same manner as their respective donors (Nissen *et al.*, 1965, Fjerdingstad *et al.*, 1965).

We interpreted these experiments to mean that two agents had developed over the course of donor training. The most powerful agent was common to all the extracts prepared from donors, regardless of which stimulus the donors were trained to respond to, and was responsible for the net positive transfer effect. The second agent, less effective than the first, was specific to the particular training of the donors and was responsible for the measurable negative transfer effect. How could the training of the donors have been responsible for producing two agents with such divergent effects?

An interpretation was suggested by the design of the experiment itself, in that the donors had been thoroughly overtrained on their task, and that some recipients had been trained to the reversed discrimination after injection. When rats are given varying degrees of practice on a visual discrimination and are then trained to the reversed discrimination, it is often reported that subjects with little training on the first task perform the second task much more inefficiently than they would have if they had had no training whatsoever on the first task. But with further practice on the first task, their performance on the reversed discrimination improves markedly, and, with extensive overtraining on the first task, becomes even better than that of subjects given no practice on the first task. This phenomenon, called the overtraining reversal effect, has been extensively reviewed by Mackintosh (1965) and has emerged, despite experimental difficulties, in a robust condition (Siegel, 1967). All attempts to explain the overtraining reversal effect have involved the assumption that during initial training the subject is learning something specific to the particular discrimination task, such as to which stimulus he must respond; while during overtraining, he learns something of a more general nature, such as the relevance of the stimulus dimension to which he must attend (Fig. 3). In the Danish experiment (Nissen

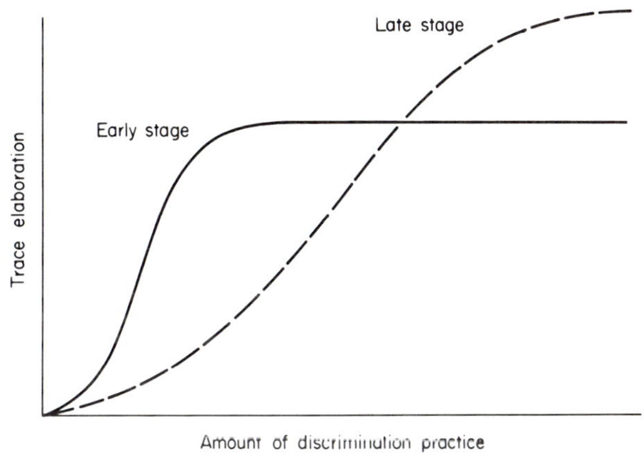

Figure 3 A two-stage model of discrimination learning. Learning in the second stage is assumed to explain the overtraining reversal effect. (Trace elaboration represents both initial encoding and redundancy of the trace.)

et al., 1965), we felt that agents resulting from both kinds of learning had been extracted from the overtrained donors.

As a partial test of this hypothesis, we designed an experiment reported in detail later in this volume (Rucker and Halstead, 1969). Briefly, the donors for this experiment were given different amounts of practice on a visual discrimination task. The recipients were first tested for stimulus preference without reward in order to get an independent estimate of the effect of the early developing factor; we had thought that the late developing factor might not operate if the stimulus dimension was not made relevant through reward. After preference testing, the same recipients were trained, with reward, on the discrimination which was the reverse of the task practiced by the donors. We had anticipated that the number of errors made during such training would simply decrease in proportion to the amount of training given to the various donors, and would give us an estimate of the late developing transfer agent.

Analysis of the results indicated that both the factor developed early in donor training and the factor developed later in donor training were operating in both tests (in addition to a third factor unrelated to discrimination training). Although the effects of the first two factors were additive in both tests, interpretation of the training results was complicated by an interaction which was manifested by the cancellation of both effects in recipients receiving extracts prepared from donors given an intermediate amount of training. Further, the effect of the rapidly developed factor was in the direction of donor training during preference testing (positive transfer), but was in the opposite direction during discrimination training (negative transfer). Whatever information had been transferred by these two agents had in fact been processed by the recipients in an entirely different manner in the two tests, a manner which could not have been predicted simply by knowing how the donor had used the same information.

Since the early- and late-developing transfer agents were found to be differentially sensitive to various chemical treatments, it would appear that we will soon be able to separate different but related transfer agents and eventually to characterize these agents in terms of the many molecular subspecies which must contribute to their net effect. Perhaps the most interesting fractionation of transfer agents will result from the use of different behavioral tests for recipients receiving the same extracts. In this setting, as in no other, the rules of biological intelligence which govern the processing of macromolecular information may become definable.

> What has been found by computer studies in the last ten years thus suggests that one central problem of the "higher" functions in the brain is to discover its methods for processing information. The methods we know today tend to be inefficient to "astronomical" degrees: the living brain may well know better [Ashby, 1966].

REFERENCES

Albert, D. J. The effects of polarizing currents on the consolidation of learning. *Neuropsychologia*, 1966, **4**, 65–77.

Ashby, W. R. Mathematical models and computer analysis of the function of the central nervous system. *Ann. Rev. Physiol.*, 1966, **28**, 89–106.

Barondes, S. H. Studies with an RNA polymerase from brain. *J. Neurochem.*, 1964, **11**, 663–669.

Barondes, S. H. and Jarvik, M. E. The influence of actinomycin-D on brain RNA synthesis and on memory. *J. Neurochem.*, 1964, **11**, 187–195.

Batkin, S., Woodward, W. T., Cole, R. E., and Hall, J. B. RNA and actinomycin-D enhancement of learning in the carp. *Psychon. Sci.*, 1966, **5**, 345–346.

Beach, G. and Kimble, D. P. Activity and responsivity in rats after magnesium pemoline injections. *Science*, 1967, **155**, 698–701.

Daniels, D. The effect of TCAP on acquisition of discrimination learning in the rat. *Psychon. Sci.*, 1967, **7**, 5–6.

Davis, R. E. and Agranoff, B. W. Stages of memory formation in goldfish: evidence for an environmental trigger. *Proc. Nat. Acad. Sci.*, 1966, **55**, 555–559.

Eccles, J. C. An electrical hypothesis of synaptic and neuromuscular transmission. *Ann. N. Y. Acad. Sci.*, 1946, **47**, 429–455.

Essman, W. B. Effect of tricyanoaminopropene on the amnesic effect of electroconvulsive shock. *Psychopharmacologia* (Berl.), 1966, **8**, 426–433.

Feldman, R. S. and Green, K. F. Antecedents to behavior fixations. *Psych. Rev.*, 1967, **74**, 272–289.

Fjerdingstad, E. J., Nissen, T., and Røigaard-Petersen, H. H. Effect of ribonucleic acid (RNA) extracted from the brain of trained animals on learning in rats. *Scand. J. Psychol.*, 1965, **6**, 1–6.

Flexner, L. B., Flexner, J. B., and Roberts, R. B. Memory in mice analyzed with antibiotics. *Science*, 1967, **155**, 1377–1383.

Gaze, R. M. Growth and differentiation. *Ann. Rev. Physiol.*, 1967, **29**, 59–86.

Gelber, B. Studies of the behavior of *Paramecium aurelia. Animal Behavior*, Supp. 1, 1965, **13**, 21–29.

Gerard, R. W. Some aspects of neural growth, regeneration, and function. In P. Weiss (Ed.), *Genetic Neurology: Conference of the International Union of Biological Science.* Chicago: University of Chicago Press, 1950. Pp. 199–207.

Glasky, A. J. and Simon, L. N. Magnesium pemoline: enhancement of brain RNA polymerases. *Science*, 1966, **151**, 702–703.

Goldsmith, L. J. Effect of intracerebral actinomycin D and of electroconvulsive shock on passive avoidance. *J. comp. physiol. Psych.*, 1967, **63**, 126–132.

Halstead, W. C. Preliminary analysis of grouping behavior in patients with cerebral injury by the method of equivalent and non-equivalent stimuli. *Am. J. Psychiat.*, 1940, **96**, 1263–1294.

Halstead, W. C. *Brain and Intelligence.* Chicago: The University of Chicago Press, 1947.

Halstead, W. C. Brain and intelligence. In L. A. Jeffress (Ed.), *Cerebral Mechanisms in Behavior: The Hixon Symposium.* New York: Wiley, 1951. Pp. 244–272.

Halstead, W. C. Thinking, imagery, and memory. In Field (Editor-in-chief), *Handbook of Physiology, Section I: Neurophysiology*, Vol. III. Washington, D.C.: Am. Physiol. Soc., 1960.

Halstead, W. C. and Rucker, W. B. Memory: a molecular maze. *Psych. Today*, 1968, **2** (No. 1), 38–41 and 66–67.

Halstead, W. C., Rucker, W. B., and McMahon, J. P. Memory. *Ann. Rev. Med.*, 1967, **18**, 1 14.

Hebb, D. O. *The Organization of Behavior.* New York: Wiley, 1949.

Jacobson, A. L., Fried, C., and Horowitz, S. D. Planarians and memory: I. Transfer of learning by injection of ribonucleic acid; II. The influence of prior extinction on the RNA transfer effect. *Nature*, 1966, **209**, 599–601.

John, E. R. *Mechanisms of Memory.* New York: Academic Press, 1967. Pp. 67–91.

Kamin, L. J. Retention of the incompletely learned avoidance response: some further analysis. *J. comp. physiol. Psych.*, 1963, **56**, 713–718.

Katz, J. J. and Halstead, W. C. Protein organization and mental function. *Comp. Psychol. Monogr.*, 1950, **20** (serial no. 103), 1–38.

Klüver, H. Eidetic phenomena. *Psych. Bull.*, 1932, **29**, 181–203.

Lashley, K. S. In search of the engram. *Symp. Soc. exp. Biol.*, 1950, **4**, 454–482.

Lodish, H. F. and Zinder, N. D. Replication of the RNA of bacteriophage f2. *Science*, 1966, **152**, 372–378.

Lorente de Nó, R. Analysis of the activity of the chains of internuncial neurons. *J. Neurophysiol.*, 1938, **1**, 207–244.

McConnell, J. V. Comparative physiology: learning in invertebrates. *Ann. Rev. Physiol.* 1966, **28**, 107–136.

McCulloch, W. S. Modes of functional organization of the cerebral cortex. *Fed. Proc., Amer. Soc. exp. Biol.*, 1947, **6**, 448–452.

Mackintosh, N. J. Selective attention in animal discrimination learning. *Psych. Bull.*, 1965, **64**, 124–150.

Marshall, R. E., Caskey, C. T., and Nirenberg, M. Fine structure of RNA codewords recognized by bacterial, amphibian, and mammalian transfer RNA. *Science*, 1967, **155**, 820–826.

Morris, N. R., Aghajanian, G. K., and Bloom, F. E. Magnesium pemoline: failure to affect *in vivo* synthesis of brain RNA. *Science*, 1967, **155**, 1125–1126.

Mosél, J. Phrase structure and short-term memory. In three parts to be submitted to *J. verb. Learning*, 1970.

Neuman, J. von and Morgenstern. O. *Theory of Games and Economic Behavior*. Princeton: Princeton University Press, 1947.

Nissen, T., Røigaard–Petersen, H. H., and Fjerdingstad, E. J. Effect of ribonucleic acid (RNA) extracted from the brain of trained animals on learning in rats. II. Dependence of RNA effect on training conditions prior to RNA extraction. *Scand. J. Psych.*, 1965, **6**, 265–272.

Olds, J. Operant conditioning of single unit responses. *Excerpta Medica International Congress Series*, 1965, **87**, 372–380.

Pauling, L. Theory of the structure and process of formation of antibodies. *J. Am. Chem. Soc.*, 1940, **62**, 2643–2657.

Penfield, W. Memory mechanisms. *Arch. Neurol. Psychiat.* (Chicago), 1952, **67**, 178–191.

Petard, H. A contribution to the mathematical theory of big game hunting. *Am. Math. Monthly*, 1938, **45**, 446–447.

Pitts, W. and McCulloch, W. S. How we know universals: the perception of auditory and visual forms. *Bull. Math. Biophys.*, 1947, **9**, 127–147.

Rosenblatt, F. and Miller, R. G. Behavioral assay procedures for transfer of learned behavior by brain extracts (in two parts). *Proc. Nat. Acad. Sci.*, 1966, **56**, 1423–1430 and 1683–1688.

Rucker, W. B. and Halstead, W. C. Memory: antagonistic transfer effects. In W. Byrne (Ed.), *Molecular Approaches to Learning and Memory*. New York: Academic Press, 1969.

Schneider, A. M. Control of memory by spreading cortical depression: a case for stimulus control. *Psych. Rev.*, 1967, **74**, 201–215.

Shannon, C. E. and Weaver, W. *The Mathematical Theory of Communication*. Urbana: University of Illinois Press, 1949.

Siegel, S. Overtraining and transfer processes. *J. Comp. physiol. Psych.*, 1967, **64**, 471–477.

Singh, J. *Great Ideas of Modern Mathematics: Their Nature and Use*. New York: Dover Publications, Inc., 1959.

Smith, R. G. Magnesium pemoline: lack of facilitation in human learning, memory, and performance tests. *Science*, 1967, **155**, 603–605.

Ungar, G. and Irwin, L. N. Transfer of acquired information by brain extracts. *Nature*, 1967, **214**, 453–455.

Wiener, N. *Cybernetics*. New York: Wiley, 1948.

Yngve, V. H. A model and an hypothesis for language structure. *Proc. of the Am. Phyl. Soc.*, 1960, **104**, 444–446.

Zinkin, S. and Miller, A. J. Recovery of memory after amnesia induced by electroconvulsive shock. *Science*, 1967, **155**, 102–103.

THE ROLE OF CONSOLIDATION IN MEMORY*

Murray E. Jarvik

Department of Pharmacology
Albert Einstein College of Medicine
Bronx, New York

A human lifetime consists of a sequence of experiences and an accumulation of memories. Scientists have a pretty good idea how most bodily functions are brought about, but the physiological basis of memory is still shrouded in mystery. We know that the learning process comprises three stages: *registration* during which environmental stimuli reach the brain via the senses, *retention* during which these messages are somehow transformed and stored, and *retrieval* during which something elicits the information which has been stored. Simple creatures like flies and frogs may go through a lifetime of reflex action without learning anything. When the interval between a stimulus and its response is lengthened from seconds to minutes, days or years, then something generally thought of as memory must intervene to keep the memory of the stimulus alive. For example, a child bitten by a dog may cry when it sees the same animal weeks later. Presumably the unpleasant experience changed the central nervous system so that subsequent reactions were changed, perhaps forever. What we would like to determine is the course of events in the brain, during and following the learning experience, that result in a memory trace.

Because the brain is so hard to reach, especially in an intact functioning animal, we must use indirect approaches to learn about the memory mechanism. One approach involves studying the conditions necessary for the formation of a memory trace. Another, and one which we have used a good deal, involves studying the destruction of a memory trace. Both clinical and

*Supported by MH 01225 and MH 05319 grants.

experimental results place certain restrictions on the models we build of mechanisms underlying memory.

The phenomenon of retrograde amnesia following injury to the head has been noted for at least a hundred years and probably longer. Ribot, in his "Disorders of Memory" (1885), describes several cases of loss of memory for events immediately preceding a head injury. Subsequently, a few clinical observations have been made on the types of retrograde amnesia encountered with head injury. Russell and Nathan (1946) indicated that most cases of concussion are accompanied by some degree of retrograde amnesia.

There is some question whether retrograde amnesia can occur as a natural consequence of interference or forgetting. Most of the evidence to date indicates that retroactive interference is not time-dependent; that is, it does not selectively impair the most recent memories.

The modern view of many verbal learning theorists that memory traces are as firm at the moment of laying down as they ever will be, is in contrast to some of the older views. For example, Mueller and Pilzecker (1900) proposed a consolidation perseveration theory which assumed that after a stimulus was perceived or registered it produced some type of continuing activity which strengthened the association and prevented interference. Thus, increasing resistance to interference with the passage of time, was analagous to increasing resistance to disruption by amnesic treatments which are the focus of my own interest. Such modern learning theorists as McGeoch and Irion (1952), Melton (1963), and Postman (1964) have all indicated that there is no evidence that the temporal point of interpolation makes any difference in retroactive interference in human verbal learning. There are many animal experiments, on the other hand, which indicate that physical or chemical treatments will produce effects which are inversely proportional to the time elapsing between training and treatment. Thus, there appears a serious discrepancy between the results of human verbal interference experiments, and experiments or observations with either humans or animals in which certain types of physical disruptive treatments are applied. It must be kept in mind that not only are the subjects different, but the types of material learned and the types of interfering treatment differ as well.

The distinction should be examined between three phenomena which are very similar in many respects and yet appear to be formally different. These phenomena are amnesia, forgetting, and extinction. Amnesia literally means without memory, but retrograde amnesia refers to the loss of memory proceeding backwards in time. This is the type of memory loss which interests us most because the temporal gradient must reflect some changing process within the organism. Retrograde amnesia is generally produced by some physical or chemical stimulus and must be contrasted with retroactive interference which is produced by material resembling that which is learned and which enters the organism through a sensory apparatus. Forgetting is a decrease in the probability of retrieval, produced either by the simple passage of time or the impact of sensory experiences during the passage of time. Most verbal learning theorists today claim that forgetting is due to either retroactive or proactive interference, and more often the latter. Extinction, on the other hand, refers to the loss of the response, which has previously been associated with a reinforcement, through lack of this reinforcement. Extinction requires that a conditioned stimulus be repeated many times, unaccompanied by its formerly associated unconditioned stimulus. Unlike forgetting, which may involve unobservable extinction trials, extinction involves readily observable trials. This is why forgetting is said to occur more slowly during sleep (Jenkins and Dallenbach, 1924), during inactivity (Minami and Dallenbach, 1946), or with anesthesia (Summerfield and Steinberg, 1957). Some experiments (*e.g.*, Andjus *et al.*, 1956) indicated that memory may remain quite intact even when cerebral activity is reduced to a very low level by cooling; these experiments, however, utilized material which was already well consolidated by the time the treatment was applied.

Electroconvulsive Shock (ECS)

The treatment which has provided more information than any other concerning retrograde amnesia is without question ECS. In the early 1930's chemical convulsions, insulin, camphor, and Metrazol were introduced for the treatment of psychoses. A few years later Bini and Cerletti (1937) introduced electroconvulsive therapy. Reports of amnesia following convulsive therapies were quite common, but Zubin and Barrera (1941) were the first investigators to attempt to verify experimentally the clinical reports of retrograde amnesia from physicians using this treatment. They found that recently learned material was more adversely affected by ECS than material learned long before the shock. Although some studies have subsequently failed to demonstrate retrograde amnesia in humans following ECS (Worchel and Narciso, 1950), others have succeeded in demonstrating a memory impairing effect (Mayer–Gross, 1943; Williams, 1950; Cronholm and Molander, 1964).

Shortly after, and as a direct result of Zubin's experiment with humans, Stone, (1946) initiated a series of investigations on the effects of ECS upon learning and behavior in animals, primarily in studies of proactive effects of ECS. The first investigator to demonstrate a temporal gradient interpretable as *retrograde* amnesia was C. P. Duncan (1949). Duncan demonstrated that if rats were trained in a shuttle box to avoid punishing shock, learning could be impaired by following each single daily trial with ECS. Furthermore, impairment was greater if the convulsion followed closely after the learning trial. It was later pointed out by Coons and Miller (1960) that punishing effects of ECS could be expected to show the same kind of interfering gradient as amnesia in an active avoidance situation.

Subsequently, Brady (1952) demonstrated a prolonged temporal gradient of effectiveness of multiple ECS treatments upon the conditioned emotional response. Now, however, the time interval was measured in days, and the closer the treatments were to the learning period, the bigger were the effects; however, spontaneous recovery of the habit occurred after the passage of several weeks. A series of experiments by Thompson and collaborators (1955, 1958) showed that ECS could produce retrograde amnesia in a variety of testing situations.

On the other hand, Lewis and Maher (1965) and their students (Misanin and Smith, 1964) found that ECS-induced amnesia is closely associated with the place where the ECS is given. They felt, therefore, that ECS has the properties of a conditioned or unconditioned stimulus. In our laboratory, however, the objections that ECS might be producing its effects exclusively through punishment was overcome with the use of a passive avoidance procedure (Pearlman *et al.*, 1961; Kopp *et al.*, 1966). With such a procedure, punishment would tend to improve performance and not to impair it, whereas ECS was shown conclusively to impair performance.

There is still considerable disagreement concerning the ability of ECS to produce retrograde amnesia in animals. Chorover and Schiller (1965) for example, felt that true amnesia is produced by ECS only up to 10–20 seconds following a trial. On the other hand, Kopp *et al.* (1966) have shown that ECS can produce gradients of amnesia extending over several hours. Since the latter studies used mice and the former used rats, it is possible that species differences could account for the discrepancies. There may also have been subtle differences in procedure both in training and in the application of electricity to the brain.

There have been very few attempts to localize the amnesic effects of electrical stimulation in the brain. Mahut (1964) found that electrical stimulation of the nonspecific thalamic nuclei during the reward period impaired learning of the Hebb Williams maze. Glickman (1958) found that stimulation of the midbrain reticular formation immediately following a one-trial learning procedure impaired subsequent performance of animals. Unfortunately, neither investigator carried out parametric time studies. Thompson (1958) found that the stimulation of the caudate nucleus interfered with performance of cats in an alternation task. Similar stimulation of the midbrain tegmentum did not have a retroactive effect. Cianci (1965) found that stimulation of limbic and lower extrapyramidal areas in the brain seemed to produce retrograde amnestic effects upon delayed response in monkeys.

Electroconvulsive shock produces multiple effects and these may sometimes work in opposite directions. For example, the disinhibition may actually facilitate performance in an active avoidance situation (Vanderwolf, 1963), whereas the amnesia will impair performance in this same type of situation. Proactive effects of ECS can be seen both as impairment of registration and retrieval in a passive avoidance situation (Kopp *et al.*, 1967, 1968).

Anoxia

Hayes (1953) demonstrated that asphyxia could produce retrograde amnesia in rats. It is not certain whether the effect was due to oxygen lack or carbon dioxide excess. Thompson and Pryer (1956) showed that anoxia could impair the retention of a discrimination task. Taber and Banuazizi (1966), however, found that anoxia was without effect in a passive avoidance test with mice.

Temperature

Cerf and Otis (1957) found that heat narcosis produced retrograde amnesia in goldfish. More recently, Agranoff (1965) found that cooling goldfish could prolong the time during which ECS was effective. Mrosovsky (1963) and Andjus *et al.* (1956) failed to find an amnesic effect of cooling on a well-learned task. Gerard (1955) reported that cooling hamsters will prolong the time during which ECS is effective in producing retrograde amnesia, a finding similar to and antedating that of Agranoff (1965).

Anesthesia

Some investigators have reported that anesthesia with barbiturates or diethyl ether will produce retrograde amnesia if it follows a learning trial (Leukel, 1957; Pearlman *et al.*, 1961), whereas others failed to find deficits in retention from postlearning with barbiturate or ether anesthesia (Russell and Hunter, 1937; Ransmeier and Gerard, 1954; Weissman, 1967). Herz *et al.* (1966) and Alpern and Kimble (1967) have indicated that depth of anesthesia may play an important role in ability to produce amnesia. Anesthesia induced amnesia has also been reported in birds (Gutekunst and Youniss, 1963; Cherkin and Lee–Teng, 1965).

Carbon Dioxide

Taber and Banauzizi (1966), Paolino *et al.* (1966), and Quinton (1966) have all found that carbon dioxide administered after learning in a passive avoidance test will produce retrograde amnesia. Animals do not appear to have a convulsion during the period of unconsciousness induced by this agent, but the mechanism is unexplained.

Spreading Depression

Bureš and Burešová (1963) have demonstrated that spreading cortical depression can produce retrograde amnesia for a passive avoidance test. Similar findings were also reported by Pearlman (1966). Subconvulsive electric shock through the brain appears to yield a state resembling that induced by spreading depression and produces amnesia (Jarvik and Kopp, 1967). Similarly, McGaugh and Alpern (1966) have indicated that passage of currents through the brain without convulsions can cause amnesia, and possibly a spreading depression may be induced in this state.

Brain Puncture

Bohdanecka *et al.* (1967) have shown that needle puncture of the brain following a passive avoidance learning trial can produce a gradient of retrograde amnesia. Again, the state of the animal suggests that spreading cortical depression may be induced by the puncture. More recently, Dorfman *et al.* (in prep.) have indicated that the degree of amnesia induced by brain puncture may be a function of either the locus or the size of the lesion. Lesions penetrating the hippocampus were more effective than those merely entering the cortex. Cerebellar punctures were without any amnesic effect, showing that mere trauma or pain was not responsible for the loss of memory.

Antibiotics

Katz and Halstead (1950) originally suggested that macromolecular structures might be responsible for carrying information and for storing memory. Subsequently, Hyden and collaborators indicated that changes in the base composition of RNA could be induced by learning experiences (Hydén and Egyházi, 1962). During the past decade, there has been a good deal of speculation and controversy concerning the role of RNA in learning. Barondes and Jarvik (1964) and Cohen and Barondes (1966) found that actinomycin D injections into the brains of mice in doses sufficient to markedly inhibit RNA synthesis did not interfere with learning in these animals. More recently,

Agranoff *et al.* (1967) have indicated that actinomycin D could, in fact, impair retention in goldfish. It appears that the discrepancy may be due to differences in the time when retention was measured.

Flexner and his colleagues (1963) indicated that intracerebral injection of puromycin could produce a gradient of retrograde amnesia extending over several days. Agranoff *et al.* (1965) found that puromycin administered intracerebrally to goldfish could also produce a retrograde amnesia resembling that induced by ECS. Barondes and Cohen (1966) found that puromycin could impair learning and produce rapid forgetting in mice. Neither Barondes and Cohen (1966) nor Flexner *et al.* (1963) could produce retrograde amnesia with ECS in the spatial T-maze they employed.

Cholinergic Systems and Amnesia

Since the turn of the century, anticholinergic drugs, such as scopolamine, have been credited with producing amnesia (Steinbuchel, 1902), but the evidence is not very convincing that the effect is on retention rather than on registration or retrieval. Almost all the work to date indicates that although pretrial injections of anticholinergic drugs are very effective in preventing learning, posttrial injections have relatively little effect (Bohdanecky and Jarvik, 1967). However, Deutsch and Hamburg (1966) have indicated that posttrial injections of scopolamine produce an amnesic effect which is greatest if the injection occurs one to three days after learning, and the animals are retested 24 hours later. The effect is less if the injection is made very shortly after learning, or delayed for seven days. The temporal curve for amnesia produced by cholinergic blocking agents appears to be a mirror image of that which is obtained with injections of anticholinesterases, such as DFP and physostigmine. As a matter of fact, under some circumstances the anticholinesterases appear to produce facilitation.

Drug Facilitation of Learning

McGaugh (1966) and colleagues have demonstrated that certain analeptic and stimulant drugs administered either before or after the learning trial can facilitate learning. Strychnine, pentylenetetrazol, and picrotoxin, as well as physostigmine have been shown by these authors to facilitate learning. Furthermore, the action of these drugs seems to be greatest if they are administered soon after the learning trial. Cook *et al.* (1963) have reported that yeast RNA extracts will produce facilitation of learning of an active avoidance task. Bovet and colleagues (1963) have indicated that nicotine can produce facilitation of learning under certain circumstances.

SUMMARY

The existence of consolidation seems fairly well established although the generality of the phenomenon is not. On the other hand, there is strong evidence that many types of memories successfully survive amnesic treatments or are not facilitated. One would have to conclude that these memories are consolidated rapidly, perhaps instantly, whereas the susceptible memories are consolidated slowly. The nature of the processes involved in consolidation remains a mystery. Inhibition of protein synthesis and possible of RNA synthesis will impair memory. Consolidation may depend upon changes in electrical activity, in receptor sensitivity, or in macromolecular synthesis, or may be a combination of all of these. Hopefully, the emergence of drugs with specific actions will throw more light on these questions.

REFERENCES

Agranoff, B. W. Molecules and Memories. *Perspectives in Biol. and Med.*, **9**, 13–22, 1965.

Agranoff, B. W., Davis, R. E., and Brink, J. J. Memory fixation in the goldfish. *Proc. Nat. Acad. Sci.*, **54**, 788–793, 1965.

Agranoff, B. W., Davis, R. E., Casola, L., and Lim, R. Actinomycin D blocks formation of memory of shock-avoidance in goldfish. *Science*, **158**, 1600–1601, 1967.

Alpern, H. and Kimble, D. P. Retrograde amnesic effects of diethyl ether and bis (trifluoroethyl) ether. *J. comp. physiol. Psychol.*, **63**, 168–171, 1967.

Andjus, R. K., Knopfelmacher, F., Russell, R. W., and Smith, A. U. Some effects of severe hypothermia on learning retention. *Quart. J. exp. Psychol.*, **8**, 15–23, 1956.

Barondes, S. H. and Cohen, H. D. Puromycin effect on successive phases of memory storage. *Science*, **151**, 594–595, 1966.

Barondes, S. H. and Jarvik, M. E. The influence of actinomycin D on brain RNA synthesis and on memory. *J. Neurochem.*, **11**, 187–195, 1964.

Bohdanecka, M., Bohdanecky, Z., and Jarvik, M. E. Amnesic effects of small bilateral brain puncture in the mouse. *Science*, **157**, 334–336, 1967.

Bohdanecky, Z. and Jarvik, M. E. Impairment of one trial passive avoidance learning in mice in scopolamine, scopolamine methylbromide, and physostigmine. *Int. J. Neuropharmac.*, **6**, 217–222, 1967.

Bovet, D., Bignami, G., and Robustelli, F. Psychopharmacologie. Action de la nicotine sur le conditionement a la reaction d'evitement chez le rat. *C. R. Acad. Sci.*, **256**, 778–780, 1963.

Brady, J. V. The effect of electroconvulsive shock on a conditioned emotional response: the significance of the interval between the emotional conditioning and the electroconvulsive shock. *J. comp. physiol. Psychol.*, **45**, 9–13, 1952.

Bureš, J. and Burešová, O. Cortical spreading depression as a memory disturbing factor. *J. comp. physiol. Psychol.*, **56**, 268–272, 1963.

Cerf, J. A. and Otis, L. S. Heat narcosis and its effect on retention of a learned behavior in the goldfish. *Fed. Proc.*, **16**, 20–21, 1957.

Cerletti, U. and Bini, L. L'Elettroshock. *Archivio Generale di Neurologia, Psichiatria, e Psicoanalisi*, **19**, 266–268, 1938.

Cherkin, A. and Lee-Teng, Evelyn. Interruption by halothane of memory consolidation in chicks. *Fed. Proc.*, **24**, 328, 1965.

Chorover, S. L. and Schiller, P. H. Short-term retrograde amnesia in rats. *J. comp. physiol. Psychol.*, **59**, 73, 1965.

Cianci, S. N. Effects of cortical and subcortical stimulation on delayed response in monkeys. *Exp. Neurol.*, **11**, 104–114, 1965.

Cohen, H. D. and Barondes, S. H. Further studies of learning and memory after intracerebral actinomycin D. *J. Neurochem.*, **13**, 207–211, 1966.

Cook, L., Davidson, A. B., Davis, D. J., Green, H., and Fellows, E. J. Ribonucleic acid: effect on conditioned behavior in rats. *Science*, **141**, 268–269, 1963.

Coons, E. E. and Miller, N. E. Conflict versus consolidation of memory traces to explain "retrograde amnesia" produced by ECS. *J. comp. physiol. Psychol.*, **53**, 524–531, 1960.

Cronholm, B. and Molander, L. Memory disturbances after electroconvulsive therapy. *Acta. psychiat. scand.*, **40**, 212–216, 1964.

Deutsch, J. A. and Hamburg, M. D. Anticholinesterase-induced amnesia and its temporal aspects. *Science*, **151**, 221–223, 1966.

Dorfman, L., Bohdanecka, M., and Jarvik, M. E. Effects of intracerebral injections on memory in mice. (In prep.)

Duncan, C. P. The retroactive effect of electroshock on learning. *J. comp. physiol. Psychol.*, **42**, 32–44, 1949.

Flexner, J. B., Flexner, L. B., and Stellar, E. Memory in mice as affected by intracerebral puromycin. *Science*, **141**, 57–59, 1963.

Gerard, R. W. Biological roots of psychiatry. *Science*, **122**, 225–230, 1955.

Glickman, S. E. Deficits in avoidance learning produced by stimulation of the ascending reticular formation. *Canad. J. Psychol.*, **12**, 97–102, 1958.

Gutekunst, R. and Youniss, J. Interruption of imprinting following anesthesia. *Percept. mot. Skills*, **16**, 348, 1963.

Hayes, K. J. Anoxic and convulsive amnesia in rats. *J. comp. physiol. Psychol.*, **46**, 216–217, 1953.

Herz, M. J., Peeke, H. V. S., and Wyers, E. J. Amnesic effects of ether and electroconvulsive shock in mice. *Psychon. Sci.*, **4**, 375–376, 1966.

Hydén, H. and Egyházi, E. Nuclear RNA changes of nerve cells during a learning experiment in rats. *Proc. Nat. Acad. Sci.*, **48**, 1366–1373, 1962.

Jarvik, M. E. and Kopp, R. Transcorneal electroconvulsive shock and retrograde amnesia in mice. *J. comp. physiol. Psychol.*, **64**, 431–433, 1967.

Jenkins, J. G. and Dallenbach, K. M. Obliviscence during sleep and waking. *Am. J. Psychol.*, **35**, 605–612, 1924.

Katz, J. J. and Halstead, W. C. Protein organization and mental function. *Comp. Psychol. Monogr.*, **20**, 1–38, 1950.

Kopp, R., Bohdanecky, Z., and Jarvik, M. E. A long temporal gradient of retrograde amnesia for a well discriminated stimulus. *Science*, **153**, 1547–1549, 1966.

Kopp, R., Bohdanecky, Z., and Jarvik, M. E. Proactive effect of a single ECS on step-through performance of naive and punished mice. *J. comp. physiol. Psychol.*, **64**, 22–25, 1967.

Kopp, R., Bohdanecky, Z., and Jarvik, M. E. Proactive effect of a single electroconvulsive shock (ECS) on one-trial learning in mice. *J. comp. physiol. Psychol.*, 1968, **65**, 514–517.

Leukel, F. P. A comparison of the effects of ECS and anesthesia on acquisition of the maze habit. *J. comp. physiol. Psychol.*, **50**, 300–306, 1957.

Lewis, D. J. and Maher, B. A. Neural consolidation and electroconvulsive shock. *Psychol. Rev.*, **72**, 225–239, 1965.

Mahut. H. Effects of subcortical electrical stimulation on discrimination learning in cats. *J. comp. physiol. Psychol.*, **58**, 390–395, 1964.

Mayer–Gross, W. Retrograde amnesia, some experiments. *Lancet*, **245**, 603–605, 1943.
McGaugh, J. L. Time dependent processes in memory storage. *Science*, **153**, 1351–1358, 1966.
McGaugh, J. L. and Alpern, H. P. Effects of electroshock on memory: amnesia without convulsions. *Science*, **152**, 665–666, 1966.
McGeoch, J. A. and Irion, A. L. *The Psychology of Human Learning*. New York: Van Rees Press, 1952.
Melton, A. W. Implications of short-term memory for a general theory of memory. *J. Verb. Learn. Verb. Behav.*, **2**, 1–21, 1963.
Minami, H. and Dallenbach, K. M. The effect of activity upon learning and retention in the cockroach. *Am. J. Psychol.*, **59**, 1–58, 1946.
Misanin, J. and Smith, N. F. Role of response-lined fear in the effects of a single ECS on an avoidance response. *J. comp. physiol. Psychol.*, **58**, 212–216, 1964.
Mrosovsky, N. Retention and reversal of conditioned avoidance following severe hypothermia. *J. comp. physiol. Psychol.*, **56**, 811–813, 1963.
Mueller, G. E. and Pilzecker, A. Experimentalle Beitraege zur Lehre vom Gedaechtnis. *Z. Psychol.*, **1**, 1–288, 1900.
Paolino, R. M., Quartermain, D., and Miller, N. E. Different temporal gradients of retrograde amnesia produced by carbon dioxide anesthesia and electroconvulsive shock. *J. comp. physiol. Psychol.*, **62**, 270–274, 1966.
Pearlman, C. A. Similar retrograde amnesic effects of ether and spreading cortical depression. *J. comp. physiol. Psychol.*, **61**, 306–308, 1966.
Pearlman, C. A., Sharpless, S. K., and Jarvik, M. E. Retrograde amnesia produced by anesthetic and convulsant agents. *J. comp. physiol. Psychol.*, **54**, 109–112, 1961.
Postman, L. Short-term memory and incidental learning. In A. W. Melton (Ed.) *Categories of Human Learning*. New York: Academic Press. Pp. 145–201, 1964.
Quinton, E. E. Retrograde amnesia induced by carbon dioxide inhalation. *Psychon. Sci.*, **5**, 417–418, 1966.
Ransmeier, R. E. and Gerard, R. W. Effects of temperature, convulsion and metabolic factor on rodent memory and EEG. *Am. J. Physiol.*, **179**, 663–664, 1954.
Ribot, T. *Disorders of Memory*. London: Kegan, Paul, Trench and Co., 1885.
Russell, R. W. and Hunter, W. S. The effects of inactivity produced by sodium amytal on the retention of the maze habit in the albino rat. *J. exp. Psychol.*, **20**, 426–436, 1937.
Russell, W. R. and Nathan, P. W. Traumatic amnesia. *Brain*, **69**, 280–300, 1946.
Steinbuchel, R. von. Vorlaüfige Mittheilung uber die Anwending von Skopolamin-Morphium Injectionen in der Geburtshilfe. *Zentralblatt Gynäk*, **26**, 1304–1306, 1902.
Stone, C. P. Effects of electroconvulsive shock on daily activity of albino rats in revolving drums. *Proc. soc. exp. Biol. Med.*, **61**, 150–151, 1946.
Summerfield, A. and Steinberg, H. Reducing interference in forgetting. *Quart. J. exp. Psychol.*, **9**, 146–154, 1957.
Taber, R. I. and Banuazizi, A. CO_2-induced retrograde amnesia in a one-trial learning situation. *Psychopharmacologia*, **9**, 382–391, 1966.
Thompson, R. The effect of intracranial stimulation on memory in cats. *J. comp. physiol. Psychol.*, **51**, 421–426, 1958.
Thompson, R. and Dean, W. A further study on the retroactive effect of ECS. *J. comp. physiol. Psychol.*, **48**, 488–491, 1955.
Thompson, R., Haravey, F., Pennington, D. F., Smith, J. Jr., Ganon, D. and Stockwell, D. Analysis of the differential effects of ECS on memory in young and adult rats. *Canad. J. Psychol.*, **12**, 83–96, 1958.
Thompson, R. and Pryer, R. S. The effect of anoxia on the retention of a discrimination habit. *J. comp. physiol. Psychol.*, **49**, 297–300, 1956.

Vanderwolf, C. H. Improved shuttle-box performance following electroconvulsive shock. *J. comp. physiol. Psychol.*, **56**, 983–986, 1963.

Weissman, A. Drugs and retrograde amnesia. *Int. Rev. Neurobiol.*, **10**, 167–198, 1967.

Williams, M. Memory studies in electric convulsion therapy. *J. Neurol. Neurosurg. Psychiat.*, **13**, 30–35, 1950.

Worchel, P. and Narciso, J. C. Electroshock convulsions and memory: interval between learning and shock. *J. Abnorm. Soc. Psychol.*, **45**, 85–98, 1950.

Zubin, J. and Barrera, S. E. Effect of electric convulsive therapy on memory. *Proc. Soc. exp. Biol.*, **48**, 596–597, 1941.

QUESTIONS AND ANSWERS

QUESTION: (Dr. A. S. Kulkarni, Dow Human Health Research and Development Labs) From the findings presented here and earlier in the literature, is it now generally accepted that short-term memory can be present even though long-term consolidation may have been disrupted? And that consolidation does not necessarily start at the end of short-term process?

ANSWER: I favor the view that consolidation is really part of the formation of the long term process and that the short term process simply decays with the passage of time. This is based mainly on the findings of Dr. Anne Geller that the amnesic effects of ECS appear to require an incubation time and do not become evident for several hours after treatment. A similar phenomenon has been described by Dr. Samuel Barondes with antibiotics.

QUESTION: (Dr. A. S. Kulkarni, Dow Human Health Research and Development Labs) It seems that most of the evidence in favor of the consolidation process comes from the reports in which procedures to disrupt the memory were used. Haven't there been any attempts to seek evidence for the process of consolidation where enhancement of the memory rather than its disruption is employed?

ANSWER: Dr. James McGaugh and many of his followers have produced evidence which indicates that administration of certain drugs following learning will facilitate memory, with a greater effect occurring if the drugs are given sooner after learning. These results are quite compatible with a consolidation explanation.

QUESTION: (Lawrence Kunstadt, New York University) Please comment on the problem of recall-performance as a measure of the recall of learning: nonperformance is not necessarily a measure of forgetting.

ANSWER: While it is true that poor performance does not necessarily indicate loss of information and that, indeed, reminiscence studies demonstrate just such preservation, performance is unfortunately the best measure we have. If, in spite of all our past efforts it becomes impossible to re-elicit a learned performance, then I don't see how we can assume that it still exists in some latent form. However, the possibility naggingly remains. Of course, some day we will be able to examine the anatomical or biochemical substrate of the memory trace and then we will know for sure.

QUESTION: (Dr. Roger M. Morrell, The Upjohn Company) Experience in the clinical management of patients who have sustained concussions of apparently equal severity indicates that the state of retrograde amnesia is not fixed, but that "memory" may be variably regained. In some cases recovery seems absolute. In others, exhaustive testing reveals subtle defects which are always retained. Can you put these observations in the perspective of your discussion of consolidation, forgetting, and extinction?

ANSWER: I don't see why it is necessary to assume that retrograde amnesia is an all-or-nothing affair or, for that matter, that the retention it produces is necessarily irreversible. It is quite conceivable that the memory traces might be deformed rather than destroyed. Something like this must occur in the cases of reminiscence following concussion and/or traumatic experiences. With present testing methods we can never be sure that memory loss is permanent.

SOME CRITICAL VARIABLES IN STUDIES OF THE EFFECT OF INHIBITORS OF PROTEIN SYNTHESIS ON MEMORY*

Samuel H. Barondes

Departments of Psychiatry and Molecular Biology
Albert Einstein College of Medicine
Bronx, New York

Memories are probably stored by a process which results in the facilitation of synaptic connections. This might be achieved by protein synthesis in the neurons involved (Barondes, 1965). If this were so, inhibition of cerebral protein synthesis should impair learning or memory. Studies with inhibitors suggest that cerebral protein synthesis is necessary for "long-term" memory storage, but not for learning or "short-term" memory storage. Although such experiments are conceptually simple, proper interpretation of the results requires careful attention to experimental details (Barondes, 1966). The major purpose of this paper is to call attention to some critical variables which must be controlled or considered in such work. Studies from my laboratory will be described to illustrate this.

STUDIES WITH PUROMYCIN: THE IMPORTANCE OF "SIDE-EFFECTS"

Mice were injected in both temporal regions of the brain with puromycin five hours before training so that maximal inhibition of cerebral protein synthesis was established at the time of training. They were trained to escape shock by choosing the left limb of a one-choice maze to a criterion of nine

*The work cited was supported in part by Career Development Award K3-MH 18232 and Research Grant MH 12773 from the National Institute of Mental Health.

out of ten consecutive correct responses (Barondes and Cohen, 1966). The puromycin-injected mice learned this task in the same number of trials as saline-injected mice. Retention 15 minutes after training was normal. However, in the ensuing few hours, retention became markedly impaired and three hours afterward, profound amnesia was observed. The amnesia persisted thereafter. Since puromycin markedly inhibited cerebral protein synthesis, it was assumed that this action of the drug was responsible for the amnesia. Subsequent studies indicated, however, that puromycin was acting not only by inhibiting cerebral protein synthesis but also by producing abnormalities in cerebral electrical activity (Cohen *et al.*, 1966; Cohen and Barondes, 1967). The first indication that this might be the case came from studies with cycloheximide. Cerebral protein synthesis was inhibited with cycloheximide to a greater extent than with puromycin. Yet, it was found that mice trained in the same task after administration of cycloheximide remembered normally (Barondes and Cohen, 1967a). This suggested that the amnesic effects of puromycin might be due to some other action. For this reason, studies were made of the electrical activity of the brain in mice five hours after intracerebral injection with puromycin or cycloheximide. Cycloheximide had no effect on the amplitude or frequency of recordings from the hippocampal region of the mouse brain (Fig. 1). In contrast, puromycin produced very marked suppression of electrical activity in this region five

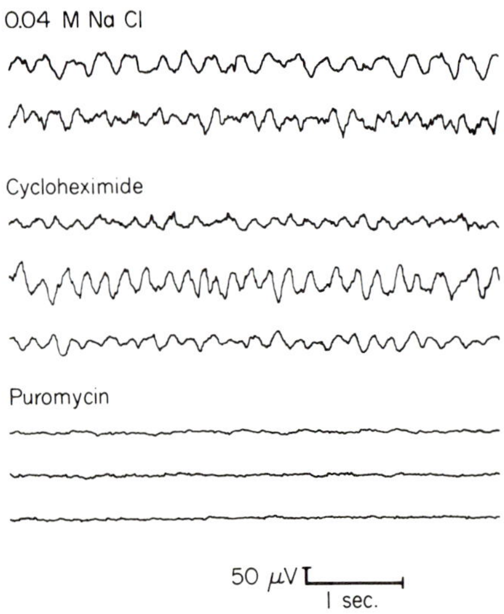

Figure 1 Electrical activity recorded from the hippocampal region of the brain. Mice were injected in both temporal regions of the brain with ten microliters of 0.04 M NaCl containing 100 μg of the indicated drug. Recordings were made in unanesthetized mice (Cohen *et al.*, 1966).

hours after injection (Fig. 1). When recordings were made from the brains of puromycin-injected mice shortly after the drug was given, frequent seizure activity was found, which was not apparent from observation of their gross behavior. Furthermore, when subconvulsive doses of pentylenetetrazol were administered to puromycin-injected animals, they had overt seizure activity, whereas acetoxycycloheximide-injected animals, whose cerebral protein was more extensively inhibited, did not (Cohen and Barondes, 1967). These experiments indicated that puromycin, or a substance formed in its presence, increased the irritability of the brain, and that the electrical disturbance recorded was probably a post-ictal phenomenon. Administration of the anticonvulsant, diphenylhydantoin, concurrently with puromycin interfered with the amnesic effect. These experiments suggest that the puromycin effect on memory appears to be due, at least in part, to disturbances of cerebral electrical activity. They demonstrate the danger of assuming that a drug's effect on memory is due to the action for which it is administered. Because of its effect on cerebral activity, puromycin is not a useful drug for studying the hypothesis that memory storage is mediated by the synthesis of protein. Since acetoxycycloheximide and cycloheximide did not produce a detectable effect on cerebral electrical activity, studies with these drugs were continued. However, the possibility that they too might have an undetected "side-effect" should not be overlooked.

STUDIES WITH ACETOXYCYCLOHEXIMIDE: THE PROBLEM OF OVERTRAINING

Initial studies with cycloheximide and acetoxycycloheximide indicated that they had no effect on learning or memory of a left-right discrimination, when training was continued to a criterion of nine out of ten consecutive correct responses (Barondes and Cohen, 1967a). This suggested that protein synthesis was not necessary for memory storage. However, it remained possible that the amnesic effect of these inhibitors was being obscured by the training procedure. Mice learn a left-right discrimination in only a few trials. In training to a criterion of nine out of ten correct responses there is extensive repetition of the correct solution. It seemed possible that the residual protein synthesizing capacity which remained after injection of cycloheximide or acetoxycycloheximide was sufficient for memory storage after overtraining in a simple task. To test this, studies were made of the effect of acetoxycycloheximide after less extensive training. Mice were trained to escape shock in a T-maze to a criterion of three out of four consecutive correct responses. Mice whose cerebral protein synthesis was inhibited more than 95% at the time of training learned as well as saline controls (Barondes and Cohen,

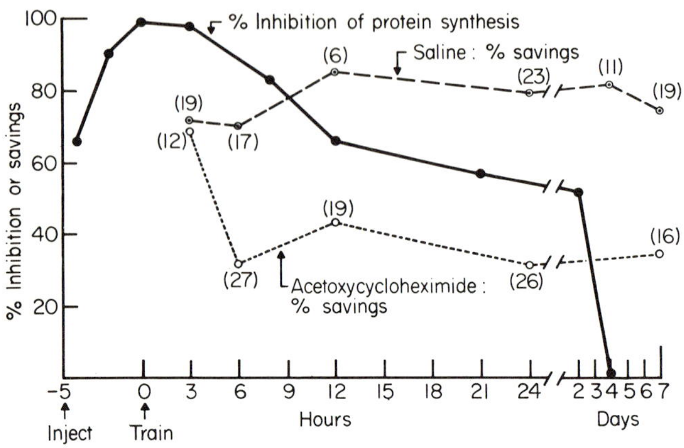

Figure 2 Effect of intracerebral acetoxycycloheximide on cerebral protein synthesis and memory. Mice were injected in both temporal regions of the brain with a total of 20 microliters of 0.15 *M* NaCl with or without 20 μg of acetoxycycloheximide. Five hours later they were trained to escape shock by choosing the correct limb of a one choice maze to a criterion of three out of four consecutive correct responses. Retention was determined at the times indicated. Protein synthesis inhibition was estimated in three—six mice at the times indicated. There was no significant difference in the saving of the two groups three hours after training but thereafter acetoxycycloheximide-injected mice had significantly poorer savings ($P < 0.05$ or less) than saline-injected mice (Barondes and Cohen, 1967b).

1967b). When tested three hours later, their memory was normal. However, retention six hours after training or thereafter was markedly impaired (Fig. 2). This suggests that overtraining had indeed obscured the amnesic action of the protein synthesis inhibitor. Similar results were found when a more difficult task was studied. Mice were trained to escape shock by choosing the lighted limb of a T-maze to a criterion of five out of six or nine out of ten consecutive correct responses (Cohen and Barondes, 1968a). This task is considered more difficult than the left-right discrimination since more training is required for the mice to achieve any given criterion of correct performance. Acetoxycycloheximide-injected mice learned this task as well as controls. Memory was normal three hours after training, but was markedly impaired six or more hours after training. Therefore, the difficult task is more susceptible to the amnesic effect of acetoxycycloheximide.

FURTHER STUDIES WITH ACETOXYCYCLOHEXIMIDE

The studies with acetoxycycloheximide supported a possible role of protein synthesis in "long-term" memory. To test this conclusion, several variations were made in the experimental design. These included studies of the effects of subcutaneous administration of the drug and of its effect on an appetitive task. Studies with subcutaneous injections were performed, since it seemed possible that the increased intracranial pressure produced by intracerebral injections or the high local concentration of the drug at the injection site might be related to the amnesic effect. Subcutaneous administration of large doses of acetoxycycloheximide produced the rapid onset of inhibition of 90–95% of cerebral protein synthesis. Injection of the drug five or more minutes prior to training had no effect on learning or on memory three hours after learning. However, memory was markedly impaired six hours, seven days (Fig. 3), and six weeks after learning (Barondes and Cohen, 1968). Therefore, the amnesic effect of acetoxycycloheximide is not related to intracerebral injections. Administration of the drug immediately after training had a less marked but significant amnesic effect (Fig. 3). This suggests that some of the protein synthesis apparently required for "long-term" memory occurs within minutes after training. Since injections 30 minutes after training had no effect (Fig. 3), the diarrhea and dehydration which appears several hours after the drug is given is probably not related to the subsequent memory impairment.

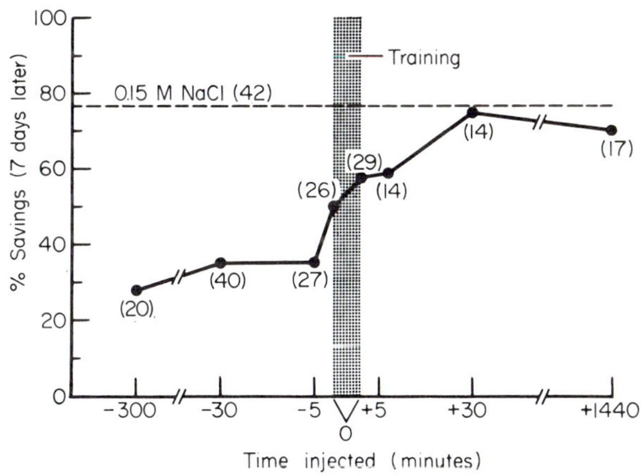

Figure 3 Effect of subcutaneous administration of acetoxycycloheximide at various times before or after training on memory. Mice were injected subcutaneously with 240 μg of acetoxycycloheximide at the indicated time relative to training. They were trained to escape shock by choosing the lighted limb of a T-maze to a criterion of five out of six correct responses. Training took an average of eight minutes. Approximately 90% of cerebral protein synthesis was inhibited within 10–15 minutes of subcutaneous injection of acetoxycycloheximide. All mice were tested for retention seven days after training, long after they had recovered from the drug. The mice injected before or within five minutes after training all had significantly less savings ($P < 0.05$ or less, Mann–Whitney U test) than those injected immediately after training. Injections 30 minutes or more after training had no effect on memory (Barondes and Cohen, 1968).

To determine the generality of the acetoxycycloheximide effect, studies have recently been made in a task motivated by thirst rather than escape from shock. The results in the appetitive task are identical with those in the aversive task (Cohen and Barondes, 1968b).

CONCLUSION

This series of experiments suggests the following conclusions: (1) Learning and "short-term" memory (for at least three hours after training in the situations studied) are apparently independent of cerebral protein synthesis. Presumably, some other mechanism is responsible for memory storage for this period. (2) It is difficult to interpret the effects of puromycin on memory since this agent not only inhibits cerebral protein synthesis but also increases cerebral irritability and the propensity to occult seizures which could interfere with memory. (3) Prolonged training in a simple task may obscure the effects of inhibition of cerebral protein synthesis on memory, presumably because the small amount of residual protein synthesizing capacity may be sufficient to mediate memory storage under this condition. (4) Acetoxycycloheximide, an agent which does not apparently interfere with intracerebral electrical activity, impairs "long-term" memory. This suggests that protein synthesis is required for "long-term" memory storage. However, the possibility that the drug produces amnesia by some other mechanism cannot be excluded. (5) The protein synthesis apparently required for "long-term" memory storage occurs during learning and/or within minutes after learning. (6) Since the amnesic effect of acetoxycycloheximide is not observed until six hours after training, the "short-term" mechanism which is apparently independent of protein synthesis may survive until this time and temporarily obscure the effect of the inhibitor on the "long-term" process.

REFERENCES

Barondes, S. H. (1965). Relationship of Biological Regulatory Mechanisms to Learning and Memory. *Nature*, **205**, 18–21.

Barondes, S. H. (1966). The Effect of Inhibitors of Protein and RNA Synthesis on Learning and Memory, Proc. Vth Int. Congr. C.I.N.P. *Excerpta Medica*, Series **129**, 208–211.

Barondes, S. H. and Cohen, H. D. (1966). Puromycin Effect on Successive Phases of Memory Storage. *Science*, **151**, 594–595.

Barondes, S. H. and Cohen, H. D. (1967a). Comparative Effects of Cycloheximide and Puromycin on Cerebral Protein Synthesis and Consolidation of Memory in Mice, *Brain Research*, **4**, 44–51.

Barondes, S. H. and Cohen, H. D. (1967b). Delayed and Sustained Effect of Acetoxycycloheximide on Memory in Mice. *Proc. Nat. Acad. Sci.*, **58**, 157–164.

Barondes, S. H. and Cohen, H. D. (1968). Memory Impairment after Subcutaneous Injection of Acetoxycycloheximide. *Science*, **160**, 556–557.

Cohen, H. D., Ervin, F., and Barondes, S. H. (1966). Puromycin and Cycloheximide: Different Effects on Hippocampal Electrical Activity. *Science*, **154**, 1557–1558.
Cohen, H. D. and Barondes, S. H. (1967). Puromycin Effect on Memory May be Due to Occult Seizures. *Science*, **157**, 333–334.
Cohen, H. D. and Barondes, S. H. (1968a). Acetoxycycloheximide Effect on Learning and Memory of a Light-Dark Discrimination. *Nature*, **218**, 271–272.
Cohen, H. D. and Barondes, S. H. (1968b). *Comm. Behav. Biol.*, **1**, 337–340.

QUESTIONS AND ANSWERS

The following was prepared to clarify the relationship between the work reported and that of others which could not be covered in the brief presentation at the symposium. The questions were raised at the symposium or in ensuing discussions.

QUESTION: Flexner has reported that administration of puromycin into the temporal region of the brain up to *several days* after training produces amnesia. Yet, your experiments indicate that acetoxycycloheximide is effective only if given before or to some slight extent immediately after training. How do you explain this difference?

ANSWER: Acetoxycycloheximide differs from puromycin in its biochemical mechanism of action and in its effects on cerebral electrical activity and on behavior. Both Flexner and we have found that administration of acetoxycycloheximide hours or days after training has no effect on memory despite the fact that this drug (in doses used) inhibited cerebral protein synthesis even more extensively than puromycin. Therefore, it appears that puromycin, in the experiments which you describe, may be working not by inhibiting the synthesis of protein required for memory but, rather, by some other mechanism. Flexner and Flexner (*Proc. Nat. Acad. Sci., U. S.*, **57**, 165, 1967) have recently provided support for this conclusion. They have shown that the amnesic effect of puromycin produced by injections days after training can be subsequently *reversed* by intracerebral injections of saline. This suggests that the puromycin, given long after training, does not interfere with memory storage but possibly with "retrieval." Thus far, only experiments in which inhibitors are given before or immediately after training have implicated protein synthesis in the formation of "long-term" memory. Although the action of puromycin up to several days after training may have nothing to do with the participation of protein synthesis in memory storage, this finding remains of great interest. This is particularly true since temporal injections of puromycin are no longer effective if given one week after training (J. B. Flexner, L. B. Flexner, E. Stellar, *Science*, **141**, 57, 1963). We have confirmed this in our laboratory. This suggests that some aspect of the storage process continues to change over a period of days. Whatever its mechanism, this phenomenon merits extensive study.

QUESTION: Do intracerebral injections of saline have any effect on the acetoxycycloheximide induced amnesia?

ANSWER: When acetoxycycloheximide is administered intracerebrally or subcutaneously, prior to training, there is amnesia observed six or more hours after training. Injections with saline into the brain sites in which acetoxycycloheximide had been given either one day or one week after acetoxycycloheximide has no effect on memory measured one day or one week after the saline injections.

QUESTION: You find that injections of acetoxycycloheximide immediately after training have only a slight effect whereas Agranoff and his colleagues, studying goldfish, find a more prominent effect which is demonstrable even if acetoxycycloheximide is administered 30 minutes after training. What is your feeling about this difference?

ANSWER: Whole mouse cerebral synthesis is not maximally inhibited until more than three hours after intracerebral injections of acetoxycycloheximide. When intracerebral injections of acetoxycycloheximide are given immediately after training, we found no memory impairment. It is only with subcutaneous injections which inhibit whole cerebral protein synthesis more rapidly (presumably because the whole brain is promptly perfused by the drug rather than being only slowly affected by diffusion from the intracerebral injection sites) that we have seen an effect of acetoxycycloheximide administered immediately after training. The goldfish studies of Agranoff and his colleagues (reported in this volume) may differ in several regards. (1) Intracranial injections of acetoxycycloheximide may more rapidly diffuse through the small goldfish brain and produce inhibition of protein synthesis more rapidly than is found with the intracerebral injections in mice. (2) Even if this were true, the period during which the "relevant" protein synthesis occurs, seems to be much longer in goldfish than in mice. Perhaps these processes are slower in goldfish brain because it is at room temperature (25°C) rather than 37°C. (3) The time when this protein synthesis must occur may vary with the behavioral situation studied.

RECENT STUDIES ON THE STAGES OF MEMORY FORMATION IN THE GOLDFISH*

Bernard W. Agranoff

Mental Health Research Institute
Department of Psychiatry and Biological Chemistry
University of Michigan
Ann Arbor, Michigan

This paper will review briefly our major findings on the formation of memory of shock avoidance in the goldfish and describe some recent experiments. Finally, these findings will be related to our present concept on how permanent memory is formed in the brain.

In our basic experiment, we give fish 20 one-minute training trials. They have 20 seconds following the onset of a light stimulus to swim over a barrier in a shuttle box, in order to avoid a punishing electric shock. Three days later, they receive ten additional trials in which we measure memory of the initial training (Agranoff and Klinger, 1964; Agranoff *et al.*, 1966). We have varied the apparatus and timing schedule and find that each paradigm gives generally similar results (Agranoff and Davis, 1968a). The apparatus we have been using for the past year has a clear plastic gate which the fish must deflect in order to swim to the opposite side of the shuttle box (Agranoff *et al.*, 1967). Ten fish are run simultaneously in individual shuttle boxes and their avoidances and escapes are recorded automatically. In the present apparatus, fish show a seven–eight-fold increase in probability of avoidance in the ten testing (retraining) trials, compared with the first ten training trials. This compares favorably with a two–three-fold increase observed in an

*Supported by grants from NSF and NIH.

earlier apparatus. The ten test trials can be given immediately after the 20 training trials or as late as one month later. There is no detectable loss of avoidance responding, which is our operational measure of memory. Consequently, we have a task which goldfish can learn in minutes and can retain indefinitely.

In addition to the rapid acquisition of permanent memory, the goldfish has another advantage for us. Its brain resides in a relatively large cranial cavity into which we can easily inject drugs through the skull by means of a microsyringe. A fine needle fitted with a plastic stop assures that the needle tip does not touch the brain itself (Agranoff and Klinger, 1964). Unanesthetized fish are removed from their tanks, are injected, and returned to the water within a few seconds without gross behavioral effects.

In a typical experiment, fish are injected intracranially with antibiotic inhibitors of protein synthesis. We have used puromycin, reported to block memory (Flexner *et al.*, 1963), and acetoxycycloheximide (AXM), reported to have an amnestic effect (Agranoff *et al.*, 1966). By varying the time after training at which fish are injected, and then testing three days later, we find evidence for a time-dependent process following training in which memory changes from a susceptible to an insusceptible form. The hypothetical process by which long-term memory forms is often referred to as the consolidation or fixation of memory. Animals given 130–170 μg of puromycin or 0.1–0.2 μg of AXM immediately following training, then returned to their home tanks for 72 hours, have a profound memory loss on retesting. The same amount of these drugs injected a few hours after the training session has little or no effect on the 72-hour memory. These experiments have been taken to mean that normal levels of protein synthesis are required following training for the formation of permanent memory. In other experiments performed with isotopic precursors, we have shown that puromycin and AXM, in the doses used, produce maximal (80%) inhibition of protein synthesis in the goldfish brain of several hours duration, with recovery in about a day (Brink *et al.*, 1966). One thousand times as much puromycin as AXM is needed to produce a similar protein block. One thousand times as much puromycin is also needed to produce a similar memory block (Agranoff *et al.*, 1966, 1967). When 0.2 μg of AXM is injected intraperitoneally instead of intracranially, there is no effect on memory. The doses of puromycin and AXM used in these experiments do not affect incorporation of labeled uridine into goldfish brain RNA (Casola and Agranoff, 1967).

In other experiments, we found that electroconvulsive shock (ECS) can also provide evidence for a consolidation process in the goldfish which is complete in two hours (Davis *et al.*, 1965). The poikilothermic nature of the teleost enabled us to demonstrate that the ECS-disruptible process was temperature-dependent. Animals cooled 10°C for two hours following training remained susceptible to ECS. ECS is also accompanied by a transient decrease in protein labeling in the goldfish brain, but the magnitude of the decrease is less than is observed with doses of puromycin which produce far less memory loss (Agranoff, 1965). We conclude that ECS can block memory formation in an additional unspecified way.

Unlike ECS, the antibiotic blocking agents do not cause observable neurological or behavioral changes. Following intracranial injection, fish swim about normally. We were therefore encouraged to study learning, or acquisition of shock avoidance following injection of puromycin or AXM. We found that acquisition of shock avoidance is not impaired in injected animals, but that memory 72 hours later is lost, as with injections immediately post trial (Agranoff *et al.*, 1965). The results of these experiments suggest that memory associated with learning (short-term memory) is formed by a process which is not blocked by protein inhibitors. The fate of this short-term memory can be studied in fish injected with puromycin or with AXM immediately before or after training. We performed experiments in which groups of animals were trained and then injected immediately following trial 20 with the antibiotics, and retested at various times (Davis and Agranoff, 1966). Fish tested immediately following injection showed normal memory. Others, not given their ten test trials for 72 hours, retained little memory. By retraining other groups at intermediate times, we were able to observe the gradual disappearance of short-term memory over the three-day period.

If short-term is converted to long-term memory, we might expect the short-term memory decline to complement the memory fixation curve. Short-term memory should decay as long-term memory forms. This is not the case. As judged from experiments with either puromycin or AXM, memory fixation is complete within three hours after training, while short-term memory decay takes two–three days. Fish can be trained, injected immediately thereafter with a protein inhibitor, and yet demonstrate quite good memory 24 hours later. If the same animals had been retrained 72 hours after injection, little memory would have remained. The opportunity to form long-term memory was somehow lost. Times for long-term memory formation and short-term memory decay for actinomycin D, similar to those for puromycin and AXM have been obtained (Agranoff *et al.*, 1967). We recently have demonstrated that actinomycin D blocks RNA synthesis without measurable effect on protein synthesis for the first few hours following its injection (Agranoff *et al.*, 1967). The results with this drug indicate that the cytoplasmic translation of RNA into protein does not alone mediate the process of long-term memory formation. We do not know whether the amnestic effect of actinomycin D is related to its block of rapidly turning over messenger RNA or to some other action, such as depletion of ribosomes.

While intact protein and RNA synthesis are necessary during a critical posttrial period, these requirements are undoubtedly not sufficient for memory formation. Dr. Roger E. Davis has performed extensive studies on the effect of the posttrial environment on long-term memory formation in the goldfish (Davis and Agranoff, 1966; Davis, 1968). He finds that fish maintained in the training environment instead of being returned to the home tanks do not fix memory, but remain susceptible to blocking agents. When they are at last returned to the home tanks, long-term memory is formed as an ordinarily observed following trial 20. The fixation process has been moved forward in time. We have speculated elsewhere regarding the possible significance of the environmental dependence of the fixation process (Agranoff and Davis, 1968b).

We have recently investigated the effects of intracranial injections of six percent KCl. This treatment causes severe convulsions and incapicitates the fish for about two hours. Protein synthesis is markedly depressed but recovers within four hours. Memory formation is susceptible to KCl and remains susceptible for over 12–18 hours, a much longer time than is seen with the antibiotics. Short-term memory decay following KCl, also takes 6–12 hours. With KCl, the two processes are complementary. This result suggests that the disappearance of short-term memory and the formation of long-term memory are parts of a unitary process such as one might envision occurring within the same structural units.

Since long-term memory formation is blocked by inhibitors of protein synthesis, it was hypothesized that protein synthesis might be required for the decay of short-term memory as well. Such a relationship could account for the prolonged decay of short-term memory following injection of protein inhibitors. The antibiotics then prevent the initiation of long-term memory formation during a critical posttrial period, and the short-term memory remains until normal protein synthesis is restored. A test of this hypothesis is to see whether repeated injections of AXM can prevent the decay of short-term memory. Animals injected with this drug immediately following training and retested 72 hours later had lost three responses (see Table I). Other fish were injected with 0.2 μg of AXM immediately following training, and again at 24 hours and at 48 hours. When retested at 72 hours, these animals had lost only 1.2 responses, significantly less than the singly-injected groups. A performance decrement due to illness would be expected to decrease the difference between the two groups. This makes the observed result all the more striking. In other experiments, we injected animals with both KCl and AXM immediately following trial 20 and retested them 72 hours later. As can be seen in the Table, AXM appears to retard the KCl-mediated short-term decay, but only temporarily.

Taken together, these experiments suggest that short-term memory decay, like long-term memory formation, requires protein synthesis. The results

TABLE I

LOSS OF AVOIDANCE RESPONSES FOLLOWING IC INJECTION OF KCl (600 μg) AND AXM (0.2 μg)

Interval between Training and Injection (Hours)	Interval between Injection and Retraining (Hours)[a] Retention Score[b]			
	4–6	24	48	72
AXM (0)	+0.52	−0.20	−2.4	−3.0
AXM (0, 24, 48)				−1.2
KCl (0)	−2.0	−3.6		−3.6
KCl + AXM (0)		−2.1		−3.7

[a] N = 23–32 fish per group except KCl + AXM (72 hours) which contained 17 fish.

[b] Achieved = predicted scores for trials 21–30.

support the concept that short-term memory is *converted* to long-term memory. This new finding may explain in part the reported lack of effect of AXM and of puromycin plus AXM (Flexner and Flexner, 1966; Barondes and Cohen, 1967) in mice whose memory is susceptible to puromycin alone. The severe, prolonged block of protein synthesis obtained with large doses of AXM may have prevented short-term memory from decaying; retraining in this circumstance would not reveal an amnesic effect.

REFERENCES

Agranoff, B. W. *Perspectives in Biol. and Med.*, **9**, 13 (1965).

Agranoff, B. W. and Davis, R. E. "The Central Nervous System and Fish Behavior," D. Ingle (Ed.), Univ. Chicago Press, p. 193 (1968).

Agranoff, B. W. and Davis, R. E. "Symposium on Physiological and Biochemical Aspects of Nervous Integration," F. D. Carlson (Ed.), Prentice-Hall, p. 309 (1968).

Agranoff, B. W., Davis, R. E. and Brink, J. J. *Brain Research*, **1**, 303 (1966).

Agranoff, B. W., Davis, R. E. and Brink, J. J. *Proc. Natl. Acad. Sci., U.S.*, **54**, 788 (1965).

Agranoff, B. W., Davis, R. E., Casola, L. and Lim, R. *Science*, **158**, 1600 (1967).

Agranoff, B. W. and Klinger, P. D. *Science*, **146**, 952 (1964).

Barondes, S. H. and Cohen, H. D. *Brain Research*, **4**, 44 (1967).

Brink, J. J., Davis, R. E. and Agranoff, B. W. *J. Neurochem.*, **13**, 889 (1966).

Casola, L. and Agranoff, B. W. Proceedings of First International Meeting of the International Society for Neurochemistry, Strasburg, July 1967.

Davis, R. E. *J. comp. physiol. Psychol.*, **65**, 72 (1968).

Davis, R. E. and Agranoff, B. W. *Proc. Natl. Acad. Sci.*, **55**, 555 (1966).

Davis, R. E., Bright, P. J. and Agranoff, B. W. *J. comp. physiol. Psychol.*, **60**, 162 (1965).

Flexner, J. B., Flexner, L. B. and Stellar, E. *Science*, **141**, 57 (1963).

Flexner, L. B. and Flexner, J. B. *Proc. Natl. Acad. Sci., U.S.*, **55**, 369 (1966).

THE EFFECT OF TRAINING ON THE INCORPORATION OF RADIOACTIVE PRECURSORS INTO RNA OF BRAIN*

John W. Zemp†, Linda B. Adair,

John Eric Wilson, and Edward Glassman

Department of Biochemistry and the Genetics Curriculum
University of North Carolina
Chapel Hill

Over the past few years a number of investigations have directly or indirectly suggested a possible involvement of RNA in the learning process or in early stages of memory consolidation (Dingman and Sporn, 1961; Hydén and Egyházi, 1964; Flexner *et al.*, 1967; Zemp *et al.*, 1966, 1967; Barondes and Cohen, 1967; Agranoff *et al.*, 1967). The earliest direct studies, and probably the most influential in attracting investigators into this area of research, were those of Hydén and Egyházi (1962, 1963, 1964). They studied RNA changes in isolated neurons and glia and demonstrated increased RNA synthesis and base ratio changes associated with prolonged training experiences. Interpretation of the results of these and other experiments has varied from the one extreme of a postulation of an RNA or protein memory

*Supported by NIH Training Grant (2T01 GM-00404) to the Department of Biochemistry (L.B.A.), by NSF Postdoctoral Fellowship (46060) (J.W.Z.), by a Research Career Development Award (GM-K3-14, 1911) from the Division of General Medical Sciences NIH (E.G.), and by research grants from the U.S. Public Health Service (GM-08202; GM-07699; NB-07457), the University of North Carolina Research Council, and by an NSF Equipment Grant (GB-4577).

†Present Address: Department of Chemistry, Medical University of South Carolina, Charleston.

molecule to the other extreme of the total irrelevance of RNA changes to the phenomena of learning and memory. Some possible roles of RNA in nerve function are summarized in Fig. 1.

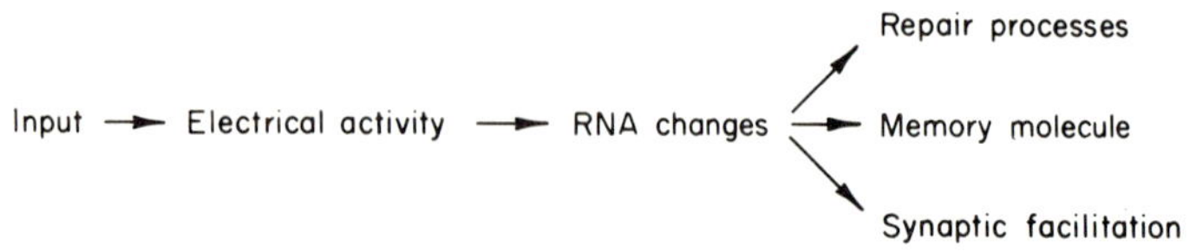

Figure 1 Possible Roles of RNA in Nerve Function (Glassman, 1967).

The first alternative proposes that observed RNA changes reflect a requirement for replacement of metabolites or enzymes lost or changed during neuronal activity. Thus, the input to the system has led to altered cellular activity which requires the mediation of RNA for a return to the normal state. The second alternative, the postulation of an RNA or protein memory molecule, is not only conceptually difficult, but also has little direct experimental proof. The third postulation of an RNA change associated with synaptic facilitation or the establishment of new synapses represents an attempt to fit the observed RNA changes within the general hypothesis relating memory storage to stable anatomical circuits.

In our studies (Zemp *et al.*, 1966, 1967) we have used a double radioactive isotope technique to study the effect of a 15-minute conditioned avoidance training experience on the incorporation of radioactive uridine into RNA of nuclear and ribosomal-polysomal (RP) fractions of mouse brain. The training apparatus consisted of a double compartment box with a continuous grid floor. In one half of the box there is a shelf (the training side), and on the other half there is no shelf (the yoked side). A buzzer and a light, which illuminates both sides of the apparatus equally, are mounted on the back of the box.

Table I describes the experimental procedure, and summarizes results obtained for the incorporation of radioactive uridine into nuclear and RP RNA isolated from liver, kidney, and brain of trained and yoked mice sacrificed immediately after training. The results are expressed as a percentage increase of radioactivity found in the trained mouse, corrected for the relative amount of radioactivity found in uridylic acid (UMP) in each organ of the two mice. The radioactivity found in UMP is not interpreted as an indication of the size of the immediate precursor pool of RNA, but represents a measurement of the relative amount of intracellular isotope in each organ. No significant differences in radioactivity were observed when RNA of liver and kidney from trained and yoked mice were compared. There was, however, a significant and substantial increase in uridine incorporation into

TABLE I

CHANGES IN INCORPORATION OF URIDINE INTO RNA FROM BRAIN, LIVER, AND KIDNEY OF TRAINED AND YOKED MICE SACRIFICED IMMEDIATELY AFTER TRAINING

Tissue	Number of Pairs	Average Percentage Increase in Trained Mouse (Corrected for Radioactivity in UMP)	
		RP RNA	Nuclear RNA
Liver	5	−2.2	−2.2
Kidney	4	1.8	1.9
Brain	10	40.2	29.2

As described previously (Zemp *et al.*, 1966), one mouse was given an intracranial injection of ^{3}H-uridine, and another mouse was given ^{14}C-uridine. Mice in which liver and kidney were used were given an additional injection intraperitoneally. After 30 minutes, one mouse was trained while the other served as a yoked animal; the mice were coded following the injection, and the animal to be trained was selected at random. *Trained* mice (avoidance conditioning I) were presented with three seconds of light and buzzer followed by the onset of a mild shock through the grid floor. The shock was terminated when the animal jumped to the shelf. No shock was presented when the mouse being trained avoided the conditioned stimuli. The *yoked* mice, therefore, received light, buzzer, and shock to the same extent as the *trained* mice. The training experience lasted 15 minutes, after which the brains were quickly removed and homogenized together. Ribosomes and nuclei were isolated, and the RNA extracted from each. The amount of ^{14}C and ^{3}H was determined. In addition, UMP was isolated from the supernatant solution after removal of nuclei and ribosomes, and the amount of ^{14}C and ^{3}H in it was determined. The percentage increase was calculated from the relative amount of radioactivity in RNA of the trained: yoked mice corrected for the relative amount of radioactivity in UMP of trained : yoked mice.

RNA from the brains of trained mice. Twenty-five pairs of mice were sacrificed immediately after training and such an increase was present in all 25 pairs. To determine if there were any differences in the pattern of radioactivity of the RNA isolated from trained and untrained mice, the nuclear and RP RNA were centrifuged on sucrose density gradients. The amount of radioactivity in nuclear and RP RNA of the trained mouse was consistently higher in all fractions from the gradient, but there was no evidence for any special type of RNA being produced.

In addition to the training procedure described in Table I, we have compared the effect of three other types of experience on the incorporation of uridine into nuclear and RP RNA. Table II describes these experiences and summarizes the results we obtained. Increases in radioactivity were observed when RNA from the brains of *trained* mice was compared with that from

TABLE II

EFFECTS OF FIFTEEN MINUTES EXPERIENCE ON THE INCORPORATION OF URIDINE INTO RNA FROM MOUSE BRAIN

Paired Experiences	Number of Pairs	Change in RNA Radioactivity
Trained vs. Yoked	18	+ (Trained)
Trained vs. Quiet	7	+ (Trained)
Yoked vs. Quiet	4	0
Shocked vs. Quiet	4	0

The biochemical procedures used and the experience of *trained* (avoidance conditioning I) and *yoked* mice are exactly as described in Table I. *Quiet mice* were kept in their home cages as undisturbed as possible for 45 minutes after injection. *Shocked mice* received 30 shocks during the 15-minute period.

quiet mice, suggesting that we are observing an increased amount of radioactivity in the *trained* mice rather than a decrease in the *yoked* mice. This suggestion is supported by the observation that the amount of radioactivity in RNA from the brains of *yoked* and *quiet* mice is the same. The lack of difference between *yoked* and *quiet* mice also indicates that light, buzzer and shock *per se* have not led to a change. Since there was no difference between *shocked* and *quiet* mice, neither that type of experience nor the random activity exhibited by the *shocked* mice could be causative factors for the changes observed.

The increases which were observed in the trained mice were obtained with whole brains. To determine whether the results represented a generalized response throughout the brain or a relatively large increase in a smaller area of the brain, the brains of the *trained* and *yoked* mice were dissected into six parts, and each part was analyzed for the relative incorporation of radioactivity into UMP and RNA. Table III identifies the areas examined and summarizes the results. The only area which showed a consistent and significant increase ($p<0.01$) was D, which contains the diencephalon, ventral hippocampus, and parts of the basal ganglia. There also appeared to be a small but consistent decrease in the amount of radioactivity in RNA in the parietal and frontal areas from the brains of trained mice.

The localization of the increased amount of radioactivity in RNA within area D does not clarify which aspect of the training situation is responsible for the increase. The difference in experience of the trained and untrained mice could be stress, attention, organized jumping, or some parameter related to learning (see Glassman, 1967). Several investigations have indicated that stress might be involved. Bryan and co-workers (1967) reported

TABLE III

CHANGES IN INCORPORATION OF URIDINE INTO RNA ISOLATED FROM PARTS OF BRAIN OF TRAINED AND YOKED MICE

Part of Brain	Average Percentage Increase in Trained Mouse (Corrected for Radioactivity in UMP)	
	RP RNA	Nuclear RNA
O	2.0	1.0
H	5.4	6.4
F	0	−12.4
P	−17.0	−3.0
D	32.4	21.6
C	5.7	0
N = 10		

The procedures were exactly as described in Table I except that the brains were dissected into six parts as previously described (Zemp *et al.*, 1967).

O = olfactory bulbs and tracts, H = dorsal hippocampus and temporal areas, F = frontal areas, P = parietal areas, D = diencephalon and anatomically associated areas, and C = cerebellar areas.

increased incorporation of ^{3}H-uridine into RNA of mouse brain following at least one hour of stress produced by taping mice to wood dowels. It is difficult, however, to compare their results with ours, since the paradigms were totally different. Other stress situations (Pevzner, 1965) have also been shown to cause both increases and decreases in RNA.

Using a water reward to minimize stress, Bowman and Strobel (1969) have reported increased incorporation of ^{3}H-cytidine into RNA of the hippocampus of rats undergoing reversal spatial discrimination in a Y-maze. Table IV describes the experimental approach and summarizes this data. There were significant increases in pool tritium (as defined in Table IV) in the experimental rats in both the hippocampus and pyriform cortex. Since the pool radioactivity includes tritium which has been lost from cytidine, as well as tritium still attached to cytidine and its metabolites (Zemp and Wilson, 1965), the pool changes are difficult to interpret. The results presented in Table IV show that the hippocampus was the only area which had a significant increase in the ratio of (tritium in RNA): (tritium in pool). Since large portions of the hippocampus are contained within the area we called D, these results may be consistent with ours.

Bowman and Strobel point out that their results could be caused by

TABLE IV

MEAN PERCENTAGE INCREASE OF POOL ^3H AND RNA ^3H OF EXPERIMENTAL COMPARED TO CONTROL ANIMALS[a]

	Pool ^3H	RNA ^3H / Pool ^3H	
		Nuclei	*Cytoplasm*
Hippocampus	+14%[b]	+23%[b]	+26%[b]
Residual Subcortex	+5	−3	+24
Pyriform Cortex	+17[c]	0	+1
Anterior Cortex	0	+5	0
Posterior Cortex	+6	−12	−16

[a] Modified from Bowman and Strobel (1969).

[b] $p < 0.01$.

[c] $p < 0.05$.

As described by Bowman and Strobel (1969), all rats, control and experimental, were overtrained to their preferred side of a Y maze for 100 trials daily for four days. On the first two days the rats were rewarded for every correct choice. On the next two days only 75% of the correct responses were rewarded in an attempt to adapt the rats to the frustration of nonreward. On the fifth day, ^3H-cytidine was injected immediately before one hour of training. The control rats were again rewarded for their preferred choices, but the experimental rats were forced to reverse their choice for reward. RNA was isolated from a cold trichloroacetic acid precipitate. The pool ^3H was estimated from the supernatant solution after trichloroacetic acid precipitation.

frustration due to nonreward, and suggest that no one experiment may be adequate to eliminate all variables. Adair *et al.* (1968) have carried out a series of experiments in which additional experiences related to our training apparatus were tested for their effects on incorporation of uridine into polysomes from the brain. Table V summarizes the results. The increases in incorporation of radioactive uridine into the polysomes from the brains of *trained* mice (avoidance conditioning I) compared to *yoked* mice agree with the changes observed when RNA was isolated from the RP pellet (Zemp *et al.*, 1966). The lack of difference between *yoked* and *quiet* mice also agrees with the earlier data. Although the data are not included in the table, no difference between *trained* and *yoked* mice was found when polysomes were isolated from liver and kidney.

TABLE V

EFFECTS OF FIFTEEN MINUTES EXPERIENCE ON THE INCORPORATION OF URIDINE INTO POLYSOMES FROM MOUSE BRAIN[a]

Paired Experiences	Number of Pairs	Change in Radioactivity of Polysomes	Learning
Avoidance conditioned I mice vs. yoked mice	36/42[b]	+	Yes
Yoked mice vs. quiet mice	6	0	No
Avoidance conditioned II mice vs. yoked mice	12	+	Yes
Classically conditioned mice vs. quiet mice	10	0	?[c]

[a]Modified from Adair (1968).

[b]There was an increase in the trained mouse in 36 pairs, no change in five pairs and a decrease in one pair.

[c]The classically conditioned mice may have undergone a type of learning, but probably not of the same type undergone by the instrumentally conditioned mice.

The injection procedures were exactly as described in Table I. Polysome preparations were isolated and analyzed as described by Adair *et al.*, (1968). The experience of the *avoidance conditioned I*, the *yoked*, and the *quiet* mice were described in Tables I and II.

The *avoidance conditioned II* mice were trained without presentation of light and buzzer. They were placed on the grid floor of the training half of the apparatus for three seconds prior to the onset of shock. The floor becomes the conditioned stimulus and the mice learned to avoid the floor.

The *classically conditioned* mice were placed in the yoked half of the training apparatus. They received three seconds of light and buzzer followed by 15 seconds of shock. They could neither escape nor avoid.

The mice trained by avoidance conditioning II learned to avoid the floor without presentation of light and buzzer. This type of instrumental conditioning also led to an increase in the radioactivity of polysomes from brain. Classical conditioning, however, was not sufficient to produce such an increase. Presumably, the *classically conditioned* mice learned to associate light and buzzer with shock, but this type of learning was not accompanied by changes in amounts of radioactivity in polysomes from brain.

Table VI summarizes the effects of prior exposure to a number of experiences on the incorporation of radioactivity into polysomes from brain. The four groups of mice were given a daily 15-minute exposure to the experiences described in Table VI for two days. On the third day the mice

TABLE VI

EFFECTS OF PRIOR EXPERIENCE ON THE INCORPORATION OF URIDINE INTO POLYSOMES FROM BRAINS OF MICE DURING TRAINING[a]

Prior Experience	Number of Pairs	Change in Radioactivity in Polysomes	Learning
Avoidance conditioning I	5	0	No
Yoked	6	+	Yes
Avoidance conditioning II	5	0	No
Classical conditioning	6	+	Yes

[a] Modified from Adair (1968).

The experimental procedures were exactly as described in Tables I and V. The experiences of the mice were described in Tables I and V. All experimental mice received a daily 15-minute exposure for two days to the indicated experience. On the third day the mice were injected and 30 minutes later subjected to 15 minutes of avoidance conditioning I (light, buzzer, and shock on the training half of the apparatus). All groups were compared to yoked mice with no previous experience in the training apparatus.

were injected with radioactive uridine and 30 minutes later subjected to 15 minutes of training (avoidance conditioning I). All mice were paired with yoked mice with no previous experience in the training apparatus.

Mice given two days of training using the avoidance conditioning I procedure showed no increases in the radioactivity of the polysomes from the brain when required to perform on the third day. There is the possibility that this lack of difference in polysomes from the brains of previously-trained mice was not related to the absence of learning, but was due to habituation to the training apparatus. In an attempt to eliminate this possibility, mice were exposed to prior experience as yoked or classically conditioned animals. They were possibly habituated to these experiences on the yoked side of the apparatus and, when trained on the third day, were able to learn to avoid light and buzzer. Both of these groups exhibited increased radioactivity in polysomes when compared to yoked mice.

To accustom mice to the training half of the apparatus, they were conditioned to avoid the floor for two days (avoidance conditioning II). On the third day an attempt was made to train them to avoid light and buzzer. However, the mice would not stay on the floor long enough for light and buzzer to be presented. Their performance of avoiding the floor was not associated with a change in the amount of radioactivity in the polysomes from the brain.

SUMMARY AND CONCLUSIONS

In this paper we have presented the effects of different experiences on the incorporation of radioactive precursors into RNA or polysomes isolated from brains. In our experiments we used a double radioactive isotope technique to compare the effect of avoidance conditioning with other experiences. We have shown that the incorporation of radioactive uridine into RNA from the brains of *trained* mice is increased when compared to RNA isolated from *yoked* or *quiet* mice. The increased radioactivity is not associated with a general biochemical response to the training experiences, since there is no change in RNA from liver or kidney. Indeed, the increased radioactivity in RNA from brain is not due to a general change throughout the brain, but is confined to an area containing diencephalon, hippocampus, and parts of the basal ganglia.

The increased radioactivity in polysomes was observed in mice trained with either light and buzzer or the grid floor as conditioned stimuli. Increases throughout the polysome region were demonstrable in mice trained to avoid light and buzzer after prior exposure to an experience as yoked mice or to a classical conditioning paradigm of light, buzzer, and shock with no possibility of escape. However, no increase occurred when classical conditioning alone constituted the experience. Performance in the training apparatus does not lead to changes in the amount of radioactivity in polysomes, since polysomes from the brains of mice previously trained, and then allowed to perform, were the same as those from the brains of yoked mice. Bowman and Strobel suggest that their data also tend to eliminate performance as a causative factor. They compared rats undergoing reversal training in a Y-maze, with rats overtrained to their preferred side of the maze. Learning to reverse a previous choice was associated with increased incorporation of radioactive cytidine into the hippocampal area of the brain.

Two major problems of interpretation of these data are apparent. It is not possible to determine which aspects of the training situation are responsible for the increased radioactivity in RNA. It is equally difficult to determine the role of the RNA being produced.

The cause of the increases appears to be associated with instrumental conditioning. Performance of the conditioned response may be eliminated as a possible cause. Stress is not completely eliminated as a factor, but such stress would have to be uniquely associated with instrumental conditioning. It is possible that learning may represent the causative agent (Dingman and Sporn, 1961), but this does not mean that learning requires RNA synthesis, especially since learning appears to occur in the absence of RNA synthesis (Cohen and Barondes, 1966; Agranoff *et al.*, 1967). Other aspects of the training experiences still must be considered.

The functional significance of the RNA produced under these conditions also remains obscure. Unless there is some unknown function for RNA in brain, the RNA changes observed are undoubtedly involved in protein synthesis. This does not clarify the situation since *protein changes* may be substituted for *RNA changes* in Fig. 1 without eliminating any alternative. It can not be decided from these or any other data whether the observed increases are part of the early steps of memory consolidation or represent simple maintenance and recovery processes associated with neuronal functioning. The latter interpretation is suggested by results reported by Appel *et al.*, (1967) who found increased amounts of polysomes in the visual cortex of rats exposed to 15 minutes of light following three days of darkness. This brief, presumably nonlearning, situation points toward RNA involvement in simple neuronal functioning, and indicates that caution must still be exercised in interpreting the significance of RNA and protein changes in learning and memory.

REFERENCES

Adair, L. B., Wilson, J. E., Zemp, J. W., and Glassman, E., 1968. *Proc. Nat. Acad. Sci., U.S.*, **61**, 606.

Adair, L. B., 1968. Doctoral dissertation, University of North Carolina.

Agranoff, B. W., Davis, R. E., Casola, L., and Lim, R., 1967. *Science*, **158**, 1600.

Appel, S. H., Davis, W., and Scott, S., 1967. *Science*, **157**, 836.

Barondes, S. H. and Cohen, H. D., 1967. *Proc. Nat. Acad. Sci., U.S.*, **58**, 157.

Bowman, R. E. and Strobel, D. A., 1969. *J. Comp. and Physiol. Psychol.*, **67**, 448–456.

Bryan, R. N., Bliss, E. L., and Beck, E. C., 1967. *Fed. Proc.*, **26**, 709.

Cohen, H. D. and Barondes, S. H., 1966. *J. Neurochem.*, **13**, 207.

Dingman, W. and Sporn, M. B., 1961. *J. Psychiat. Res.*, **1**, 1.

Flexner, L. B., Flexner, J. B., and Roberts, R. B., 1967. *Science*, **155**, 1377.

Glassman, E., 1967. "Molecular Approaches to Psychobiology." Belmont, Calif.: Dickenson Publishing Co.

Hydén, H. and Egyházi, E., 1962. *Proc. Nat. Acad. Sci., U.S.*, **48**, 1366.

Hydén, H. and Egyházi, E., 1963. *Proc. Nat. Acad. Sci., U.S.*, **49**, 618.

Hydén, H. and Egyházi, E., 1964. *Proc. Nat. Acad. Sci., U.S.*, **52**, 1030.

Pevzner, L. Z., 1965. "Macromolecules and Behavior." J. Gaito (Ed.) New York: Applegate –Century–Croft.

Zemp, J. W., Wilson, J. E., Schlesinger, K., Boggan, W. D., and Glassman, E., 1966. *Proc. Nat. Acad. Sci., U.S.*, **55**, 1423.

Zemp, J. W., Wilson, J. E., and Glassman, E., 1967. *Proc. Nat. Acad. Sci., U.S.*, **58**, 1120.

Zemp, J. W. and Wilson, J. E., 1965. *Fed. Proc.*, **24**, 664.

QUESTION AND ANSWER

QUESTION: (Dr. H. Røigaard–Petersen, University of Copenhagen) What do you mean when you say: intracranial injection?

ANSWER: Injections were made bilaterally into the frontal lobes, approximately 2 mm from the midline, 3 mm posterior to the olfactory bulbs, and to a depth of approximately 3 mm. Similar injections with trypan blue showed that the dye was distributed throughout the sub-arachnoid space almost immediately after injection; radioautography of brain slices from mice sacrificed 30 minutes after injection showed labelling of nuclei in areas remote from the site of injection.

COMMENTS

Robert E. Bowman

University of Wisconsin
Department of Psychology
Madison, Wisconsin

Dr. Zemp and his colleagues are to be congratulated on a careful series of studies, in which the use of numerous control groups has considerably narrowed the number of possible explanations of their data. Biochemically, it seems plausible that the RNA effect they note in the *trained* mice arises from increased RNA synthesis. Of course, an increased ratio of uridine per molecule of synthesized RNA could contribute to this effect, along the lines suggested by data of Hydén and Egyházi (1964). The use of two isotopic precursors simultaneously in each mouse would be of interest in testing this possibility.

It is, at least, clear from the data that more uridine is being incorporated into RNA in the *trained* mice than in the *quiet*, *yoked*, and *shocked* mice. Zemp *et al.* suggest that this difference could result from "stress, attention, organized jumping, or some parameter related to learning." Their own controls seem to me to make stress one of the less likely possibilities. To illustrate, the *trained* mice were likely to have been intermediate between the *quiet* and *shocked* mice in terms of amount of stress, thus requiring that uridine incorporation into RNA be an inverted U-shaped function of amount of stress, if stress were to be the determining factor. Although this is possible, the matter is complicated by the necessity (for the stress theory) of then having to assume that the *yoked* mice were either like the *quiet* mice or the *shocked* mice in amount of stress, rather than like the *trained* mice which received the same number of shocks as the *yoked* mice did. This problem might be further resolved by measurements of blood corticosterone as indicants of stress for these various conditions, or by determining the curve of uridine incorporation into RNA as a function of number of shocks received.

The possibility that the uridine incorporation might be affected by differences in motor activity, such as "organized jumping," seems certainly possible from the first two studies described by Zemp *et al. Shocked* and *yoked* mice might be expected to have exhibited considerable "freezing" behavior during the test (*i.e.*, crouching and remaining motionless) They might thereby have resembled the *quiet* mice in motor activity more than they resembled the actively avoiding *trained* mice. This could account for the *trained* mice showing the most uridine incorporation. However, the experiments of Adair appear to decrease somewhat the likelihood of this explanation. Although her data cannot be adequately evaluated until presented more fully, they indicate that mice on the third day of training on this task do *not* exhibit higher uridine incorporation into brain polysomes. These data would bear stronger on this point if we knew what differences in motor activity existed between the first and third days of training, and in particular if the mice were quieter on the third day.

Adair's data (1968) would further seem to rule out generalized attentional factors as accounting for the uridine incorporation, since she failed to find increased incorporation in mice trained on a classical conditioning procedure similar to the type described by Hunt (1965) for training a conditioned emotional response (or CER). Of course, this finding could be embarassing to the memory consolidation hypothesis of this effect, since by a number of learning criteria (EEG changes, development of partially adaptive responses or of a conditioned emotional response, etc.), CER trained animals must be said to be learning. However, the learning of instrumental responses, no doubt, differs from CER learning in several ways. As one example, avoidance trained mice, but not CER trained mice, would experience a large reinforcement change consequent upon the correct response. This could result in different emotional states as well as in a difference in stimulus-response-reinforcement discrimination processes in the brain (I will return to this below).

Our own experiments (Bowman and Strobel, 1969) compliment those of Zemp *et al.* in a number of ways. As indicated in the description of our experiment by Zemp *et al.* in this symposium, we obtained an effect similar in a number of ways to their results, using a different RNA precursor (cytidine) in rats trained to an appetitive reward (water) rather than to a noxious stimulation. The RNA effect therefore has some generality across species and behavioral tasks. We also localized our effect to the hippocampus, although some of the other subcortical structures, which were all pooled for analysis, might have been found to exhibit this effect if assayed separately. In addition to an increase of about 25% in cytidine incorporation into both nuclear and cytoplasmic RNA, we obtained a 14% greater amount of total RNA in the cell nuclei of the hippocampus of our trained rats, although no change in total cytoplasmic RNA. From all measures taken together, it seems likely that we were dealing with greater amounts of RNA synthesis in trained rats.

Our experimental design would seem to rule out sensory and motor factors as contributing to the RNA effect, since both control and learning rats performed the same type of response in the same maze and received the same number of water reinforcements. The only operational difference was that the control rats performed an already learned spatial discrimination, whereas the experimental rats were trained for an average of five spatial discrimination reversals during a 60-minute experimental session following intravenous injection of tritiated cytidine. The only differences of note between these experimental conditions would appear to be in the reversal *learning* of the experimental rats, and in the possible *frustration* effects these rats may have experienced when failing to get water reward for errors of choice during learning (reinforcements missed in the choice arm of the maze were presented immediately upon the return of the rat to the start box, to equate water intake between the groups).

It is interesting to speculate that localization of altered RNA metabolism to the hippocampus by us, and to a subcortical fraction containing ventral hippocampus by Zemp *et al.*, may prove to be a most important clue as to the meaning of the effect. For example, a number of studies, cited by Green (1964), are suggestive that the hippocampus may be involved in memory processes. More specifically, Flexner *et al.* (1963) found that puromycin injections in the vicinity of the hippocampus 24 hours following training blocked retention subsequently. Recent studies by Deutsch and collaborators (1966, 1967) have found retention deficits following hippocampal injections of DFP (an anticholinesterase) and of scopolamine (an anticholinergic). The hippocampal EEG, particularly the theta rhythm (4–7 cps), has been shown by Adey and co-workers (1962) to vary in systematic fashion with discrimination learning, and these investigators have traced some of the neural relationships between the hippocampus and other structures during learning. Finally, Erickson and Chalmers (1966) have shown that when hippocampal theta rhythm was sustained via injections of an anticholinergic drug, rats failed to make lever avoidance responses to which they had already been conditioned. But the rats did retain a pronounced orientation behavior toward the CS, a behavior which was characteristic of a classically conditioned CER (in their avoidance conditioning, the rat appeared to learn a CER preliminary to acquiring the lever avoidance response). This suggests that the hippocampus may not be important in learning the CER, but may be integrally involved in the instrumental avoidance response. This possibility, and the assumption that the RNA effect of Zemp and co-workers occurred mainly in the hippocampus, could explain why Adair's data showed no RNA effect in rats trained in a CER paradigm. It would be interesting to know whether classical conditioning which included discriminative stimuli (*e.g.*, foot shock following a 5000 cps tone but not a 10,000 cps tone) would produce an RNA effect, or whether the opportunity for either escape or avoidance responses in the situation would be necessary for the RNA effect. The theory of Douglas and Pribram (1966) that the hippocampus is involved in an "error-evaluate" process would suggest in fact that responses might be necessary to the RNA effects, since this theory indicates the hippocampus may be involved only in situations where errors can occur.

It is still to early to tell whether the above measured changes in RNA synthesis might be related to emotional changes consequent on reinforcement, or to maintenance of neuronal function (as in some error-evaluate function of the hippocampus) or to consolidation of memory, or to some combination of these. However, the recent report of Agranoff *et al.* (1967) that retention of avoidance conditioning by the goldfish was impaired by an inhibition of RNA synthesis would probably be consonant only with the last of these possibilities, since the RNA inhibition was produced by actinomycin D injected only after training was concluded. Thus, although much work remains to be done to settle this issue, the possibility remains viable that RNA synthesis is integrally involved somehow in memory processes.

REFERENCES

Adey, W. R., Walter, D. O., and Lindsley, D. F. *Arch. Neurol.*, 1962, **6**, 94–207.

Agranoff, B. W., Davis, R. E., Casola, L., and Lim, R. *Science*, 1967, **158**, 1600–1601.

Bowman, R. E. and Strobel, D. A. Brain RNA metabolism in the rat during learning. *J. comp. Physiol. Psychol.*, 1969, **67**, 448–456.

Deutsch, J. A. and Rocklin, K. W. *Nature*, 1967, **216**, 89–90.

Deutsch, J. A., Hamburg, M. D., and Dahl, H. *Science*, 1966, **151**, 221–223.

Douglas, R. J. and Pribram, K. H. *Neuropsychologia*, 1966, **4**, 197–220.

Erickson, C. K. and Chalmers, R. K. *Arch. Int. Pharmacodyn.*, 1966, **163**, 70–78.

Flexner, J. B., Flexner, L. B., and Stellar, E. *Science*, 1963, **141**, 57–59.

Green, J. D. *Physiol. Rev.*, 1964, **44**, 561–608.

Hydén, H. and Egyházi, E. *Proc. Nat. Acad. Sci., U.S.*, 1964, **52**, 1030–1035.

Hunt, H. F. *Trans. N. Y. Acad. Sci.*, 1965, **27**, 923–945.

TIME COURSES OF EFFECTS OF DIFFERENTIAL EXPERIENCE ON BRAIN MEASURES AND BEHAVIOR OF RATS*

Edward L. Bennett†, Mark R. Rosenzweig‡, and Marian C. Diamond§

University of California, Berkeley

For a number of years we have been investigating the cerebral and behavioral changes induced by differential experience in hopes of shedding light on brain mechanisms of learning and memory storage. In previous publications we have shown that significant cerebral effects occur when littermate rats are exposed to enriched or impoverished environments for a period of 80 days following weaning (Bennett *et al.*, 1964; Rosenzweig, 1966; Diamond, 1967). We have also demonstrated effects of differential experience on problem-solving behavior when animals were kept in the enriched or impoverished environments for 30 days following weaning (Krech *et al.*, 1962). After reviewing these studies briefly, we will turn to results of recent experiments in which we have varied both the duration and the starting age of exposure to differential experience. This survey is based in part on published studies, in part on as yet unpublished research, and in part on work in progress; it is possible to give only some of the highlights of the current research of our group. The discussion will include a section on problems of extrapolation from studies of infra-human animals to human behavior.

*This research was supported by National Science Foundation Grant GB–5537, and has also received support from the United States Atomic Energy Commission.

†Associate Director, Laboratory of Chemical Biodynamics, Lawrence Radiation Laboratory.

‡Professor of Psychology, Department of Psychology.

§Associate Professor of Neuroanatomy, Department of Physiology-Anatomy.

ENRICHED AND IMPOVERISHED BEHAVIORAL CONDITIONS

In most of the experiments to be reported here, littermate male rats were assigned to the enriched condition (EC) or the impoverished condition (IC). In the enriched condition, rats are housed in groups of 10 to 12 in a large cage that is provided with "toys" such as ladders, wheels, boxes, platforms, etc. The cage arrangement is changed daily, with some toys being exchanged for objects from a large set. To enrich further their experience, the rats are taken out each day in a group for a half-hour exploratory session in a 3-ft. square field with a pattern of barriers that is also changed daily. Each EC rat has a littermate assigned to the impoverished condition in which animals live in individual cages with solid walls so that the animals cannot see or touch one another. These isolation cages are placed, in most experiments, in a separate, quiet, dimly-lit room. In other experiments the lighting and sound have been the same for enriched and deprived groups, with no apparent change in the EC-IC cerebral effects. Littermates are assigned to groups randomly, with the restriction that littermates differing widely in body weight are excluded and that total body weight of groups be substantially equal; the groups are then assigned randomly to experimental conditions. All of the experiments reported in this paper have employed only male rats. Unless otherwise noted, the subjects were of the Berkeley S_1 strain, but other strains have also been used to cross-check many of these results. After rats have been maintained in the differential conditions for the experimental period, they are sacrificed, and various anatomical and biochemical measures are made of their brains.

CEREBRAL EFFECTS OF DIFFERENTIAL EXPERIENCE

When animals have been kept for the 80 days after weaning in the differential conditions, we have found a number of differences in the cerebral cortex between enriched and impoverished rats; these differences include changes in tissue weight, in enzymatic activity, and in a number of anatomical measures. In 13 replications of this experiment, the most consistent result has been that the enriched-environment rat has a heavier cerebral cortex than the littermate kept in the isolated condition. Weight refers to wet weight of tissue samples, taken by our standardized dissection methods (Rosenzweig *et al.*, 1962). The consistency of the EC-IC weight differences is shown in Fig. 1, which presents results of experiments run from 1960 through 1965. Each experiment included from 10 to 12 EC-IC littermates. In the top graph the bar for each experiment indicates the percentage by which the EC group exceeded the IC group in weight of total cortex. The

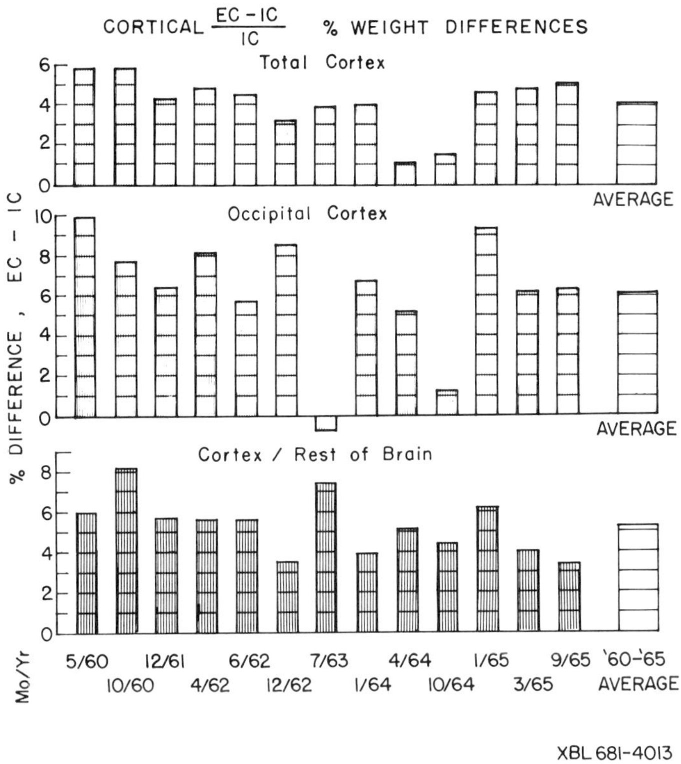

Figure 1 Percentage differences between brain weight measures of littermateS_1 rats from enriched condition (EC) and impoverished condition (IC). Stability of effects is seen in 13 experiments performed over a six-year period.

effects ranged from one to six percent, with each experiment showing a positive effect and with an overall mean EC-IC difference of slightly greater than four percent.

The effects of differential experience are larger, although somewhat more variable, in the occipital or visual region of the cortex, as is shown in the middle panel of the figure. Here it will be seen that the average effect was about a six percent difference. These weight differences in the occipital cortex have been found to be significantly larger than similar effects in the nearby somesthetic region of the cortex. Many of our histological measures have been restricted to the occipital area. Although the EC-IC effects are especially prominent in the occipital region, we do not believe that visual experience plays a major or even necessary role in producing the effects. We have found significant EC-IC differences in the occipital cortex even when we used rats that were blinded by enucleation at weaning (Krech *et al.*, 1963) or when we conducted such experiments in absolute darkness. Why the occipital region is especially susceptible to differential experience remains at present an unsolved question.

A rather stable measure of EC-IC effects has been the ratio of weight of the cortex to that of the rest of the brain. Some investigators have indicated that various cortical/subcortical ratios increase with phylogenetic development of the brain. We will show in a later section of this paper that the cortical/subcortical weight ratio is correlated with a measure of problem-solving ability. The stability of EC-IC differences in the ratio is shown at the bottom of Fig. 1.

Results for the occipital region—in tissue weight, enzymatic activity, and histological measures—are summarized in Table I. Let us now consider each of these measures and its implications briefly.

TABLE I

EFFECTS OF DIFFERENTIAL EXPERIENCE ON OCCIPITAL CORTEX (DIFFERENCES BETWEEN LITTERMATES OF S_1 STRAIN IN EC OR IC FROM 25 TO 105 DAYS OF AGE)

Measure	Diff.	P	No. EC > IC
Weight	6.1	<.001	106/141
Depth	6.3	<.001	36/41
Total protein	7.9[a]	<.001	25/32
Total AChE	2.3	<.01	83/140
Total ChE	10.2	<.001	77.5/98[b]
Total hexokinase	6.9[c]	<.01	17/21
Number of neurons	−3.1	NS	7/17
Number of glia	14.0	<.01	12/17
Perikaryon cross-section	13.4	<.001	11.5/13[b]

[a] Weight diff. 7.0% in these experiments.

[b] Fractional value indicates tied score.

[c] Weight diff. 5.5% in these experiments.

Changes in Cortical Bulk

Tissue weight, as is the case for all measures that we will discuss, is taken without knowledge of the experimental condition to which the particular animal had been exposed. A change in weight alone is not very revealing; this might be due to change in either bulk or density of the tissue; it might be due to increase of fluid or increase of formed material. The further measures serve to give more information about the ways in which the brain is changing as a consequence of differential experience. Measures of depth or thickness of the cerebral cortex indicate that it is, at least in large part, the bulk rather than the density of the tissue that is changing. It will be noted here that the increase in depth of the cortex is very similar to the increase in its weight. The increase in total protein, which closely parallels the weight difference, serves to show that our effect is not due simply to "water on the brain."

Changes in Brain Enzymes

While the foregoing changes are all closely similar in magnitude, enzymatic changes show rather different patterns. Acetylcholinesterase was chosen for study since it is an enzyme intimately concerned with transmission at many central synapses; it is found especially in neural cells. Cholinesterase, an enzyme that also splits the synaptic transmitter acetylcholine, but less specifically, is of interest because it is found primarily in the glial cells (Koelle, 1952, 1954; Giacobini, 1959). Hexokinase is an enzyme important in energy transformation, but it does not have a specifically neural role. Total acetylcholinesterase increases with enriched experience but less than does tissue weight, whereas total cholinesterase increases even more than does tissue weight (Table I). Total hexokinase increases to about the same extent as does tissue weight. The different EC-IC pattern of changes in acetylcholinesterase and cholinesterase led us to investigate whether there might be a change in relative numbers of neurons (which are rich in acetylcholinesterase) and glial cells (which are relatively rich in cholinesterase).

Histological Measures

To obtain reliable counts of neurons and glia we devised a method using composite photographs representing a large standard area of occipital cortex (Diamond *et al.*, 1966). The results of two experiments shown in Table I indicate that while differential experience does not lead to any significant change in the number of neurons, it does lead to a significant difference in the number of glia. Altman and Das (1964) independently found an increase in number of new glial cells in animals exposed to an enriched environment as compared with isolates. Finally, although the number of neurons does not increase with enriched experience, we have recently found the cross-section area of neuronal cell bodies to be greater by about 13% in EC over IC for two experiments (Diamond, 1967). If the cell body is considered to be roughly spherical in shape, then a difference of 13% in cross-sectional area indicates a difference of about 20% in volume.

Subcortical Changes with Differential Experience

Measures of weight and enzymatic activity have also been taken in the "subcortex" or "rest of brain"; we are defining these terms to mean all of the portions of brain remaining after total cortex has been stripped away. In general, the effects at the subcortex are rather small, and in some cases they are even opposite in direction from the effects found at the cortex. For example, after 80 days in the EC or IC conditions we usually find that the enriched-experience animals have slightly lesser weight of subcortex, although they have greater weight of cortex. It might also be noted that the enriched-experience animals are about nine percent less in total body weight at the end of the experiment than are their isolated brothers. A ratio of cortical to subcortical measures often gives the most reliable index of EC-IC effects.

In looking back over these cerebral effects of differential experience, we see that the brain is sensitive to experience in a variety of ways. Its weight, and the distribution of bulk between cortex and subcortex, activities of enzymes, and a number of histological measures can all be influenced significantly by the conditions under which the animals live.

DISTINGUISHING INFLUENCES OF ENRICHMENT FROM THOSE OF IMPOVERISHMENT

The results that we have given so far have been expressed entirely in terms of EC-IC differences. Are such differences to be ascribed chiefly to the enrichment of experience of the EC group or to the impoverishment of experience of the IC group, or to the joint effect of both treatments? In the absence of a baseline it is impossible to allocate the causes of these differences. Choice of an appropriate baseline is a difficult matter, but in 1964 (Bennett *et al.*) we suggested the use of the colony animal as one appropriate baseline, since the colony rat is the subject of most experimental work. (We recognize that in an ethological sense the EC environment probably approaches the "normal" environment of the rat more closely than does the SC condition.) In several experiments we have therefore taken three male animals from each litter and have randomly assigned one of the animals to remain in the standard colony (SC) condition. Fig. 2 shows results obtained when we compare EC animals with the SC littermate baseline and similarly compare IC results with the SC baseline. The figure is based on measures made on total cortex and shows results both for animals run from 25 to 105 days of age and for animals run from 105 to 185 days of age. In the case of weight of total cortex we see that the EC-SC difference is larger than the corresponding IC-SC difference; this is especially the case for the older animals.

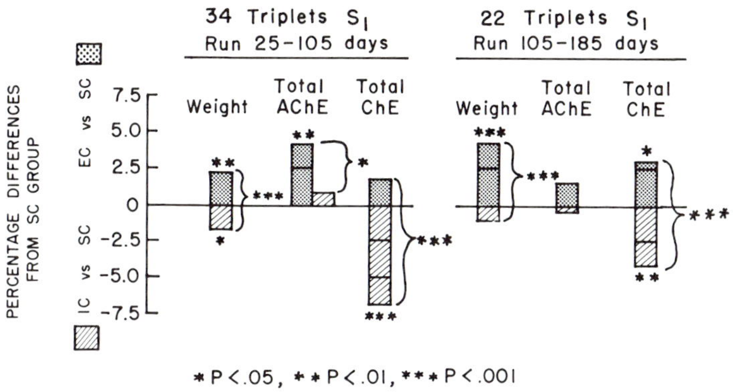

EFFECTS OF ENRICHMENT OR IMPOVERISHMENT VS. COLONY CONDITIONS ON TOTAL CORTEX MEASURES

XBL 674-1051

Figure 2 Effects of enrichment or impoverishment on total cortex measures, taken against standard colony (SC) littermates as a baseline. The left half of the figure is based on three experiments using male triplets from 34 S_1 litters, with exposure to the differential conditions extending from 25 to 105 days of age. The right half is based on two experiments, including male S_1 triplets from 22 litters, the exposure extending from 105 to 185 days of age. The differences in cortical weight and total AChE activity are seen to be larger between EC and SC than between IC and SC; that is, they are due more to enrichment than to impoverishment of experience. On the other hand, impoverishment (IC vs. SC) produces larger differences in total ChE activity than does enrichment (EC vs. SC).

In total activity of acetylcholinesterase, again the larger difference is found between EC and SC than between IC and SC. For activity of cholinesterase, the pattern is reversed, the larger difference occurring between the IC and SC littermates, rather than between EC and SC. Thus, it appears that the EC-IC differences in weight and total AChE should be attributed largely to enriched experience, whereas the differences in total ChE activity must be attributed chiefly to the restricted experience of the IC animals. This is one of several sets of results indicating that various consequences of differential experience do not necessarily covary as the type, duration, or starting age of experience are modified.

CEREBRAL EFFECTS OF VARIED DURATIONS OF DIFFERENTIAL EXPERIENCE

A few years ago we had not only found that the brain was quite plastic to effects of experience, but we had also shown that differential experience of 30 days duration after weaning led to significant differences in problem-solving behavior (Krech *et al.*, 1962). In order to test whether the cerebral and the behavioral effects of experience were critically related, we decided to vary the length of exposure to enriched or impoverished conditions and to study the effects on both brain and behavior. We anticipated that both sorts of effects would increase in magnitude with increasing duration of exposure, at least up to a ceiling. The design of such experiments in which both duration of EC-IC experience and the starting age are varied is shown in Fig. 3. For each age and duration of exposure, there are duplicate pairs

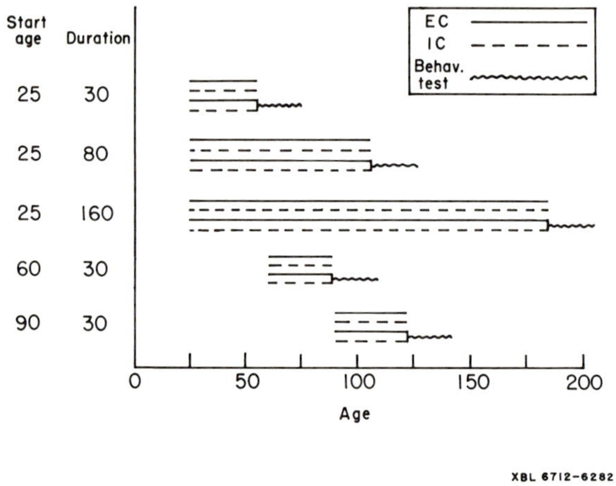

Figure 3 Design of experiments to test effects of varied durations and varied starting ages of EC or IC experience on brain and on behavior. For each starting age and duration, one pair of EC–IC groups is sacrificed for measurement of brain values at the end of the period of differential experience; the other pair is trained and tested behaviorally, then sacrificed.

of groups, one EC-IC pair being sacrificed immediately at the end of the experimental period, while the other pair is pretrained and tested on a behavioral test and then sacrificed. The groups sacrificed at the outset of behavioral testing indicate the cerebral values found at this point. It appears that pretraining and testing further alter brain measures, but this is a story that we will have to leave for another time. Let us turn first to the cerebral effects of EC-IC experience for various durations and with various starting ages. We will limit ourselves here to observations of brain weights, since data on other cerebral measures are not yet as complete.

Fig. 4 shows the increase of weight of total cortex in animals put into the enriched or impoverished situations at 25 days of age. This figure is based on three pairs of EC-IC groups, all taken from the same breeding of S_1 rats. One paired group was taken out of the EC-IC conditions after 30 days (at

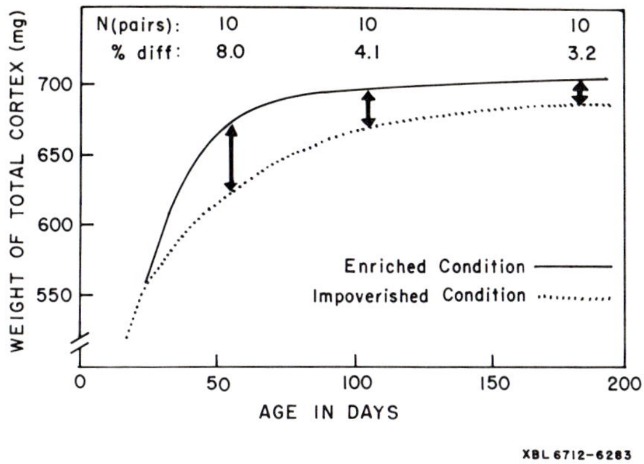

Figure 4 Growth of weight of total cortex for S_1 rats put into enriched or isolated conditions at 25 days of age. Ten EC–IC littermate pairs were sacrificed at 55 days of age, ten pairs at 105 days, and ten pairs at 185 days. The relative difference between groups was greatest after 30 days of differential experience (at 55 days of age) and declined thereafter.

55 days of age), a second pair was taken out of the differential conditions after 80 days (at 105 days of age), and the third pair was removed after 160 days of differential experience (at 185 days of age). Both the EC and IC curves show growth after 25 days of age, but the EC (solid) curve spurts up more rapidly at the start. The difference between the curves increases over the first month and then the two curves come closer together. At 55 days of age the EC-IC difference amounted to eight percent; at 105 days of age, the difference had dropped to 4.1% and at 185 days of age the difference had decreased still further to 3.2%. These results can be added to those of other experiments for the same three EC-IC durations, but where the various groups were not all taken from the same breeding (see Table II). Table II shows results not only for weight of total cortex, but also for the weight of the rest of the brain (subcortex) and for the ratio of weight of cortex to that of the rest of the brain. It will be seen that the EC-IC difference in total cortex weight decreases steadily with duration of EC-IC experience. The EC-IC difference in weight of rest of brain changes from a small positive effect at 30 days to a significantly negative effect at 180 days. The ratio of weight of cortex to the rest of the brain shows an EC-IC difference that is both highly significant and substantially constant for all three EC-IC durations.

We have done two sets of experiments in which animals were placed in the differential conditions at 60 days of age and the subgroups were removed after 15, 30, 45, or 60 days; results of these experiments are shown in Table III. Here we see that EC-IC differences for both cortex and rest of brain increased from 15 to 30 days and then declined. The differences in cortical/

TABLE II

PERCENTAGE DIFFERENCES IN BRAIN WEIGHT BETWEEN S_1 LITTERMATES PUT IN ENRICHED OR IMPOVERISHED CONDITIONS AT WEANING

EC–IC Duration	30 days	80 days	160 days
N (pairs)	32	151	22
Total cortex	7.4[c]	4.1[c]	2.5[b]
Rest of brain	1.5	−1.0[b]	−3.3[b]
Cortex/Rest	5.8[c]	5.2[c]	6.1[c]

[a] $P < .05$.

[b] $P < .01$.

[c] $P < .001$.

subcortical ratio were statistically significant and quite constant for all four durations. If we exclude the 15-day column in which the effects may not yet have become fully developed, the trends in the rest of Table III resemble rather closely those of Table II in which animals were put in the EC-IC conditions at 25 days of age and left there for various durations. Although the effects beginning at 60 days were smaller than those for experiments beginning at 25 days of age, in both cases the percentage magnitude of the cortical effect reached a maximum at about 30 days and then declined, as did the magnitude of the effect in the rest of the brain, leaving the EC-IC difference in ratio of cortex to rest of the brain substantially constant.

TABLE III

PERCENTAGE DIFFERENCES IN BRAIN WEIGHT BETWEEN S_1 LITTERMATES IN ENRICHED OR IMPOVERISHED CONDITIONS FOR VARIOUS DURATIONS BEGINNING AT 60 DAYS OF AGE

Duration	15 days	30 days	45 days	60 days
N (pairs)	21	20	21	20
Total cortex	3.7[c]	6.2[c]	3.9[c]	3.2[b]
Rest of brain	0.5	2.6[a]	0.3	−0.4
Cortex/Rest	3.2[c]	3.5[c]	3.6[c]	3.7[c]

[a] $P < .05$.

[b] $P < .01$.

[c] $P < .001$.

Varied Starting Ages

Now let us see what happens when the EC-IC duration is held fixed at 30 days but the starting age is varied; this design is shown in Fig. 3 in the top, fourth, and fifth sets of lines. The upper half of Table IV shows results obtained when the 30-day EC-IC period was started at 25, 60, or 90 days of age. It should be noted that these experiments were taken from different

TABLE IV

PERCENTAGE DIFFERENCES IN BRAIN WEIGHT BETWEEN S_1 LITTERMATES IN ENRICHED OR IMPOVERISHED CONDITIONS FOR 30 OR 80 DAYS, STARTING AT DIFFERENT AGES

	30-day EC–IC duration, starting at age		
	25 days	60 days	90 days
N (pairs)	32	63	10
Total cortex	7.4[c]	6.1[c]	4.6[a]
Rest of brain	1.5	2.1[c]	4.0[a]
Cortex/Rest	5.8[c]	3.9[c]	0.5
	80-day EC–IC duration, starting at age		
	25 days		105 days
N (pairs)	151		45
Total cortex	4.1[c]		5.4[c]
Rest of brain	−1.0[b]		1.7[a]
Cortex/Rest	5.2[c]		3.7[c]

[a] $P < .05$.
[b] $P < .01$.
[c] $P < .001$.

brccdings. When the start of the experimental period was delayed beyond wcaning at 25 days of age, the animals were kept in standard colony conditions (three to a cage) prior to being assigned to the EC or IC condition. Starting the EC-IC period at different ages appears to have differential effects depending on whether one is examining total cortex, rest of brain, or the ratio of cortex to rest of brain. The EC-IC effects on weight of total cortex were largest when the treatment was initiated at 25 days of age, slightly less when they began at 60 days of age, and apparently smaller at 90 days of age. (The 90-day results should be considered with caution, since they are based on only a single experiment with ten pairs of animals.) For the subcortex the results go in the opposite direction; the EC-IC differences were small and not statistically significant for experiments started at 25 and 60 days of age, but for the one experiment started at 90 days of age they reached four percent ($P < 0.05$). Since the results for cortex and subcortex went in opposite directions, the ratio of cortex to the rest of the brain showed clearly the largest results in experiments started at 25 days of age, smaller results at 60 days and nonsignificant results in the experiment started at 90 days of age.

The lower part of Table IV shows the effects of 80-day EC-IC, beginning either at 25 or 105 days of age. The effects seen here are rather similar to those in the upper half of the table, for 30-day EC-IC starting at three different ages. The dissimilarity is that in the 80-day experiments there was a slightly larger effect on weight of total cortex for the experiments begun at 105 days of age than for the experiments begun at 25 days of age. For rest of brain and for the ratio of total cortex to rest the results were similar to those seen in the 30-day experiments; that is, the effect for rest of brain was larger when the start of the experiment is delayed to 105 days and the effect in cortical/subcortical ratio was larger with animals started at weaning.

Distinguishing Effects of Enrichment from Normal Maturation of Brain

Is it possible that the effects of enrichment which we have been describing represent merely an acceleration of normal maturation of the brain? We believe not, for several reasons: (a) Significant increases in cortical weight are obtained even when the growth of brain in colony animals has practically ceased. (The brain of the rat never completely stops growing, but the rate of growth is slow after about 100 days of age.) Fig. 5 shows changes in cortical weight when two animals of a litter were put into the EC or IC conditions at 105 days of age, the third littermate remaining in the colony condition. The EC curve is seen to rise promptly and significantly above the SC plateau. (b) Total AChE activity of brain remains practically constant after about 80 days of age (Bennett *et al.*, 1958). Enriched experience begun after this age nevertheless leads to a significant increase in total AChE activity. (c) Perhaps the clearest evidence distinguishing the two factors is that maturation and enriched experience lead to opposite effects in depth of cerebral cortex. The cortex of the rat normally decreases progressively in thickness, at least after weaning, whereas enriched experience as we have seen in Table I leads to a significant increase in cortical depth in relation to the impoverished condition. The cortex on the dorsal aspect of the occipital area measures 1340 micra in depth in the 25-day-old S_1 rat, but only 1200 micra in the 100-day-old rat kept under standard colony conditions. (These measures were made on formalin-perfused frozen sections.) This decrease in cortical thickness with age apparently occurs as the cortex, which

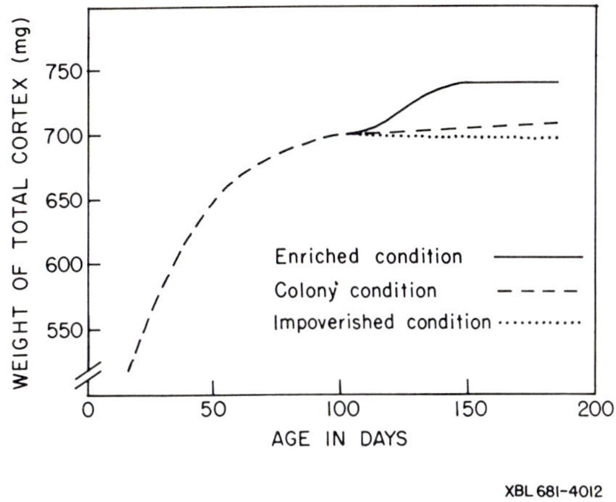

Figure 5 Growth of weight of total cortex for S_1 rats put into enriched or isolated conditions at 105 days of age. A third animal in each litter remained in the colony condition. All animals were sacrificed at 185 days of age. The 185–day values were based on three experiments including three male rats from each of 34 litters. The initial 105-day point and the intermediate values are estimated from other experiments.

is adding weight more slowly than the rest of the brain, is stretched thinner to cover the subcortex. There is an anterior-posterior cortical gradient in both the age and the experience effects, such that the posterior or occipital region shows both the greatest decrease of depth as a function of age and the greatest increase of depth as a function of environmental enrichment.

SUMMARY OF CEREBRAL EFFECTS OF DIFFERENTIAL EXPERIENCE

Let us emphasize six of the results brought out in the foregoing discussion:

(a) The EC-IC cortical weight differences are to be attributed more to enrichment than to restriction of experience (Fig. 2).

(b) Increasing the length of the EC-IC experience, at least beyond 30 days, does not enhance the EC-IC differences found in measures of weight of brain tissue (Fig. 4 and Tables II and III).

(c) This sensitivity of the brain to experiential influence is not restricted to weanlings, for significant changes in brain weights are found when animals are put into the experimental conditions at 60, 90, or 105 days of age (Figs. 2 and 5, Table III).

(d) For 30-day EC-IC experience, the largest effects in weight of cortex and in the cortical/subcortical ratio are found when the animals are put into the experimental conditions at weaning; for the 80-day duration this appears clearly only for the cortical/subcortical ratio (Table IV).

(e) Effects of differential experience can be distinguished from changes in the brain due to maturation (see preceding paragraph and Fig. 5).

(f) Several weight measures yield different temporal courses as the animals are kept for varying periods in the experimental conditions (Table II). Weight of total cortex shows a transitory component in the EC-IC effect, the difference first increasing to 30 days and then decreasing. On the other hand the cortical/subcortical ratio shows an EC-IC difference that appears to be relatively constant as the duration of the experiment is varied. Having found these differences in time courses, we are now examining time courses of EC-IC effects on brain chemistry and on histological measures; we would expect to find a variety of time courses among these measures as well.

EFFECTS OF DIFFERENTIAL EXPERIENCE ON PROBLEM-SOLVING BEHAVIOR

Now let us turn to experiments performed to determine whether effects of differential experience on problem-solving behavior might be related to the effects produced on the brain. We had earlier reported (Krech *et al.*, 1962) that when littermate animals were put into EC or IC environments at 25 days of age and left there for one month, the EC animals were significantly better in solving visual reversal discrimination problems in the Krech Hypothesis apparatus. Since then we have also found, but not published, that the same period of EC-IC experience would differentiate animals in performance on the Lashley III and Dashiell Checkerboard mazes; in both cases the EC animals performed significantly better than their IC littermates. In all of this earlier behavioral work, the EC-IC period was restricted to the 30 days following weaning. We therefore decided to vary the duration and starting age of EC-IC experience preceding behavioral tests, as described for brain and behavior in Fig. 3, and in 1967 we ran ten behavioral experiments of this sort.

At the end of the EC-IC period, the rats were removed from the differential environments and were recaged in individual colony cages. The experimenters who tested the rats did not know from what condition any individual animal came, so that the behavioral tests were run "blind," just as were the chemical and anatomical determinations.

The Visual Reversal Discrimination Test

In all of the 1967 behavioral work we used the visual reversal discrimination problem in the Krech Hypothesis apparatus. Reversal discrimination tasks, as Warren (1965) has pointed out, have certain characteristics required of a test of animal intelligence in that performance is correlated with both phyletic status and maturation. The Krech apparatus is a linear maze with four successive Y-shaped choice points. At each choice point the right or the left alley is illuminated, and the experimenter can make either of the alleys correct on a given trial by changing the position of a swinging door at the end of the unit. In a ten-day pretraining period, the animals were accustomed to the 24-hour food deprivation schedule. They were pretrained in a straight runway to leave the start box, run through doorways, and enter the goal box. Following the pretraining period the nine-day testing session began; on each day there were ten trials in the four-unit apparatus, thus giving 40 choices. On each trial the pattern of lights and doors followed a different preset order. The reversal discrimination schedule followed that used by Krechevsky (1932). The animals were first trained to go to all the lighted alleys. To reach criterion the animal could not make more than one error in five successive trials; *i.e.*, 19 correct choices out of 20. When an animal reached criterion, the very next trial (which might occur on the same test day) began the first reversal problem, dark correct. Training on the dark-correct problem was continued until the criterion was reached, whereupon the problem was switched to light correct. During the nine days most animals solved four or more reversal problems.

The use of an automatic recording system made scoring somewhat more reliable than in our previous work. In the earlier work observers had some difficulty in accurately scoring choices, especially on trials when an animal darted back and forth, making "vicarious trials and errors." For the automatic scoring system, segments of the floor of the apparatus were painted with electrically conductive paint. Entry of the rat into either alley at a choice point completed a circuit and lighted a light on a panel. Holding circuits kept the record of each of the four choices until the experimenter copied them down and reset the device at the end of each trial.

Behavioral Results

Results of five experiments, all done with S_1 rats from a single breeding, are shown in Fig. 6. In experiments A, B, and C, the animals were put in EC or IC at 25 days of age and were kept in the differential conditions for 30, 80, or 160 days, respectively. (These EC-IC periods coincided with those of the rats which furnished data for the curves of cortical growth in Fig. 4.) In experiments D and E, 30-day EC-IC periods started at 60 days and 90 days, respectively.

In each experiment, animals solved the initial light-correct problem (L-1) with relatively few errors. The next problem, dark-correct (D-1), involved

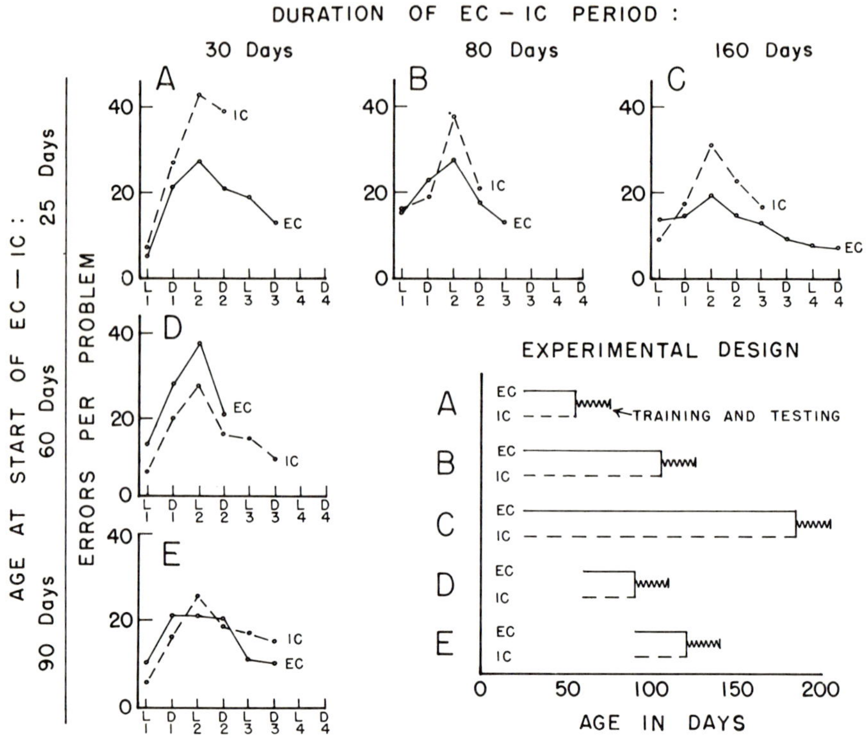

XBL 681–4020

Figure 6 Errors per problem on successive visual reversal discrimination problems, for groups exposed to enriched or isolated conditions for various durations and from various starting ages. The design of the experiment is shown in the lower right section of the figure. The successive problems are designated as L–1 (first light-correct problem), D–1 (first dark-correct), L–2 (second light-correct), and so on. Each curve extends through the last problem solved by at least half the members of the particular group. Errors increase over the first two reversals (through L–2), and then decrease. Relative errors of the groups, depending upon the particular EC–IC conditions, are discussed in the text.

more errors. Re-reversing to light-correct (L-2) was harder still. Thereafter, the animals reversed more readily, and errors were reduced. Each curve is drawn only to the last problem that was solved by at least half the animals in the group. Thus, in experiment A, at least half the EC rats solved the third dark-correct problem (D-3), while the last problem solved by at least half the ICs was D-2. In experiments A, B, and C—the groups assigned to the experimental conditions at 25 days of age—the EC curve lies below the IC curve at most points, and the EC curve stretches further out to the right.

In experiment D—where EC-IC lasted from 60 to 90 days of age—the results are reversed; the EC rats made more errors per problem and solved fewer problems than their IC littermates. In experiment E (EC-IC from 90 to 120 days of age), neither group was clearly superior to the other. Where differences exist between groups, they tend to be largest, both absolutely and relatively, on problem L-2. Thereafter the curves tend to converge, so that differences in scores on this test as a function of earlier experience do not persist.

The five other experiments run in 1967 replicated experiments A, D, and E and added two groups that were kept in EC-IC from 25 to 85 and from 60 to 120 days of age. Data for all ten experiments are presented in Table V.

TABLE V

RELATIVE PERFORMANCE ON VISUAL REVERSAL DISCRIMINATION AFTER ENRICHED (EC) OR IMPOVERISHED (IC) EXPERIENCE OF VARIOUS DURATIONS AND DIFFERENT STARTING AGES (S_1 RATS)

Period	Terminal Date	N (litters)	EC/IC Mean Number Problems Solved	P	EC/IC Mean Errors Per Reversal	P
25–55 (A)[a]	7/67	10	1.45	<.01	0.64	<.01
25–55	11/67	7	1.30	<.10	0.66	<.05
25–85	12/67	10	1.10	—	0.93	—
25–105 (B)	9/67	10	1.36	<.10	0.78	—
25–185 (C)	11/67	10	1.27	<.05	0.66	<.01
60–90 (D)	8/67	10	0.63	<.01	1.60	<.05
60–90	12/67	10	0.88	—	1.31	<.10
60–120	4/67	10	1.13	—	0.83	—
90–120	4/67	10	0.86	—	0.93	—
90–120 (E)	9/67	10	0.89	—	1.15	—

[a]The letters correspond to the group designations in Fig. 6.

The results are expressed in two ways: (a) The EC/IC ratio of mean numbers of problems solved during the nine days. If the animals that had been in EC solve more problems than their IC littermates, then the EC/IC ratio on this measure will be significantly greater than 1.0. (b) The EC/IC ratio of mean number of errors made per reversal problem. If the EC animals make fewer errors than their IC littermates in solving an average problem, then the EC/IC ratio on this measure will be significantly less than 1.0.

Looking at the upper half of the table, for all experiments in which the animals were put into the differential conditions at about 25 days of age (weaning age), we see that the expectation of superior performance on the part of the EC animal appears to be justified. The EC animals in general solved significantly more problems and made significantly fewer errors per reversal problem than the ICs. Analyses of variance on the scores confirm these impressions: in number of reversals completed, the actual scores did not differ significantly among the five experiments, and the overall EC-IC difference was highly significant ($P < 0.001$). There was not a significant interaction between experiment and EC-IC effect, so all experiments contributed similarly to the overall effect. The other measure—errors per reversal completed—was not as stable, since actual scores varied significantly among experiments ($P < 0.01$). Nevertheless this measure also showed both a highly significant overall EC-IC effect ($P < 0.001$) and no significant difference among effects in the various experiments. Thus, when S_1 rats were put into EC or IC at 25 days of age, the difference in problem-solving ability in favor of the ECs could not be distinguished according to whether the animals remained in the experimental conditions for 30, 60, 80, or 160 days. In this connection, it might be recalled that one of the brain weight measures—the ratio of weight of cortex to that of the rest of the brain—also showed EC-IC differences that were significant and of about the same magnitude regardless of length of treatment, when this started at 25 days of age.

Now let us look at the lower parts of Table V which show behavioral results when animals were put into the EC or IC conditions at either 60 or 90 days of age. It will be recalled that prior to being put into the experimental conditions, all animals were held in standard colony (SC) conditions, three animals to a cage.

The replication of 60–90-day experiment reproduced the results of the first, in which EC solved fewer problems and made more errors per reversal problem than did the ICs. Taken together, the two experiments show the ECs to be significantly worse on both counts ($P < 0.01$). For the 60–120-day experiment, the EC and IC groups did not differ significantly. The two 90–120-day experiments did not show significant differences either, although the EC solved somewhat fewer problems in both cases.

Considering all of the experiments of 30 days' duration, we see a puzzling sequence as the starting age is moved from 25 to 60 days, and from 60 to 90 days. In the 25–55-day experiments, the EC groups performed significantly better than the ICs. Surprisingly, in the 60–90-day experiments, the order was reversed; the ECs made more errors and the ICs made less errors than in the 25–55-day experiments. In 90–120-day experiments, neither replication showed a clear difference between the groups. These data give a striking indication of a critical period for beneficial behavioral effects of environmental enrichment. Before we rely upon them, however, we will want to see both further replications of these experiments and extensions to other behavioral tests. Such replications and extensions are planned for 1968.*

Just as in the case of the cerebral differences, one can ask whether the behavioral EC-IC differences found among animals put in the conditions at weaning are due chiefly to enrichment or chiefly to impoverishment of experience. As in the case of the cerebral measures, we have used the standard colony (SC) condition as a baseline against which to measure effects of enrichment or of impoverishment of experience. In the three experiments placed at the top of Table V we actually used not two but three animals from each litter. One of the animals was kept in SC while its brothers were put in EC or IC. Table VI shows the relative performance on the visual reversal discrimination tasks of the SC and IC littermates. It will be seen that on both measures for each of the first two experiments the SC animals were significantly better than their IC brothers. In fact, when SC/IC ratios of Table VI are compared with the corresponding EC/IC ratios of Table V, it will be seen that in five of six comparisons the SC animals were somewhat better than their EC brothers. Thus, it appears that performance on the visual reversal discrimination task can be significantly impaired if an animal is placed in isolation immediately after weaning, but that the performance on this task cannot be improved by giving an animal enriched experience

TABLE VI

RELATIVE PERFORMANCE ON VISUAL REVERSAL DISCRIMINATION AFTER COLONY (SC) OR IMPOVERISHED (IC) EXPERIENCE (S_1 RATS)

Period	Terminal Date	N (litters)	SC/IC Mean Number Problems Solved	P	SC/IC Mean Errors Per Reversal	P
25–55	7/67	10	1.36	<.01	0.62	<.01
25–55	11/67	7	1.42	<.05	0.64	<.05
25–85	12/67	10	1.36	<.10	0.80	—

*In a recent experiment, we have used the Lashley III maze to test rats kept in EC or IC conditions either from 25 to 55 or from 60 to 90 days of age. Both initial and repetitive entries were scored; initial errors are first entries into a given cul-de-sac on a trial, while repetitive errors are all subsequent entries on the trial. For both starting ages, the EC animals made significantly fewer initial errors than their IC littermates. The EC animals also made fewer repetitive errors in both cases, but this difference was significant only for the 25–55-day groups. In no case were the IC animals superior on the Lashley maze, as they had been in the 60–90-day group on the visual reversal discrimination test. This emphasizes the importance of using a variety of tasks in order to test the generality of behavioral effects.

over and above the colony experience. This may also help to explain why IC behavior was not inferior when animals were placed in the differential conditions only at 60 or 90 days of age. Apparently, once an animal has been in the colony condition for a month after weaning, it is protected against the deleterious effects of restriction of experience.

The behavioral results on this test thus indicate that the difference between the EC and IC animals should be attributed to early restriction of experience and not to enrichment of experience. In this sense the behavioral results resemble somewhat those found with ChE activity, in distinction to the cerebral results found with AChE activity or with wet weight. It will be recalled that in ChE activity of the cortex the main difference was found between the IC and SC groups, whereas in both AChE activity and weight of cortex the main differences were found between the EC and SC groups. Having found that the visual reversal discrimination task is not sensitive to enrichment of experience, we are now trying other behavioral tasks to see whether the same will be true of them or whether certain tasks will show effects of enrichment, rather than of impoverishment of experience.

CORRELATIONS BETWEEN BRAIN WEIGHTS AND BEHAVIORAL MEASURES

The behavioral data can fruitfully be considered from another aspect—correlations between brain measures for individual rats and individual behavioral scores. Several years ago we had reported a significant correlation between the cortical/subcortical ratio of brain weights and the mean errors per reversal problem solved (Krech *et al.*, 1962). The correlation was negative, meaning that the greater the ratio of cortex to the rest of the brain in an individual rat, the fewer the errors that animal tended to make. This result was obtained for rats kept in the enriched environment from 25 to 55 days and tested over 18 days. For a concurrent IC group, no significant correlation was obtained. We suggested that the brain values of rats from EC would probably be less affected by training and testing than would rats that had been in IC. ". . . such end-point determinations presumably reflected rather accurately the relatively stable cerebral values of the EC group during the problem-solving phase and less accurately the changing values of the IC group, thus permitting higher correlation of brain measures with behavior for the EC than for the IC group" (1962, 807).

We then tested the generality over strains of the relation between brain weight ratio and behavior, giving all rats EC experience (Rosenzweig *et al.*, 1967). Two hundred fifty-two rats of ten strains were tested. The period of testing was reduced in this work to nine days, since scores had tended to convergence in the latter half of the longer period employed previously. These experiments yielded a significant mean correlation of -0.27 ($P < 0.001$) between the brain weight ratio and errors per reversal problem. Each strain was tested in two experiments, and 17 of the 20 individual coefficients were negative; nine of the ten strains gave overall negative correlations, thus indicating that the relationship is indeed general over strains. For the S_1 strain, the correlation was -0.32 ($P < 0.1$, $N = 33$). Some additional data and scatterplots based on 299 rats have been published (Bennett and Rosenzweig, 1968).

The behavioral experiments run in 1967 and described earlier in this paper allow us to test further the reliability of this brain-behavior relation. The results also permit us to determine whether this relation appears most clearly in rats that had been in the enriched condition prior to behavioral testing. For the three experiments that included EC, SC, and IC groups, the respective correlations between the cortical/subcortical ratio of brain weight and mean errors per reversal were the following: EC, -0.42 ($P < 0.05$); SC, 0.06 (NS); and IC, 0.23 (NS). For all five experiments in which the rats were assigned to conditions at 25 days of age, the correlations were these: EC, -0.30 ($P < 0.05$), and IC, 0.08 (NS). Four of the five EC groups had negative correlations, the exception being the 25–185-day group which yielded a slightly positive coefficient (0.06). Thus the present EC correlations reproduce quite closely the S_1 value of -0.32 reported in the 1967 paper (all of those experiments having been done in 1966 or earlier). Furthermore, in the present study only animals from the enriched environment yielded a significant relationship, as was true in the original study reported in 1962.

Now let us look at the results for animals kept in EC from 60 to 90 days of age. Here the within-group brain-behavior relation departs from that of the groups started at 25 days, just as the 60–90-day error-score relation was reversed between the EC and IC groups. The two 60–90-day EC groups yielded positive correlations of 0.19 and 0.44, the combined coefficient not being significantly different from zero, but certainly departing from the -0.30 found for the groups put into EC at 25 days. Curiously, the one 60–120-day EC group yielded a correlation in the "usual" range (-0.34). The two 90–120-day EC groups gave negative brain-behavior correlations of -0.18 and -0.34, the combined coefficient not being significant but resembling closely that of the groups started in EC at 25 days. Thus the relation between the cortical/subcortical weight ratio and ability to solve reversal problems has been abundantly confirmed for groups put into the enriched environment at weaning, but, for reasons as yet unknown, such a relation was not obtained with rats kept in EC from 60 to 90 days of age.

In the experiments reported in the previous section in the present paper, it will be recalled that no consistent relation was found between group means of cerebral weight measures and mean scores on the reversal discrimination task. In line with results reported in earlier papers, when animals were assigned to EC or IC at weaning, the EC animals were superior to the IC littermate group in problem-solving scores and the EC group had the larger cortical/subcortical weight ratios; this was true regardless of the duration of the EC-IC period. Use of an additional group of SC animals now provided a critical test. The EC animals also had larger cortical/subcortical weight ratios than their littermates, yet the ECs were not superior to the SCs in behavioral scores. The groups assigned to differential environments at 60 or 90 days of age also negated the hypothesis that greater cortical/subcortical weight ratios are generally related to superior problem-solving ability. In the 60 to 90-day experiments, the EC rats were actually significantly poorer than the ICs in behavioral scores, although the ECs had the higher cortical/subcortical ratios. In the 90 to 120-day experiments, the ECs did not differ significantly from the ICs in behavior, although again the ECs had the higher brain ratio.

The results obtained with differential experience from 25 to 55 days—both for within-group correlations and for between-group differences—are not upset by the results of the 60 to 90-day experiments, but our ability to generalize on the basis of the earlier results is now considerably restricted. An overall relation between the cerebral and behavioral effects of differential experience has not so far emerged from the range of conditions studied.

DISCUSSION

The discussion will be in two parts—a section on brain plasticity and a section on problems of attempting to extrapolate from animal studies to human behavior. The section on brain plasticity will be brief because one of us has just extensively reviewed this subject elsewhere (Rosenzweig and Leiman, 1968).

Brain Plasticity

It has long been speculated that training and stimulation might lead to changes in the sizes of parts of the brain, in the interconnections of its cells, and in its chemical composition. Until very recently, however, the evidence seemed to show that the brain was very stable in its anatomy and composition. Now we and other investigators are finding that many aspects of the brain can be altered readily by treatments that involve only changing the experience of the subjects.

The variety of cerebral changes with experience is great, ranging from relatively gross measures such as weight or depth of cortex, to refined measures of enzymatic activity and cross-section areas of perikarya and their nuclei. These changes are not large in percentage terms, but they are consistent and reproducible.

Even more delicate measures are being undertaken. For example, Cragg (1967) has been studying possible changes in synaptic size and number with experience. He suggested that a reasonable first approach was to look for synaptic changes after first exposure of dark-reared animals to light, since differences in visual experience have been found to give rise to morphological effects in the visual cortex. We had shown that raising rats in the dark reduces the weight of visual cortex (Krech *et al.*, 1963). Gyllensten *et al.* (1959, 1965) reported that mice raised in the dark have smaller nuclei in cells in the upper half of the visual cortex and less internuclear material per nucleus. Cragg therefore exposed dark-reared rats to light for various periods and compared them with littermates kept in the dark. The results suggested "that exposure to light promotes the formation of new synapses of small diameter in the deeper layers of the visual cortex, while the synapses already formed in the superficial half of the cortex grow slightly larger."

When we began our work, we supposed that it would be difficult to demonstrate cerebral responses to differential experience. It was for this reason that the experience was begun at weaning because we thought that the adult brain might not show any changes due to experience. Furthermore, we chose the relatively long duration of experience of 80 days, and we exposed the animals to the experimental situations for 24 hours a day. The experiments described above have shown that the brain is more readily affected than we had supposed, and we can now add further information along this line. The first requirement that we found could be relaxed was the age at onset of the differential experience. Rats did not have to be placed in the conditions at weaning; quite similar effects could be obtained when rats were started in experiments as young adults at 105 days of age. Then, also as reported above, it was found that reducing the duration of the EC-IC period to 30 days did not diminish the cerebral effects. Indeed, for weight and total enzymatic activity of total cortex, the results of 30-day experiments were greater than those of 80-day duration. Recently, in experiments being reported elsewhere (Rosenzweig *et al.*, 1968), we have found that exposing rats to two hours of EC per day for either 30 or 60 days brings about as large cerebral changes as does exposing them for 24 hours a day for the same period. Furthermore, in recent experiments with two hours of daily exposure, the IC animals were caged in the same room and same type of cages as their EC littermates. The treatment of the two groups differed only in the two-hour EC period, yet significant brain differences were produced.

The fact that the brain is plastic in many ways to environmental demands has thus been clearly demonstrated, and detailed evidence is now coming in from a number of laboratories. How to relate this plasticity to learning and storage of memory remains a problem on which less progress has as yet been made. It represents the major challenge before us.

Problems of Extrapolating from Animal Studies to Human Behavior

It may be tempting to try to extrapolate from our studies on the rat—or from other animal experiments—to human behavior. People often ask us questions such as these: "What do your experiments suggest about the importance of early enrichment of children's experience for adult intelligence?" "Can later enrichment compensate for early neglect?" Extrapolation to man might be justified if either of two conditions held: (a) One condition would be that a variety of infra-human mammalian species has been tested adequately, and similar effects of enrichment or impoverishment of experience on behavior have been found for all species. For example, it might be found that prolonged early restriction of experience permanently impairs later learning ability. In this case, it might be hypothesized that an effect found generally in all infra-human mammals will also hold true for man. (b) In the second possible case, a variety of species has been tested adequately, and a phylogenetic progression has been found in the effects of enrichment or impoverishment of experience. For example, it might be found that effects of early restriction of experience are progressively more severe as one goes from rodents to carnivores and from carnivores to infra-human primates. In this case, it might be hypothesized that effects of impoverished experience would be especially grave in man.

A survey of the animal studies indicates that they are not yet either full enough or consistent enough to permit extrapolation to man, although they do provide many suggestive leads for research. For the species most investigated in this regard—the rat—the data are still incomplete on important points, and studies of different investigators are often inconsistent. For carnivores and primates, studies on effects of experience on later problem-solving are still relatively few. No clear relations have yet emerged that would allow extrapolation along a phylogenetic line to man. A brief review of the literature will help to highlight the complexities and remaining uncertainties in this work.*

EXPERIMENTS WITH RATS

In the case of the rat, Hebb (1949, 298) was probably the first to report that animals reared in a complex environment were better at problem-

*Some of these references were furnished to us by Dr. Jeanne Cejka La Torre whose doctoral thesis (Cejka, 1967) includes a review of this question.

solving than were colony-reared animals. This study led to several others in the early 1950's in which "free-environment" rats were reported to be superior to "restricted-environment" rats in solving various mazes (Bingham and Griffiths, 1952; Hymovitch, 1952; Forgays and Forgays, 1952). Further experiments have shown that many factors are involved in determining whether test performance of the rat will be affected by earlier differential experience. Let us note some of these factors and a few pertinent studies.

Types of differential experience. Most experiments of this sort have compared the performance of two groups of animals, one having been placed for a period in individual isolation cages and often with further restrictions of sensory input, the other having been placed in a large cage with other animals and usually with a variety of stimulus objects present. Attempts have been made to vary the types of stimulation available, and it has been shown that there may be interaction between the type of stimulation and the type of test used (Forgus, 1955). It has usually been difficult to determine whether difference should be ascribed principally to enrichment or principally to restriction of experience, since few investigators have compared enriched-experience and restricted-experience animals to colony-reared animals. One clear example of this sort also involved strain differences; this will be taken up under the next heading.

Strain differences. Cooper and Zubek (1959) tested McGill Bright and Dull strains in the Hebb–Williams apparatus. Some animals of each strain were raised in colony conditions, some in enriched conditions, and some in isolation in a restricted environment. Cooper and Zubek reported that the scores of the bright strain were not improved by environmental enrichment but were impaired by environmental restriction. Conversely, their dulls were benefited by enrichment but were not harmed by restriction. Only when coming from the usual colony condition did their two strains differ; coming from the enriched environment, the dulls caught up with the brights, and coming from the restricted environment, the brights sank to the level of the dulls. Our work reported in this paper was with the Berkeley S_1 strain (descendants of the Tryon Maze-Bright animals). It will be recalled that in this strain, as in the McGill Bright strain, early impoverishment impaired problem-solving, while enrichment did not produce better results than colony living. In the one experiment, in which we have used EC, SC, and IC groups of the Berkeley S_3 strain (descendents of Tryon's Maze-Dull animals), we have found a contrary effect: The enriched-experience group performed significantly better in visual reversal discrimination than the other two groups; the colony and isolated animals were indistinguishable in their performance. Thus, our results to date with the S_1 and S_3 strains are concordant, respectively, with results for the McGill Bright and Dull strains. Further work with the S_1 and S_3 strains is clearly indicated. One complication is that on the visual reversal discrimination test, the S_3 animals are superior to the S_1 animals.

Age at which differential experience is given. Following a preliminary finding of Hymovitch (1952) that enriched experience must occur before maturity if it is to improve problem performance, Forgays and Read (1962) examined this question in greater detail. They used a 21-day period of differential experience, starting either at birth, at day 22 (weaning), or at days 44, 66, or 88. The restricted-experience animals lived in small cages. Testing of all groups began at 114 days of age. No differential effects were found in errors in a Y-maze, but there were significant differences in errors on the Hebb–Williams test. The 21 days immediately following weaning were found to be more effective in reducing error scores than were either earlier or later periods. The results of Nyman (1967) differ from those of Forgays and Read, since Nyman found that benefits for later spatial discrimination learning were most marked when exposure to the free environment was given from 50 to 60 days of age rather than either from 30 to 40 or 70 to 80 days. Significant effects of enriched environment were found both on an alternation maze and on the Hebb–Williams test; no effects were found in a visual discrimination task in a T-maze. It will be recalled that our only experiments showing impairment of the restricted-environment animals were those in which the restriction was begun at 25 days of age (weaning).

Behavioral test employed. Many investigators have reported finding differences between enriched and impoverished animals with one test but not with another. One of the first of these reports was that of Bingham and Griffiths (1952) who found no differences in behavior of the groups in visual discrimination with the Lashley jumping-stand apparatus, but who found significant effects in two mazes. Hymovitch (1952) found significant effects when he used the Hebb–Williams test, but did not find group differences on a multiple T-maze. As was mentioned above, Forgays and Read (1962) did not find differences on a Y-maze, but found differences with the Hebb–Williams test which presents a series of problems successively within the same enclosure. In the case of our visual discrimination reversal problem, the first light-correct problem can be considered as a simple dark-light visual discrimination task. On this task the enriched and impoverished animals did not differ significantly, although they did on the succeeding reversal problems.

Wilson *et al.* (1965) have pointed out that it is paradoxical that several investigators have found significant effects of experience on performance in the Hebb–Williams test, since performance on this test does not improve with age nor with the phyletic status (Warren, 1965a). They suggest that performance on the Hebb–Williams test may benefit specifically from experience in exploring open fields which is characteristic of most "free-environment" conditions.

Is reactivity of the restricted animal the main factor? Woods *et al.* (1960) believed that the restricted-experience animals have a greater tendency to explore than enriched-experience animals, and that on many tests exploratory activities will be counted as errors against the previously isolated rats. Actually, whether enriched or restricted animals explore more is a function of the testing situation (Forgus, 1954; Lore and Levowitz, 1966).

Myers and Fox (1963) have cited an increase in fear-like responses as a cause of poor maze performance of their isolated rats. Such responses have not been noticed with our extremely tame strains. It will clearly be important to determine whether differences in problem-solving after differential experience should be ascribed to motivational or intellectual changes.

Are the differences induced by differential experience transitory or enduring? Most investigators using rats as subjects have not continued to test them long enough to determine whether differences in problem-solving ability following exposure to enriched or deprived conditions are transitory or enduring. Hebb (1949, 298–299), in the initial report on this question, stated that the rats reared as pets improved more during testing than the cage-reared animals. He concluded that there was a permanent effect of early experience on problem solving at maturity: "The . . . richer experience of the pet group during development made them better able to profit by new experiences at maturity—one of the characteristics of the 'intelligent' human being." Forgays and Forgays (1952) gave their animals two successive tests on the standard series of Hebb–Williams problems. Both enriched and restricted animals improved to the same extent, keeping the difference constant, and these authors concluded on the basis of a rather short period of testing, that the differences were enduring. In our study, it will be recalled that the differences between enriched and impoverished animals tended to disappear after the first few visual reversal problems.

EFFECT OF DIFFERENTIAL EXPERIENCE IN CARNIVORES

Work on this problem with dogs began almost as soon as the work with rats, but there have been far fewer studies using dogs than rats. Clarke *et al.* (1951) found pet-reared Scotties to be superior to restricted Scotties on a variety of learning and problem-solving tasks. This work with Scotties was replicated and extended by Thompson and Heron (1954). Melzack (1962) reported that pet Beagles were superior to restricted animals in solving a simple black-white discrimination test, and there was an even greater difference in reversal problems. Fuller (1966) also found restricted Beagles to be inferior in reversal learning, but only on the first five reversals; thereafter there was no significant difference between restricted-experience animals and pets.

In cats, Wilson *et al.* (1965) found a significant difference in favor of enriched-experience animals on the Hebb–Williams maze, but no significant difference between enriched and control animals in a simple position learning problem or in reversals of this problem or in avoidance conditioning. As mentioned above, they questioned whether the differences are to be considered as being chiefly of intelligence if they appear in the Hebb–Williams test but not in reversal discrimination.

EXPERIMENTS WITH MONKEYS

Harlow and his associates, who have carried out well-known studies on effects of isolation on social behavior of monkeys (Harlow and Harlow, 1965), have also begun to test in the present decade for effects of isolation on learning ability. An initial report from the Wisconsin laboratory by Mason and Fitz-Gerald (1962) indicated that a single monkey isolated for its first 90 days performed rather normally in both formation of learning sets and form discrimination. Rowland (1964) reported no significant difference in development of learning sets between monkeys severely isolated for their first six months and partially restricted controls. The isolated animals did show more performance problems than the controls, and the isolated animals were inferior to the controls over the first 200 problems—a far longer series than has been used in testing other species of animals. Griffin and Harlow (1966) reported that monkeys isolated for their first three months showed severe withdrawal and behavioral peculiarities when removed from the isolation chambers. Although the animals isolated for six months in Rowland's study were crippled in social behavior, the three-month isolates showed no social abnormalities, nor were they deficient in learning performance. Commenting on their failure to find effects on learning ability, Griffin and Harlow comment, "The apparent contradiction between rhesus monkeys and other species on learning after isolation may be an artifact of the types of tasks employed. Tasks that do not extensively adapt Ss or do not discard data indicating problems in adaptation will probably show performance deficits" (1966, 546).

STUDIES WITH CHILDREN

A number of investigators have studied institutionalized children in order to determine both emotional and intellectual effects of restrictions of early care in comparison with effects of normal care or enhanced stimulation and maternal responsiveness. Skeels (1966) has presented striking evidence from retesting 25 individuals who had varied experiences 21 years ago. One group of children who initially tested better were kept in an orphanage, while a second group, thought to have been mentally retarded, were transferred from the orphanage to a school for mentally retarded children and were later adopted. The adults who had spent their early life in the orphanage were found to have lost an average of 26 I.Q. points on retesting, whereas the children who were adopted made an average gain of 29 I.Q. points and had made better social adjustments than the first group. Haywood (1967) has given a full review of the literature concerning how intellectual growth in children may be stunted by deprived environments. Mussen (1967), in reviewing studies of effects of maternal care on emotional and intellectual development, pointed out that ". . . dramatic difficulties and maladjustments characterized children who encountered exceptional deprivation of intimate, personal, or warm maternal care in their earliest years." He went on to comment, "Apparently a minimum amount of such care is required for adequate development and adjustment . . ." but he noted that ". . . it is not possible to define this minimum precisely." Finally, Mussen stated that it is not yet known if there are beneficial effects of increased amounts of maternal care above the required minimum.

TENTATIVE CONCLUSIONS CONCERNING EFFECTS OF DIFFERENTIAL EXPERIENCE ON ANIMAL AND HUMAN INTELLIGENCE

From our survey, two tentative conclusions appear to be supported by the bulk of the data: (a) Inadequate early experience leads to transitory impairment of ability in animals but to permanent intellectual impairment in children. Although some investigators have reported permanent effects in animals, such effects have been difficult to demonstrate, and the test series have usually been too short to support such a claim. The longest series of animal tests have been those performed with monkeys, and these studies have not shown enduring intellectual deficits in the isolated animals, even though there were persistent abnormalities of social behavior. (b) Enrichment of experience beyond adequate care has not been demonstrated to improve later ability in either animal or man.

The first conclusion—that animals and children differ in the permanence of effects of early deprivation—cannot as yet be considered as firmly established, for several reasons: First, the methods employed for testing human and animal intelligence are not strictly comparable. Most of the human tests involve language ability, and it is possible that this ability is more critically affected by inadequate experience than are other sorts of ability. Nonverbal intelligence tests with humans might fail to show enduring deficits due to restricted experience. Secondly, tests of human intelligence are given in a single session, whereas animal tests require many sessions to teach the subjects the tasks to be performed. The tests employed to measure possible effects of experience with monkeys have involved formation of learning sets over hundreds of successive problems. Perhaps human subjects from very restricted backgrounds are not properly prepared to take intelligence tests. In that case, it is possible that prolonged teaching and testing of human subjects would overcome the effects of early deprivation in them as well as in animals. Tested in this way, human deficits might also not be permanent. Thirdly, to the extent that human intelligence has a verbal factor not paralleled in animals, the possibility of compensating for early impoverishment by later enrichment cannot be predicted from animal studies. If there is a critical period for development of verbal ability, as much evidence suggests, then later training would not be able to make up for inadequate early experience of children.

The second conclusion—that enrichment of experience does not improve later performance—also seems to us not to be sufficiently well founded to provide a basis for social application. However, it would be tragic to see the development of opposition to enrichment of education of children, simply because clear support cannot be found in either animal or human studies. The risk of underdeveloping children's ability is too great to abandon efforts at enrichment, even if the cost is not at present justified by available research results.

DIRECTIONS FOR FURTHER RESEARCH ON EFFECTS OF DIFFERENTIAL EXPERIENCE ON ANIMAL AND HUMAN INTELLIGENCE

The difficulty to date in extrapolating from animal to human behavior does not imply, in our opinion, that research in this area should be reduced or abandoned. Rather, we believe that more intensive work is needed to develop knowledge and understanding in a complex field of research. For application to educational and social problems, direct work with children is needed. This is the level at which practical "pay-offs" can occur.

At the same time, research with animal subjects should be pursued, for two main reasons. First, fuller understanding of effects observed with human subjects may be reached on the basis of comparative behavioral studies. Furthermore, greater range, and often greater rigor of behavioral experimentation is possible with animal subjects. On the basis of research to date, one can certainly conclude that inadequate experience of the young will produce undesirable effects. This is true whether one inspects results from man, rat, or monkey. (While experiments with monkeys have not shown a lower final level of problem-solving in deprived animals, they have demonstrated slower and more erratic early learning and also persisting abnormalities of social behavior.) On the other hand, there is little evidence now available to support the hypothesis that enrichment of experience, beyond the level of adequate care, will benefit intellectual ability. Nevertheless, we feel that experimental attempts to find behavioral effects of enrichment should be continued, with improved experimental designs and with better measuring instruments. There have been relatively few studies on effects of enrichment, and no overall conclusion is as yet warranted.

Secondly, when the relations of human to animal behavior have been worked out, then relations between brain and behavioral effects in animals can be extrapolated to man. Many cerebral measures are feasible only in animal subjects, yet the similarities of human and animal brain structures and processes are so great that it has often proved possible to apply findings made on the animal brain to the human brain. In the present case, the hope will be to relate effects of experience on behavior in man to effects on brain processes in animals; this will require the intermediate steps of defining relations between relevant human and animal behavior and between animal behavioral and cerebral events. Encouragement for pursuing these studies can be gained from the now abundant findings that the brain of the rat is modified chemically and anatomically by differential experience. The fact that this link between animal behavioral and cerebral events has been forged gives hope that the rest of the chain can be completed.

SUMMARY

Differential experience leads to significant changes in both brain and behavior of the rat. The cerebral cortices of rats from enriched conditions exceed those of littermates from impoverished conditions in tissue weight, depth, total protein, total activities of the enzymes acetylcholinesterase, cholinesterase and hexokinase, number of glial cells, and cross-section areas of neural cell bodies and their nuclei. When littermates in colony conditions are used as a baseline, the effects in cortical weight and total acetylcholinesterase are seen to be attributable mainly to enrichment of experience while the difference in total cholinesterase is due chiefly to impoverishment. Anatomical comparisons between environmental manipulations and the baseline colony condition are in progress. The effects of experience can be distinguished from those of normal maturation of the brain, since they occur in adult animals, and since enriched experience increases cortical depth while maturation (at least after weaning) leads to thinning of the rat's cortex. Increasing the duration of differential experience beyond 30 days does not increase the cerebral differences between animals with enriched and impoverished experience. The effect on cortical weight declines with periods of over 30 days, whereas the difference in ratio of cortical weight to that of the rest of the brain remains approximately constant.

Behavioral effects of differential experience were measured by scores on a visual reversal discrimination task. When Berkeley S_1 rats were assigned to the enriched (EC), colony (SC), or isolated (IC) conditions at weaning, IC scores were inferior to EC or SC, but the latter two groups were equivalent. When rats were in EC or IC from 60 to 90 days of age, the effect was reversed—EC scores were inferior to IC. When the EC-IC period extended from 90 to 120 days of age, there was no consistent difference between EC and IC scores. Where differences were found between EC and IC groups in errors per problem, these differences tended to disappear after the first few problems. Significant correlations were obtained between behavioral scores and the ratio of weight of cortex to that of the rest of the brain. Confirming previous studies, these indicated for most EC groups that the greater the cortical/subcortical weight ratio, the fewer the errors the animal tended to make. Here, too, the 60 to 90-day EC groups differed from the other experiments. Each of these results is based on at least two experiments, and more are being run. No general relation could be found between cerebral and behavioral measures that held over the whole range of conditions studied.

Studies on behavioral effects of differential experience in man and in several infra-human mammalian species were reviewed. Difficulties in attempting to extrapolate from animal studies to human behavior were discussed, and desirable directions for further research were indicated.

ACKNOWLEDGMENTS

We would like to acknowledge the diligent and skillful aid of co-workers in many phases of this research: in the maintenance of animals in the experimental conditions—Clarence Turtle and Donald Gassie; in the behavioral testing—Clarence Turtle, Su-Yu Chang, Donald Gassie, Yoram Jaffe, and John Frazee; in the brain dissections—Marie Hebert; in the chemical analyses—Hiromi Morimoto; in the histological analyses—Alma Raymond, Bernice Lindner, and Lennis Lyon; in statistical analyses—Ann Muto, Anne Betancourt, and Diane Rein; and in administrative assistance and preparation of the manuscript—Jessie Langford. Finally we would like to thank our students and co-workers for their many comments and suggestions on an earlier draft of this paper.

REFERENCES

Altman, J. and Das, G. D. Autoradiographic examination of the effects of enriched environments on the rate of glial multiplication in the adult rat brain. *Nature*, 1964, **204**, 1161–1163.

Bennett, E. L., Diamond, M. C., Krech, D. and Rosenzweig, M. R. Chemical and anatomical plasticity of brain. *Science*, 1964, **146**, 610–619.

Bennett, E. L. and Rosenzweig, M. R. Brain chemistry and anatomy: Implications for theories of learning and memory. In C. Rupp (Ed.), *Mind as a tissue.* New York: Hoeber Medical Div., Harper and Row, 1968. Pp. 63–86.

Bennett, E. L., Rosenzweig, M. R., Krech, D., Karlsson, H., Dye, N. and Ohlander, A. Individual, strain, and age differences in cholinesterase activity of the rat brain. *J. Neurochem.*, 1958, **3**, 144–152.

Bingham, W. E. and Griffiths, W. J. The effect of different environments during infancy on adult behavior in the rat. *J. comp. physiol. Psychol.*, 1952, **45**, 307–312.

Cejka, J. A. Environmental effects on the anatomy, chemistry, and histology of the mouse brain. Unpublished doctoral dissertation, McGill University, 1967.

Clarke, R. S., Heron, W., Fetherstonhaugh, M. L., Forgays, D. G. and Hebb, D. O. Individual differences in dogs: preliminary report on the effects of early experience. *Canad. J. Psychol.*, 1951, **5**, 150–156.

Cooper, R. M. and Zubek, J. P. Effects of enriched and restricted early environments on the learning ability of bright and dull rats. *Canad. J. Psychol.*, 1958, **12**, 159–164.

Cragg, B. G. Changes in visual cortex on first exposure of rats to light: effect on synaptic dimensions. *Nature*, 1967, **215**, 251–253.

Diamond, M. C., with technical assistance by B. Lindner and A. Raymond. Extensive cortical depth measurements and neuron size increases in the cortex of environmentally enriched rats. *J. comp. Neurol.*, 1967, **131**, 357–364.

Diamond, M. C., Law, F., Rhodes, H., Lindner, B., Rosenzweig, M. R., Krech, D. and Bennett, E. L. Increases in cortical depth and glia numbers in rats subjected to enriched environment. *J. comp. Neurol.*, 1966, **128**, 117–125.

Forgays, D. G. and Forgays, J. W. The nature of the effect of free environmental experience in the rat. *J. comp. physiol. Psychol.*, 1952, **45**, 322–328.

Forgays, D. G. and Read, J. M. Crucial periods for free-environmental experience in the rat. *J. comp. physiol. Psychol.*, 1962, **55**, 816–818.

Forgus, R. H. The effect of early perceptual learning on the behavioral organization of adult rats. *J. comp. physiol. Psychol.*, 1954, **47**, 331–336.

Forgus, R. H. Early visual and motor experience as determiners of complex maze-learning ability under rich and reduced stimulation. *J. comp. physiol. Psychol.*, 1955, **48**, 215–220.

Fuller, J. L. Transitory effects of experiential deprivation upon reversal learning in dogs. *Psychonom. Sci.*, 1966, **4**, 273–274.

Giacobini, E. The distribution and localization of cholinesterase in nerve cells. *Acta physiol. Scand.*, 1959, **45** (Suppl. 156), 1.

Griffin, G. A. and Harlow, H. F. Effects of three months of total social deprivation on social adjustment and learning in the Rhesus monkey. *Child Develpm.*, 1966, **37**, 533–548.

Gyllensten, L. Postnatal development of the visual cortex in darkness (mice). *Acta morph. Neerl.-Scand.*, 1959, **2**, 331–345.

Gyllensten, L., Malmfors, T. and Norrlin, M. L. Effect of visual deprivation on the optic centers of growing and adult mice. *J. comp. Neurol.*, 1965, **124**, 149–160.

Harlow, H. F. and Harlow, M. K. The affectional systems. In A. M. Schrier (Ed.), *Behavior of nonhuman primates*, Vol. 2. New York: Academic Press, 1965.

Haywood, H. C. Experiential factors in intellectual development: the concept of dynamic intelligence. In J. Zubin and G. Jervis (Eds.), *Psychopathology of mental development.* New York: Grune and Stratton, 1967.

Hebb, D. O. *The organization of behavior.* New York: Wiley, 1949.

Hymovitch, B. The effects of experimental variations on problem-solving in the rat. *J. comp. physiol. Psychol.*, 1952, **45**, 313–321.

Koelle, G. B. The histochemical localization of cholinesterases in the central nervous system of the rat. *J. pharmacol. exp. Therap.*, 1952, **106**, 401.

Koelle, G. B. The histochemical localization of cholinesterases in the central nervous system of the rat. *J. comp. Neurol.*, 1954, **100**, 211.

Krech, D., Rosenzweig, M. R. and Bennett, E. L. Relations between brain chemistry and problem-solving among rats raised in enriched and impoverished environments. *J. comp. physiol. Psychol.*, 1962, **55**, 801–807.

Krech, D., Rosenzweig, M. R. and Bennett, E. L. Effects of complex environment and blindness on rat brain. *Arch. Neurol.*, 1963, **8**, 403–412.

Krechevsky, I. Antagonistic visual discrimination habits in the white rat. *J. comp. Psychol.*, 1932, **14**, 263–277.

Lore, R. K. and Levowitz, A. Differential rearing and free versus forced exploration. *Psychonom. Sci.*, 1966, **5**, 421–422.

Mason, W. A. and Fitz-Gerald, F. C. Intellectual performance of an isolation reared monkey. *Percept. mot. Skills*, 1962, **15**, 594.

Melzack, R. Effects of early perceptual restriction on simple visual discrimination. *Science*, 1962, **137**, 978–979.

Mussen, P. Early socialization: learning and identification. *New directions in psychology*, Vol. 3. New York: Holt, Rinehart and Winston, 1967. Pp. 51–110.

Myers, R. D. and Fox, J. Differences in maze performance of group- vs. isolation-reared rats. *Psychol. Rep.*, 1963, **12**, 199–202.

Nyman, A. J. Problem solving in rats as a function of experience at different ages. *J. genet. Psychol.*, 1967, **110**, 31–39.

Rosenzweig, M. R. Environmental complexity, cerebral change, and behavior. *Amer. Psychol.*, 1966, **21**, 321–332.

Rosenzweig, M. R., Bennett, E. L. and Diamond, M. C. Effects of differential environments on brain anatomy and brain chemistry. In J. Zubin and G. Jervis (Eds.), *Psychopathology of mental development*. New York: Grune and Stratton, 1967. Pp. 45–56.

Rosenzweig, M. R., Krech, D., Bennett, E. L. and Diamond, M. C. Effects of environmental complexity and training on brain chemistry and anatomy: a replication and extension. *J. comp. physiol. Psychol.*, 1962, **55**, 429–437.

Rosenzweig, M. R. and Leiman, A. L. Brain functions. *Ann. Rev. Psychol.*, 1968, **19**, 55–98.

Rosenzweig, M. R., Love, W. and Bennett, E. L. Effects of a few hours a day of enriched experience on brain chemistry and brain weights. *Physiol. and Behav.*, 1968, **3**, 819–825.

Rowland, G. L. The effects of total social isolation upon learning and social behavior in rhesus monkeys. Doctoral dissertation, University of Wisconsin, 1964.

Skeels, H. M. Adult status of children with contrasting early life experiences. *Monogr. soc. Res. child Develpm.*, 1966, **31**, 1–65.

Thompson, W. R. and Heron, W. The effects of restricting early experience on the problem-solving capacity of dogs. *Canad. J. Psychol.*, 1954, **8**, 17–31.

Warren, J. M. Comparative psychology of learning. *Ann. Rev. Psychol.*, 1965a, **16**, 95–118.

Warren, J. M. Primate learning in comparative perspective. In A. M. Schrier, H. F. Harlow, and F. Stolnitz (Eds.), *Behavior of nonhuman primates: modern research trends*. New York: Academic Press, 1965b. Pp. 349–382.

Wilson, M., Warren, J. M. and Abbott, L. Infantile stimulation, activity, and learning by cats. *Child Develpm.*, 1965, **36**, 843–853.

Woods, P. J., Ruckelshaus, S. I. and Bowling, D. M. Some effects of "free" and "restricted" environmental rearing conditions upon adult behavior in the rat. *Psychol. Rep.*, 1960, **6**, 191–200.

QUESTIONS AND ANSWERS

QUESTION: (Robert V. Crow, Midwest Research Institute, Kansas City) Are there changes in catecholamine levels in the cortex of rats as a function of experience?

ANSWER: Our group has not measured catecholamines, but differences in catecholamines have been reported by Geller, Yuwiler, and Zolman from UCLA (In Variation in Chemical Composition of the Nervous System, G. B. Ansell (Ed.), Pergamon Press, N.Y., 1966). They found increased norepinephrine in the cortex of isolated animals, and even larger increases in the caudate nucleus.

QUESTION: (S. E. Locke, Cornell University) The application of terms "enriched" or "impoverished" to the nature of the environment is rather subjective; isn't it possible that the complexities of the so-called "enriched" environment are, in fact, producing a stress-created anxiety state responsible for the observed changes? (Certainly, there is ample evidence to show the effect on neuroendrocrine function from stress situations to cause one to question the causal relationship implied by your report.)

ANSWER: This is a question that is frequently asked and on which we have tried to make some observations. Both our enriched and impoverished animals remain easy to handle throughout the experiments, although some investigators have reported increased aggressiveness of isolated rats of other strains. Whatever signs we find of stress suggest the isolation condition is more likely to be stressful than is enrichment. We have found the isolated rats to have somewhat larger adrenals, but only in proportion to their larger body weight. Following a suggestion of Dr. Guy Everett, we observed piloerection, a sign of circulating norephinephrine and presumably of stress, in the IC animals but not in their EC littermates. Dr. Walter H. Riege, in our laboratories, is now comparing effects of stress with those of enriched, isolated, and standard colony conditions. Further research of Dr. Reige has shown both that the effects of stress on brain weights are smaller than those of differential environments and that they do not resemble closely the pattern caused by either enrichment or impoverishment.

MACROMOLECULAR SYNTHESIS IN SYNAPTOSOMES*

Stanley H. Appel†,

Departments of Medicine and Biochemistry
Duke University Medical Center
Durham, North Carolina

Lucila A. Autilio,

Department of Neurology
Hospital of the University of Pennsylvania
Philadelphia, Pennsylvania

Penelope Pettis

Department of Neurology
Hospital of the University of Pennsylvania
Philadelphia, Pennsylvania

Within recent years advances in molecular biology have focused attention on the many processes by which information storage may be regulated. Synthesis of nucleic acids and proteins, alteration of enzyme activity by feedback inhibition or activation, regulation of metabolic pathways by hormones and small molecules, and translocation of substrates or products, may all contribute to the way in which cellular activity is expressed. Their potential relevance to neural function and behavior is being actively

* This investigation was supported in part by grants from The National Multiple Sclerosis Society and the Public Health Service.

† Career Development Awardee of the Public Health Service.

investigated in many laboratories. One approach is the use of antibiotics and antimetabolites in mice in the experiments of Flexner and Flexner (1966) and Barondes and Cohen (1966), and in goldfish in the experiments of Agranoff *et al.* (1965).

These studies implicate *de novo* protein synthesis as vital to information processing by demonstrating that antibiotics which inhibit protein synthesis alter specific behavior. Such studies are always subject to the uncertainty of whether the drugs produce their effects by their stated action on protein synthesis, or by some other effect related either to other actions of the drug or to secondary repercussions of the altered protein synthesis.

It has been demonstrated recently that puromycin, an inhibitor of protein synthesis, alters neuronal mitochondria (Gambetti *et al.*, 1968). The interpretation of puromycin effects upon behavior are, therefore, obscured by possible alterations in mitochondrial function. Consequently, without knowing the full range of cellular metabolic events which are either primary or secondary consequences of the administration of any antibiotic, it becomes difficult to attribute an antibiotic's action to a specific cellular process.

An additional approach to understanding the relationship between macromolecules and behavior is to ask what changes occur in the rate or the characteristics of macromolecular synthesis, and what cellular structures are involved when behavior of the organism has been altered. The experiments of Zemp *et al.* (1966) suggest that newly synthesized RNA, especially in the subcortex, is associated with a change in mouse behavior. The experiments of Bennett *et al.* (1964) focus on the marked change in glial cells with environmental enrichment. Our own experiments indicate that alterations in polysomes and the capacity for protein synthesis occur in response to environmental changes (Appel *et al.*, 1967). Many of these experiments pose the question whether the observed alterations in macromolecular synthesis affect macromolecular constituents of the neural elements, and possibly influence the subsequent transfer of information.

Electrophysiological data have indicated the significant and often key regulatory and rate-limiting role which the synapse may assume. It is a site where alterations in any processes such as transmitter synthesis and release, receptor interaction, and membrane properties could affect intercellular communication.

Consequently, our specific studies have focused on the use of isolated nerve endings as an *in vitro* model for understanding the way in which alterations in macromolecular synthesis result from synaptic activity and modulate subsequent synaptic function.

Nerve endings are isolated from 18-day-old rats' cerebral cortices by modifications of the techniques of DeRobertis *et al.* (1963) and Whittaker *et al.* (1964). It has been found necessary to use discontinuous Ficoll gradients in contrast to the usual sucrose gradients for the separation of "viable" synaptosomes. These synaptosomes are identified as enriched populations by their high specific activity of acetylcholinesterase, their low activity of succinic dehydrogenase, and by electron microscopy.

At 37°C these synaptosomes incorporate radioactive leucine into lipid-free hot trichloroacetic acid precipitates at a relatively linear rate for 20 minutes. Ribonuclease and chloramphenicol inhibit less than 25%, whereas puromycin and acetoxycycloheximide inhibit greater than 70% in concentrations of 10 μg/ml.

As in the studies of mitochondrial protein synthesis, it is necessary to demonstrate that bacterial contamination does not account for the observed radioactive leucine incorporation into protein. Several different experiments provided evidence that such contamination does not contribute significantly to the observed protein synthesis. The bacteria employed in these studies represent a mixed population of 10^9 cells/ml grown from the 10^2–10^3 cells/ml present as contaminants in the synaptosome preparation.

1. Chloramphenicol was found to inhibit only 25% when present in concentrations of 50 μg/ml. With bacteria, chloramphenicol inhibited incorporation of radioactive leucine into protein approximately 96%.
2. The incorporation of leucine into synaptosome protein was linear for only 20 minutes and virtually stopped at 50 minutes, whereas the time course of incorporation into bacteria exhibited a lag and then continued for more than five hours.
3. Bacteria whose protein had been labelled did not sediment in the same fraction as the synaptosomes.
4. When the synaptosome preparation was kept at 4°C for 24 hours and checked for amino acid incorporation into protein at that point, the result was approximately 20% of the incorporation in a fresh preparation. With bacteria the incorporation demonstrated no decrease after 24 hours at 4°C.

The requirements for this incorporation into protein were found to be minimal. Only sodium and potassium appeared to affect activity. 100 m*M* sodium and 10 m*M* potassium resulted in a four-fold enhancement of activity. ATP, ADP, AMP, and glucose were without effect.

It is of interest that the level of monovalent cations is less critical than the specific ions that are present. 100 m*M* Na and 10 m*M* K gave significantly more enhancement than 100 m*M* K and 10 m*M* Na. By testing all intervening values, a curve of activation was obtained, which was practically superimposable on the known activation of the sodium-potassium-ATPase which has been described to be present in high concentration within synaptosome fractions. In addition, ouabain at 5×10^{-4} *M* was found to be markedly inhibitory, and the extent of inhibition depended on the presence of optimal Na and K. Exogenous energy was not required. Substances which interfere with oxidative phosphorylation—such as dinitrophenol, potassium cyanide, and oligomycin—were found to be markedly inhibitory.

Serotonin (10^{-4} *M*) was found to inhibit protein synthesis 40%. At similar concentrations desmethylimipramine inhibited 70%; methysergide inhibited 60%, and curare (10^{-5} *M*) inhibited 65%.

Following incorporation of amino acid into protein, the synaptosomes were subjected to hypotonic shock, and the preparation electrophoresed in polyacrilamide gels. Many protein bands were obtained, and when removed from the gel several of these possessed radioactivity. In addition, the synaptosomes that had been exposed to hypotonic shock were subjected to sedimentation and discontinuous sucrose density gradients according to the technique of Whittaker *et al* (1964). With this technique the majority of the protein synthesized was associated with membrane particulate fractions. When microsomal protein which had also been synthesized *in vitro* with leucine-^{3}H was compared to synaptosomes which had incorporated leucine-^{14}C, the pattern of incorporation of the two preparations as analyzed on a sucrose gradient appeared entirely different. This indicates that what is synthesized in the synaptosomes may be a different set of proteins than that synthesized with a microsomal preparation from the brain. The work of Rubin and Stenzel (1965) demonstrated that cell-free protein synthesizing systems from brain reflect, at least partially, the activity of glia, whereas the protein synthesized in synaptosomes predominantly reflects neuronal activity. Therefore, it is possible that the differences observed in the synaptosome and microsome incorporation reflect different populations of cells performing protein synthesis, rather than different parts of the same neuronal population.

Several interesting conclusions emerge from these studies. (1) Isolated synapses are capable of synthesizing proteins. Not only does this support the relative integrity of the nerve endings, but it also represents one of the first observations of protein synthesis in neurons, as opposed to the more usual neuronal and glial synthesis observed *in vitro*. Although it is unclear at the present time whether the protein synthesis is within the presynaptic or the postsynaptic process, it seems more likely that this is taking place within presynaptic processes. (2) The protein synthesis is modulated by membrane events and environmental ionic constituents, thereby indicating that at least, *in vitro*, neurohormones, ion fluxes, and the availability of mitochondrial endogenous energy may regulate neuronal protein synthesis. (3) Such proteins are incorporated predominantly into membrane fractions. They thereby provide a model in which alterations in intercellular communication may affect the extent of membrane protein synthesis, and conversely, a model in which alterations in the extent of membrane synthesis may affect subsequent neuronal expression and intercellular communication.

ACKNOWLEDGMENTS

We are grateful to Dr. Pier–Luigi Gambetti and Dr. Nicholas Gonatas for their elegant electron micrographs of the synaptosome fractions.

REFERENCES

Agranoff, B. W., Davis, R. E. and Brink, J. J. *Proc. Nat. Acad. Sci.* (U.S.) **54**, 788, 1965.
Appel, S. H., Davis, W. and Scott, S. *Science*, **157**, 836, 1967.
Barondes, S. and Cohen, H. D. *Science*, **151**, 594, 1966.
Bennett, E. L., Diamond, M. C., Krech, D. and Rosenzweig, M. R. *Science*, **146**, 610, 1964.
DeRobertis, E., Rodriguez, D. G., Salganicoff, L., Pellegrino Delraldi, A. and Zieher, L. H. *J. Neurochem.*, **10**, 225, 1963.
Flexner, L. B. and Flexner, J. B. *Proc. Nat. Acad. Sci.* (U.S.), **55**, 369, 1966.
Gambetti, P. L., Gonatas, N. and Flexner, L. *J. Cell. Biol.*, **36**, 379, 1968.
Rubin, A. L. and Stenzel, K. H. *Proc. Nat. Acad. Sci.* (U.S.), **53**, 963, 1965.
Whittaker, V. P., Michaelson, I. A. and Kirkland, R. J. A. *Biochem. J.*, **90**, 293, 1964.
Zemp, J. W., Wilson, J. E., Schlesinger, K., Boggan, W. O. and Glassman, E. *Proc. Nat. Acad. Sci.* (U.S.), **55**, 1423, 1966.

QUESTION AND ANSWER

QUESTION: (Dr. Roger M. Morrell, Biochemical Research, The Upjohn Company) Have you tried isolating polysomes from a synaptosomal preparation? If so, what is their protein-synthetic activity?

ANSWER: Preliminary experiments have been unsuccessful in demonstrating amino acid incorporation into protein in extracts of isolated synaptosomes.

CYTOLOGICAL CORRELATES OF FUNCTIONAL CHANGE IN CENTRAL NEURONS OF INSECTS: A SPECULATIVE SUMMARY*

Melvin J. Cohen

Department of Biology
University of Oregon
Eugene, Oregon

The neural correlates of behavioral change can be divided into two broad categories: (1) The first type consists of specific membrane permeability changes resulting in electrical activity associated with conduction along neurons and transmission at synapses. This kind of activity is generally rapid (milliseconds) and is the immediate precursor to the behavioral act. (2) A second type of neuronal function consists of metabolic changes other than those immediately coupled to the electrochemical processes of conduction and transmission. These may involve the metabolism of ribonucleic acid (RNA) and proteins and be associated with alterations in ultrastructural components such as ribosomes, endoplasmic reticulum, and other intracellular membranous organelles (Hyden, 1960; Holtzman *et al.*, 1967). Such metabolic activity is concerned with functional changes over relatively long periods of time as in maturation, use and disuse, and injury and repair. This trophic aspect of neuronal activity is the primary topic of this summary.

The thoracic ganglia of the cockroach *Periplaneta americana* has proved useful for investigating certain aspects of trophic change in neurons (Cohen, 1967a, 1967b; Cohen and Jacklet, 1967). This material has been recently

*Supported by PHS Research Grant 5 R01 NB01624 and PHS Career Development Grant 1 K03 NB04992–01.

reviewed and it is the purpose here to present only a summary of the salient points and the questions which they raise.

LEARNING

The thoracic ganglia of the cockroach can learn simple avoidance-conditioning tasks. This has been demonstrated in the relatively intact animal (Horridge, 1962) and in the isolated thoracic ganglion (Eisenstein and Cohen, 1965). Some electrophysiological correlates of avoidance conditioning have been examined (Hoyle, 1965). These studies demonstrate that the cockroach thoracic ganglia are involved in complex integrative behavior and led to subsequent work on the individual nerve cells within the ganglionic mass.

RNA CHANGES ASSOCIATED WITH NERVE INJURY AND REPAIR

Vertebrate neurons respond to injury of their axon by a cyclic change in the distribution of cytoplasmic RNA (Brattgård *et al.*, 1957). This involves a breakdown of the large basiphilic aggregates in the cytoplasm (Nissl bodies) in a process referred to as chromatolysis (Nissl, 1892; Bodian and Mellors, 1945).

The central neurons of the cockroach also show a reorganization of cytoplasmic RNA following axonal injury and subsequent regeneration (Cohen and Jacklet, 1965; Cohen, 1967a). Within 12 hours after axon injury a ring of RNA accumulates in the cytoplasm around the nuclear membrane (Fig. 1). This perinuclear RNA ring reaches a maximum at two days after injury and disappears by ten days. We have used this perinuclear ring response to injury as an anatomical marker for preparing cell maps of the cockroach metathoracic ganglion (Cohen and Jacklet, 1967). These maps identify which central cell bodies send their axons out a given peripheral nerve trunk, and in some cases, which cell body innervates a particular limb muscle. A three dimensional representation of such a map is indicated in Fig. 2. This permits the identification of individual nerve cells involved in certain specific acts of behavior.

ULTRASTRUCTURAL CHANGES

The normal insect central neuron shows a relatively uniform distribution of cytoplasmic RNA in contrast to the large RNA aggregates characteristic of vertebrate neurons (Hess, 1958; Cohen and Jacklet, 1965). This is associated

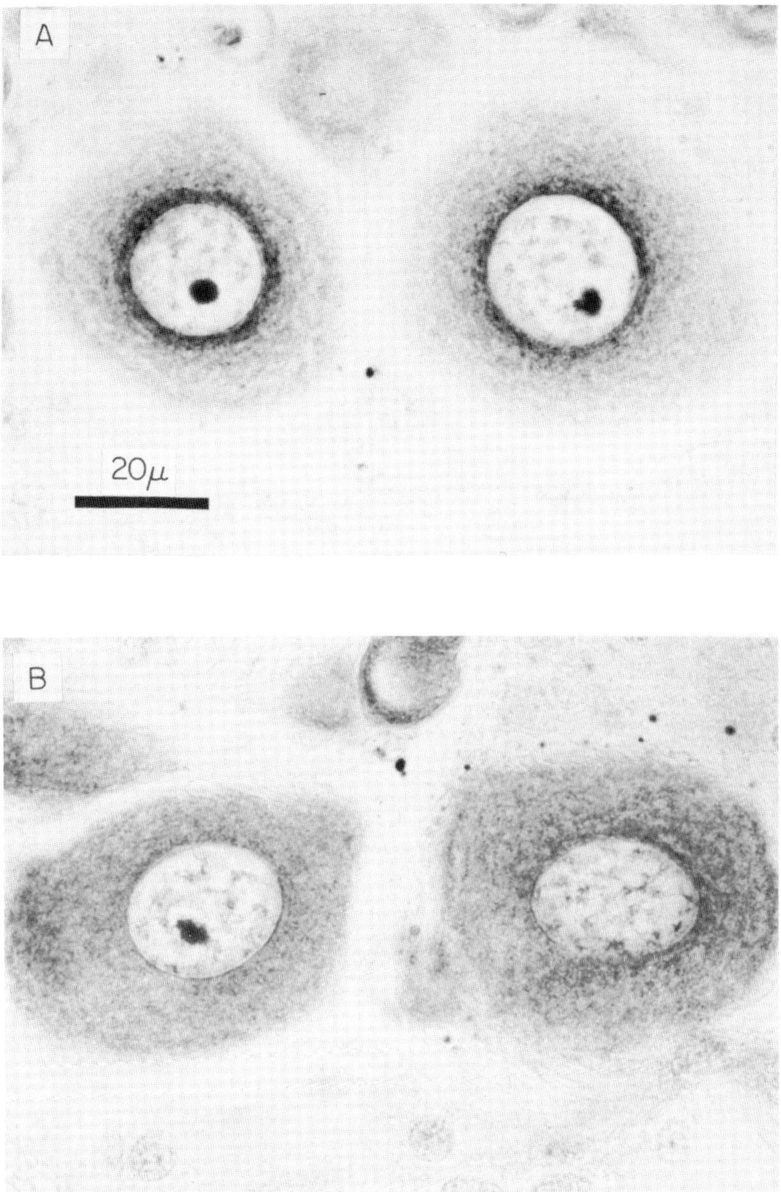

Figure 1 Cell bodies of neurons in the metathoracic ganglion stained with pyronine-malachite green. A: Cell bodies whose axons were cut two days previous to fixation. Note the dense accumulation of pyronine stain in the perinuclear cytoplasm. This has been shown to be RNA. B: The bilaterally matched cells from the opposite half of the same ganglion. The axons of the cells in B were left intact as a control. Note the lack of the dense perinuclear RNA ring in the control cells (modified from Cohen and Jacklet, 1967).

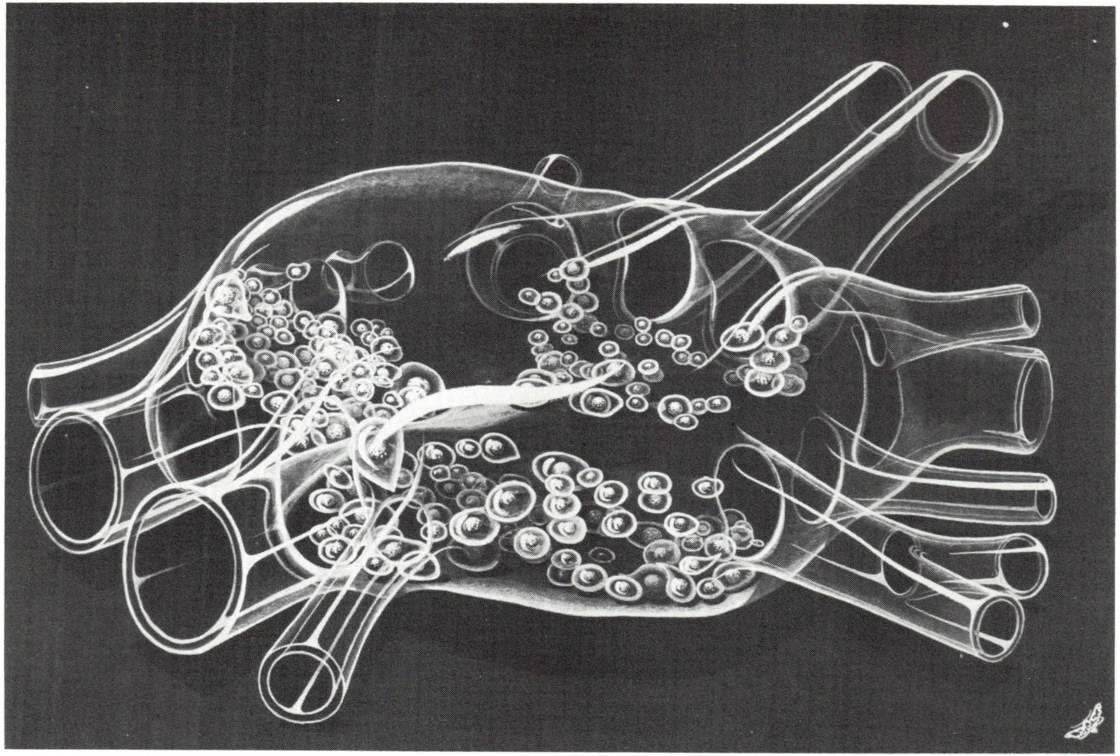

Figure 2 A three dimensional representation illustrating the distribution of motor nerve cell bodies in the metathoracic ganglion of the cockroach. The cells have been assigned numbers and can be identified from one animal to the next. Maps are available showing the particular peripheral nerve trunk through which each of these cells sends an axon to the periphery. The large pair of anterior connectives are seen at the left and the posterior connectives are directed toward the upper right corner. The ganglion is one mm. in its longest dimension (modified from Cohen and Jacklet, 1967).

with a scarcity of endoplasmic reticulum in roach neurons, and a more or less uniform dispersal of free ribosomes (Hess, 1958; Cohen, 1967a). The basiphilic aggregates of the vertebrate Nissl body are composed of layered rough endoplasmic reticulum (ER) with interspersed polysomes (Palay and Palade, 1955; Palay, 1964). Examination of injured and regenerating nerve cells in the roach indicates that the accumulation of perinuclear RNA is associated with an increase in rough ER and polysomes, *i.e.*, the injured roach neuron begins to form vertebrate-like Nissl bodies (Fig. 3).

The accumulation of perinuclear RNA occurs relatively early after injury, before the axon actually begins regeneration. From this point, the process then resembles that of vertebrate chromatolysis in that the ribosomes subsequently disperse from their scaffolding of endoplasmic reticulum and the cell begins to make the protein that is required for growth of a new axon.

When the roach neuron is injured, it shifts into a condition resembling the normal vertebrate Nissl body organization by laying down rough endoplasmic reticulum and polysomes. This is interpreted as a preparatory stage for protein synthesis. Perhaps the precise arrangement of ribosomes on a scaffolding of endoplasmic reticulum is necessary for the proper coupling between ribosomes and soluable RNA to make the new protein required for axon regeneration. Once this coupling is achieved, then the ribosomes disperse from the membrane of the endoplasmic reticulum and this is associated with growth of the new axon filaments.

The Nissl body organization in the cytoplasm of vertebrate neurons has been related to a high basal level of protein synthesis (Brattgård *et al.*, 1957). The normal insect neuron, on the other hand, with the cytoplasmic RNA in a dispersed state, might be thought to have a relatively low basal level of protein synthesis. When a high protein demand is placed upon the insect neuron, it must first shift into the rough ER organization characteristic of vertebrate Nissl bodies before it can make protein at a higher rate. It is as if the vertebrate neuron is continually primed and ready for high level protein synthesis while the insect neuron is set at a lower basal synthetic level and must first be shifted into the Nissl-type organization before it can meet the new metabolic demand. It may be that the relatively stereotyped behavior of insects, as compared to vertebrates, is related to the metabolic consequences of these differences in ultrastructural organization.

GANGLION TRANSPLANTS

A metathoracic ganglion from a donor adult cockroach can be transplanted into the coxa of another adult cockroach (Fig. 4). Bodenstein (1957)

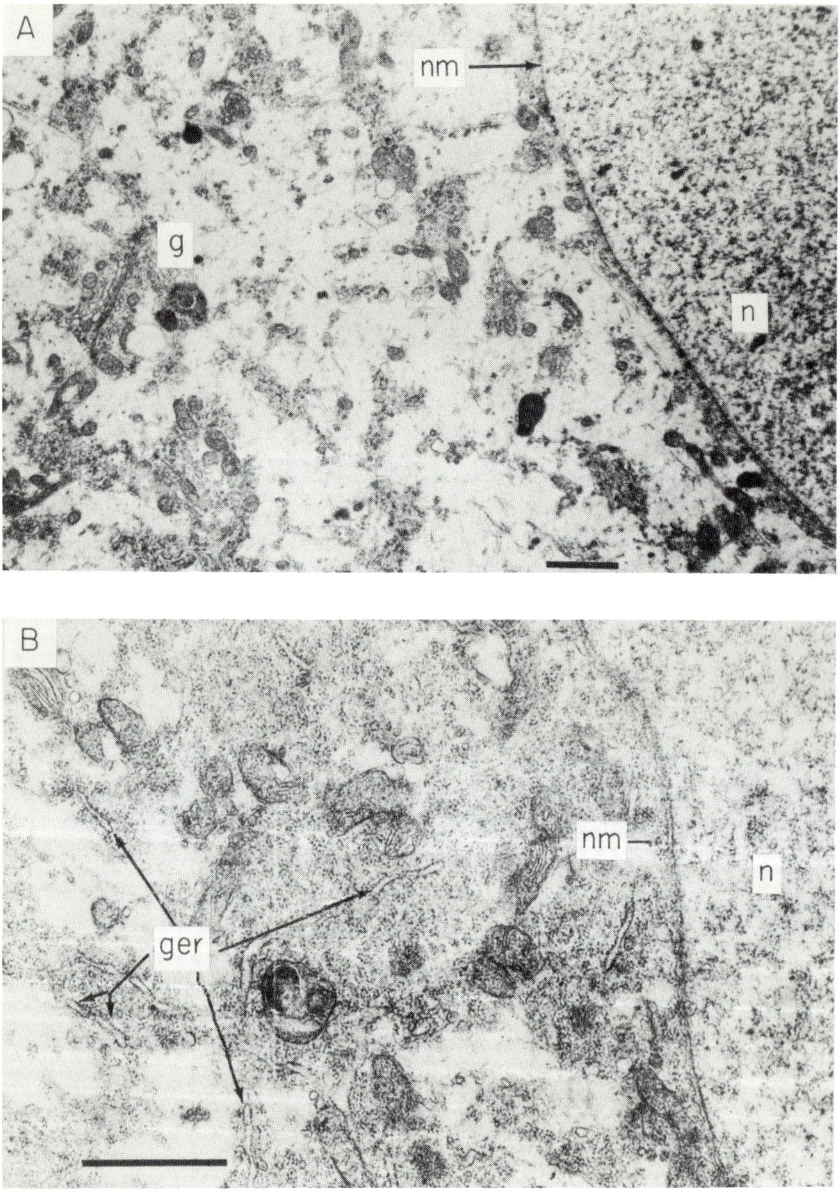
A
nm
g
n
B
nm
n
ger

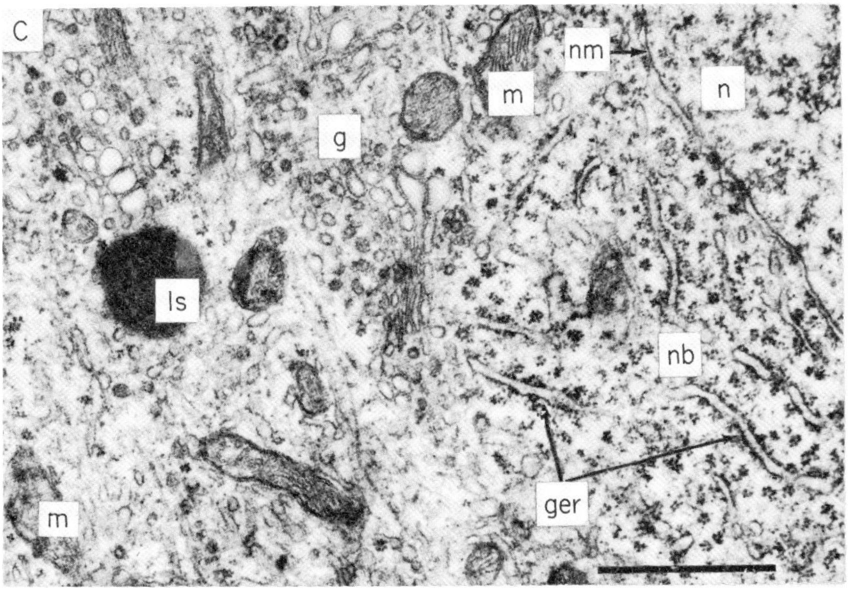

Figure 3 Electron micrographs of nerve cell bodies showing the perinuclear region of the cytoplasm. A: Normal ganglion cell from the metathoracic ganglion of the cockroach. There is relatively little endoplasmic reticulum and a low density of ribosomes. B: Cockroach ganglion cell whose axon was cut two days prior to fixation. Note the granular endoplasmic reticulum and the high density of ribosomes. C: Normal Purkinje cell from the rat cerebellar cortex in the region of a Nissl body. Note the general similarity to the injured cockroach neuron in the presence of granular endoplasmic reticulum and ribosomes. *g*, Golgi body; *ger*, granular endoplasmic reticulum; *ls*, lysosomes; *m*, mitochondrion; *n*, nucleus, *nb*, Nissl body; *nm*, nuclear membrane (from Cohen, 1967a). Micrographs in A and B taken by Dr. D. Ashhurst; C is by courtesy of Dr. S. Palay.

reported that muscles of the host limb fibrillate several weeks after transplantation of the ganglion. We have demonstrated that this fibrillation is caused by neural activity from the transplanted ganglion (Jacklet and Cohen, 1967). Axons from the transplanted ganglion will innervate only those muscles of the host limb that have been previously denervated.

The use of regenerating systems permits the study of specific junctions between excitable units while they are in the process of formation. The knowledge of nerve cell body distribution and their peripheral connections would seem to make this preparation particularly useful for investigating the question of the specificity of connections at the level of individual cells.

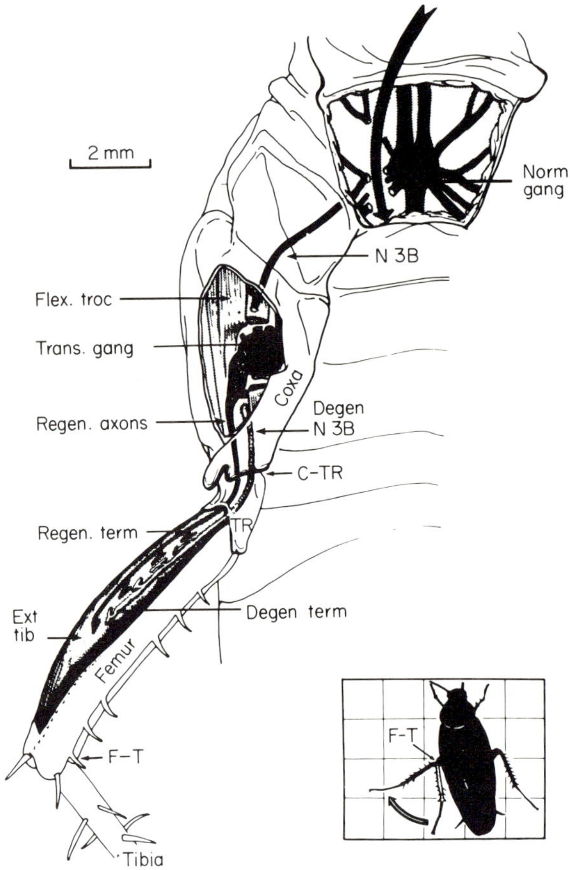

Figure 4 Diagram of the ventral surface of the metathorax and right metathoracic leg in the cockroach illustrating the ganglion transplantation procedure. A metathoracic ganglion from a donor cockroach (trans. gang.) is transplanted into the coxa of the host. Regenerating axons (regen. axons) grow from the transplant through the coxal-trochanter (c-tr) joint to innervate (regen. term.) the *extensor tibia* muscle. When clonic extensions of the tibia are seen about 30 days after the transplant, the normal metathoracic ganglion (norm. gang.) of the host is exposed and all normal nerve trunks to the operated leg are cut as indicated by the curved arrow. After this operation the clonic extensions of the host leg continue, indicating that they are related to regenerated axons from the transplanted ganglion. *Inset:* Dorsal view showing the direction of the clonic twitches in the left methoracic leg housing a transplanted ganglion. The leg moves at the femoral tibial joint (f-t) from the open position to the solid position. Grid squares, one cm. (from Jacklet and Cohen, 1967).

REFERENCES

Bodenstein, D. (1957). Studies on nerve regeneration in *Periplaneta americana*. *J. Exp. Zool.*, **136**, 89–115.

Bodian, D. and Mellors, R. C. (1945). The regenerative cycle of motoneurons, with special reference to phosphatase activity. *J. Exp. Med.*, **81**, 469–488.

Brattgård, S. O., Edström, J. E. and Hyden, H. (1957). The chemical changes in regenerating neurons. *J. Neurochem.*, **1**, 316–325.

Cohen, M. J. (1967a). Correlations between structure, function and RNA metabolism in central neurons of insects. In: *The Invertebrate Nervous System*, pp. 65–78. C. A. G. Wiersma, Ed. University of Chicago Press, Chicago.

Cohen, M. J. (1967b). Some cellular correlates of behavior controlled by an insect central ganglion. In: *Chemistry of Learning*, pp. 407–424. W. C. Corning and S. C. Ratner, Eds. Plenum Press, New York.

Cohen, M. J. and Jacklet, J. W. (1965). Neurons of insects: RNA changes during injury and regeneration. *Science*, **148**, 1237–1239.

Cohen, M. J. and Jacklet, J. W. (1967). The functional organization of motor neurons in an insect ganglion. *Phil. Trans. Roy. Soc. (London) B.*, **252**, 561–569.

Eisenstein, E. M. and Cohen, M. J. (1965). Learning in an isolated prothoracic insect ganglion. *Animal Behavior*, **13**, 104–108.

Hess, A. (1958). The fine structure of nerve cells and fibers, neuroglia, and sheaths of the ganglion chain in the cockroach (*Periplaneta americana*). *J. Biophys. Biochem. Cytol.*, **4**, 731–742.

Holtzman, E., Novikoff, A. B. and Villaverde, H. (1967). Lysosomes and gerl in normal and chromatolytic neurons of the rat ganglion nodosum. *J. Cell Biology*, **33**, 419-435.

Horridge, G. A. (1962). Learning of leg position by the ventral nerve cord in headless insects. *Proc. Roy. Soc. (London) B.*, **157**, 33–52.

Hoyle, G. (1965). Neurophysiological studies on "learning" in headless insects. In: *12th International Congress of Entomology*, London, pp. 203–232. J. E. Trehern and J. W. L. Beament, Eds. Academic Press, New York.

Hyden, H. (1960). The neuron. In: *The Cell. Biochemistry, Physiology, Morphology*. Vol. IV, pp. 215–323. J. Brachet and A. E. Mirsky, Eds. Academic Press, New York.

Jacklet, J. W. and Cohen, M. J. (1967). Synaptic connections between a transplanted insect ganglion and muscles of the host. *Science*, **156**, 1638–1640.

Nissl, F. (1892). Ueber die Veränderungen der Ganglienzellen am Facialiskern des Kaninchens nach Ausreissung der Nerven. *Allg. Z. Psychiat.*, **48**, 197–198.

Palay, S. L. (1964). The structural basis for neural action. In: *Brain Function*, Vol. II: *RNA and Brain Function: Memory and Learning*, pp. 69–108. M. A. B. Brazier, Ed. University of California Press, Los Angeles.

Palay, S. L. and Palade, G. E. (1955). The fine structure of neurons. *J. Biophys. Biochem. Cytol.*, **1**, 69–88.

THE PLANARIAN CONTROVERSY

W. C. Corning and D. Riccio

Animal Behavior Laboratories, Department of Psychology
Fordham University
New York, New York

INTRODUCTION: GREAT EXPECTATIONS

In evaluating the significance of any animal preparation currently being utilized to research the physiological bases of learning and memory, better sense might be made out of various investigations if we first ask, what is it that the experimenter *expects* from the animal? Viewed in this way, much of the controversy that frequently accompanies the development of preparations in new research areas might be better understood.

Now the topic of this symposium is molecular memory and it has brought together a number of people whose primary concern is the definition of those cellular events that accompany and maintain learning. A lesser concern is the organism that is doing the learning. What typifies many investigators in the area is their treatment of the laboratory subject as a learning machine, a preparation that will display consistent and unequivocal learning under automated and somewhat restricted conditions. Their *expectations* allow for little consideration of the organism and when the beast rebels and dirties up the data, psychologists may prefer to conclude that learning is impossible in the particular animal and those of other disciplines may begin to express impatience with the science of behavior. This is sometimes followed by a negative review paper and a return to safer pastures, moves that could increase the unemployment rate among psychologists wanting to work in biochemistry.

The psychologist in turn, has certain *expectations* with respect to what has been labeled the "engram" or memory trace. One can almost imagine a quivering, virginal macromolecule which processes the CS at one end, the US at the other, counts the interval between the two (0.5 seconds?), awaits the consequences of the US (pleasure, pain, etc.), makes certain that the intertrial interval is not too short (or too long), and then flips into an engrammatic state which solidifies over time.

These points are made since the *expectations* will determine to a certain extent *what* is looked for, and more importantly, the *way* in which the research is conducted. It is going to be difficult to convince the biochemist that the organism in his laboratory does more than wait for a psychologist to come along and give it a CS and US, and it is going to be equally difficult to convince psychologists that memory involves more than a unique molecule sequestered in chosen cells.

PLANARIANS AND THE CHEMISTRY OF LEARNING

The relevance of planarian research to the present disquisitions requires no new emphasis. The earlier findings that memory could survive regeneration (McConnell *et al.*, 1959) and that it was possible to interfere biochemically with this survival (Corning and John, 1961), were followed by even more spectacular data which suggested that a "transfer of training" could be effected by feeding trained worms to naive subjects (John, 1964; McConnell, 1962). An extensive review of invertebrate studies and the chemistry of learning is now available (Corning and Ratner, 1967).

There has followed a period of disaffection due, on the one hand, to the reported failure of some laboratories to replicate the conditioning results, and on the other hand to the unwillingness of others to assume that memory could involve molecules. However, a peremptory dismissal of the findings in this area would not only be premature (active experimentation began only nine years ago), but also unwarranted in view of the existing evidence. The planarian studies that have reported negative data need to be given serious attention, but they should not be misconstrued as having negated the entire body of planarian research. In many cases, these studies have outlined critical variables that need to be controlled. They also underscore the fact that the planarian can be a difficult beast to work with. But anyone who cares to *read* the literature, rather than *react* to only those comments that fit a priori assumptions, will find that the planarian preparation is very much alive and still occupies an influential position in current memory research. One of the problems has been that many critics writing under the guise of "objective review" are too frequently making their comments on the basis of emotional biases and in some cases they have been appallingly ignorant of the facts. We will demonstrate this by examining in detail the comments of one of the more vociferous critics.

There are two questions to be asked in considering the current status of planarian research: first, a more general question of whether learning is possible in phylum Platyhelminthes; and second, whether specific studies relevant to the biochemistry of memory have utilized adequate learning paradigms. More specifically, can we be satisfied that the retention after regeneration studies and the transfer of memory studies have demonstrated true learning? In dealing with these questions we will first review what has to be considered as one of the best demonstrations of learning at this phylogenetic level and then turn to the critical comments.

A LEARNING DEMONSTRATION

Wells (1967) has mentioned some of the usual objections that are registered against the use of the light-shock classical conditioning procedure and the T-maze discrimination task as methods for demonstrating learning in planarians. In the classical conditioning situation which uses light as a CS and shock as a US, the question of light neutrality and the unknown consequences of whole-body shock can sometimes make distinction between sensitization and true conditioning difficult. In the typical maze situation, Wells objects to the necessity of manipulating subjects by the experimenter and also points out that the type of task required of the planarian may be unnatural.

To obviate these problems, Wells turned to the procedures first described by Van Oye in 1920 (reviewed by Jacobson, 1963; McConnell, 1967a). The procedures allowed for a minimum handling of subjects, the avoidance of stimuli such as light and shock, and the use of a natural response mode of the planarian—food-seeking behavior. Generally speaking, the Van Oye technique consisted of training planarians to locate food on the end of a wire which is lowered into the container housing the worms. The motivation in this case—hunger—could be regulated by starving the subjects. Since the trials were administered in the animals' bowl, there was little danger of interference by the experimenter. Furthermore, it permitted objective assessment of response changes in the subjects; they were at the end of the rod or they were not.

These training procedures are schematized in Fig. 1. The reward (a piece of meat, etc.) is placed on the end of the wire which is lowered below the surface in a bowl containing a number of naive worms. At first, none finds it but then the wire is raised so that the food is on the surface making it easier for the animals to locate it by crawling up the side of the container,

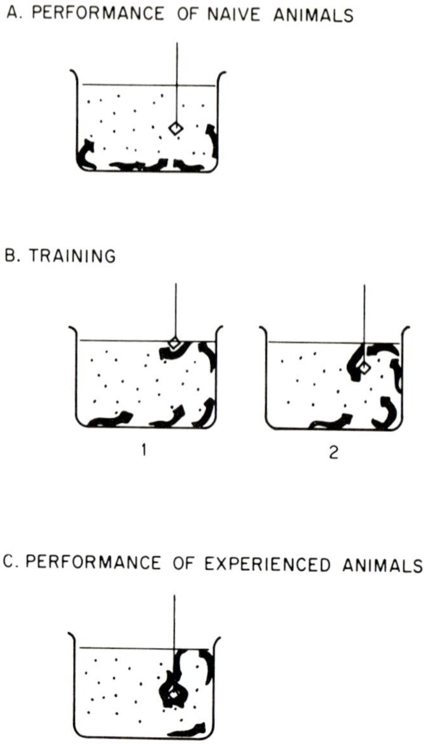

Figure 1 General schematic of the Van Oye technique.

across the undersurface of the water, and then onto the wire with its reward. By gradually lowering the wire it is possible to train planarians to reach the reward when it is several centimeters below the surface of the water.

The experiments of Wells and associates (reviewed in Wells, 1967) have further established the validity of these training procedures by the inclusion of control groups that rule out alternative explanations.

The techniques were quite similar to those of Van Oye. Several subjects were placed in each bowl and the training proceeded by lowering the food in four stages until it was 32 mm below the surface. In the control groups the baited rods were on the bottom of the container. During test sessions, any subject reaching the suspended bait within an hour was counted.

Because planarians could be reacting to stimulus gradients in this situation rather than acquiring a specific maze habit, the animals were tested with clean, empty rods in a beaker containing water mixed with meat juice. Thus, the chemical stimulus (meat juice) was present, but there was no gradient. When tested under these circumstances, trained animals still found the empty goal in greater numbers than naive animals. Further elaboration of these procedures demonstrated the possibility of stimulus association. When a light is associated with the training periods it was found that it took on conditional properties—the presentation of the light along with an unbaited rod resulted in a significantly greater number of trained subjects appearing at the end of the rod than control animals. A summary of these various investigations is presented in Table I.

TABLE I

SUMMARY OF WELLS' DATA

		Number reaching goal		
Experiment	N	Experimental	Control	p
1	60	23	3	<0.01
2	50	17	3	<0.01
3	50	15	3	<0.01
4[a]	50	12	3	<0.05
5[a]	80	17	0	<0.01
		84	12	

[a] Stimulus pairing conditions: Ss exposed to baited rod only when a light is on.

In addition to the replication of Wells, McConnell (1967a) also reports success with the technique. To facilitate location of the route to the baited rod, a colored stripe was painted along the side of the beaker. Other methodological improvements are also suggested.

From these studies it can be seen that the general question posed earlier is easy to answer; planarians do possess the capacity to learn. The Van Oye experiment has now been replicated in two independent laboratories and Wells alone has run at least five separate experiments. What is curious about all this is that these studies are carefully avoided by many reviewers who have concluded that learning in planarians is highly suspect if not impossible.

There is one dissenting report, the Bennett and Calvin (1964) paper. The procedures were generally similar to those described above: in the experimental group the food was progressively lowered below the surface while in the control group the food remained at a fixed position below the surface. Because the experimental group did not find the food with shorter latencies or with increased frequency as trials progressed, it was concluded that the methods were unsuccessful. However, the written report is somewhat confusing since the authors state:

> When the food was presented to the experimental planarians in the center of the beaker on the water surface for 45 minutes, the subjects found food in 35% of the 54 opportunities; when presented food 0.5 cm below the surface, the subjects found food in 44% of 18 opportunities. The subjects found food 1.0 cm below the surface in 30% of 30 opportunities, and food 2.0 cm below the surface was found in 23% of 30 opportunities. The control subjects found food in 5% of the trials [Bennett and Calvin, 1964, 21].

It is true that the distance to the bait was never as great for the experimental group as it was for the control group, but as far as the study went it appeared that experimental animals had a higher probability of finding the bait than controls. Two methodological changes might have improved data: The wire should have been closer to the side of the beaker and the beakers could have been smaller. In any event, this study does not refute Wells or Van Oye. On the contrary, the data tend to be more supportive than damaging and had the research been pushed further, a more unequivocal picture might have been obtained. But then, the Bennett and Calvin paper never intended to state that worms could not learn—only that they could not be trained *reliably*.

THE STRATEGY OF THE SKEPTIC

A frequently cited review of planarian learning is Jensen's 1964 Cambridge paper (Jensen, 1965) which deserves special attention because it represents the sort of emotional trap one can fall into when the *expectations* get in the way of scientific thinking. The paper takes on double importance because it serves to reinforce the biases of others who have decided independently of fact that there is not much to the planarian story. For example, R. F. Thompson in his *Foundations of Physiological Psychology* (1967) devotes extensive space to the initial classical conditioning study (Thompson and McConnell, 1955) which, although demonstrating response modification in planarians, left some questions unanswered. Ignoring the later, more systematic improvements in conditioning techniques, Thompson then turns to Jensen's review, stating:

> Jensen . . . has carefully reviewed all studies of planaria learning through 1964. In all experiments in which learning was reported, serious methodological and control problems exist. The most carefully controlled experiments generally have found negative results [598].

Apparently, Thompson is content to rely upon a review paper rather than read the literature himself because Jensen did not review all studies (less than half, to be exact) and his referral to the negative studies as "the most carefully controlled" is a gross violation of fact. There have been good and bad studies on both sides—to imply that all positive studies have serious problems and that the negative studies are the most carefully run is a good example of the application of "the strategy of the skeptic" which will be defined below.

Beginning with an overview of the paramecium learning controversy and a consideration of how confounding factors might have operated in certain studies, Jensen then outlines "strategies of inquiry" which might characterize various approaches to science:

> . . . one can be a mystic, ignoring data and accepting the teachings of tradition or the inner voice of revelation.
>
> . . . the strategy of the demonstrationist. . . . Demonstrations are searches for data which conform to and confirm a pre-determined theory.
>
> . . . one can be a scientific investigator, one who uses data to develop theory [Jensen, 1965, 12].

A fourth strategy, included in the prepublication copy of the Cambridge paper is particularly interesting:

> . . . the strategy of the skeptic, who doubts any and every view, but will not assuage his doubt or selectively apportion it on the basis of data. The skeptic deals in doubt, but not data [Jensen, 1964, 8].

We shall show how Jensen applies this strategy to the planarian literature.

The focal point of the attack is the first classical conditioning paper of Thompson and McConnell which, as Jensen points out, is important because later studies are based upon it. This study happened not to use the best pseudoconditioning control and had other more or less minor faults. Since that time, proper controls and other procedural changes have been instituted which have complied with most major objections. However, the skeptics continue to live in 1955 and use the weaknesses in the first study as a basis for rejecting all subsequent planarian studies. It would seem that if the business at hand were the question of conditioning planarians, then those seriously interested in this problem would be following current research techniques. If one is interested in applying the "strategy of the skeptic," then 1955 is a good place to stop reading. Before writing off the earlier research, however, we ought to examine whether there has ever been an adequate refutation of it.

The principal experimental results which many point to as adequate evidence against planarian learning derives from a series of investigations by Halas and associates (Halas *et al.*, 1961, 1962; James and Halas, 1964).

Detailed procedural comparisons of the McConnell and Halas studies will be presented in a later section. Basically, they differ in the criteria which will be used to determine whether learning occurs, and in the experimental methodology used to achieve learning. In the initial attempt (Thompson and McConnell, 1955), acquisition scores provided the measure; the responses of individual subjects on every trial were recorded. In the study by Halas *et al.* (1961), no attempt was made to measure acquisition—subjects were mass trained in one container and then tested individually under *extinction* conditions (CS only) in a different container. There was no way of knowing what sort of response modifications were taking place in the experimental and control groups during training. The reasons for adopting these procedures were that individual conditioning required the constant attention of the experimenter to keep the animal gliding smoothly and to keep the animal oriented properly with respect to the electrodes. To prevent the experimenter from biasing the situation, mass conditioning was used where ". . . all Ss are treated more alike and are relatively free of biasing" (James and Halas, 1964, 23). Recognizing that under mass training conditions the shock would not be the same for all animals, they raised their current intensity to a level where the subjects would react regardless of what their orientation was. In this case, Halas is properly concerned with experimenter bias, but ignores the fact that in the group training situation, the subjects may bias matters and prevent the uniform experience that Halas desired. Their procedures also ensured that subjects received a highly varied US experience. Since the shock level was raised so that all subjects would contract even when they were improperly oriented, the correct orientation resulted in a greater than threshold shock level. Such procedures can be damaging to the planarian.

In experiments at Brookhaven National Laboratories we spent considerable effort in developing a mass training apparatus and eventually arrived at the training container depicted in Fig. 2. This has an interesting history. At first, an attempt was made to mass condition animals (by pairing light and shock) with the electrodes at either end of a rectangular trough. Several things happened after the onset of conditioning trials: the subjects would climb up on the electrodes, slip behind them, and position themselves in certain places in the trough where it appeared they were minimally affected by the shock or were immobilized. If Halas claims that there is homogeneity of experience in this type of situation, he has never observed his animals very closely.

We then decided to use the circular training compartment of Fig. 2. The electrode climbing was prevented by enclosing the animals within a porous clay restraining wall (furnace core) and positioning the platinum electrodes outside it. Two pairs of electrodes were used and by randomly switching the

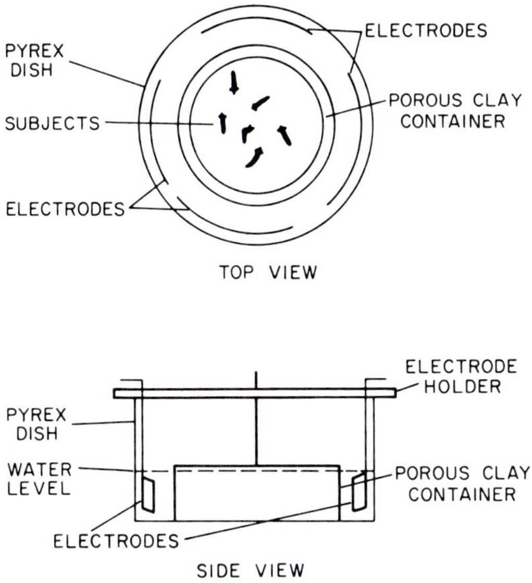

Figure 2 Mass training apparatus for planarians (Corning and Freed, 1968; Freed, 1966).

polarity we could prevent the subjects from adopting any one orientation. The current spread was also quite uniform within the compartment. With this set-up, the minimum effective shock for one animal approximated the effective shock for all other subjects in the bowl, providing sizes were approximately the same. The schematic in Fig. 3 represents the extent one can go to in order to ensure precise and automatic programming of light and shock treatments.

Our problems didn't always end at this point. On a number of occasions, subjects were observed to crawl up the sides of the porous clay restraining wall as soon as the shock treatments began and once out of the water they would, of course, no longer experience the shock. To outwit planarians that used this tactic, we placed a piece of "Saran Wrap" over the top of the inner wall, secured it tightly on the rim, and raised the level of the water over the rim. It should be mentioned here that it only took the planarian two or three days to learn how to avoid the shock but it took the experimenters two weeks to learn how to prevent this behavior.

These details are presented to demonstrate that some of the assumptions we make concerning our animals can be quite tenuous. The least that could be done in something like the mass training situation would be for the experimenter to check his subjects' performance periodically to ensure that other factors are not making the behavior "noisy." We did this at Brookhaven by removing samples of subjects at certain points in the course of

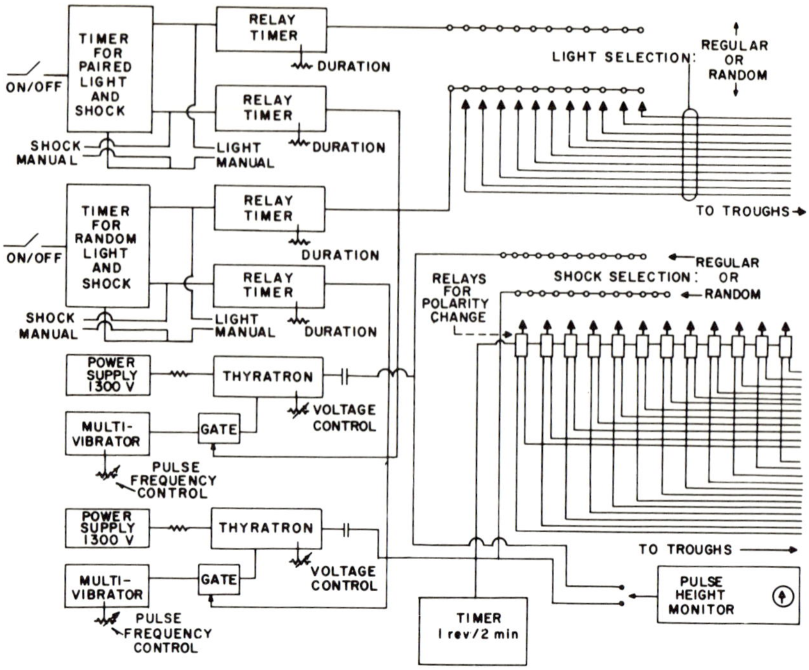

Figure 3 Schematic of apparatus used to control light and shock treatments in mass training (Freed, 1966).

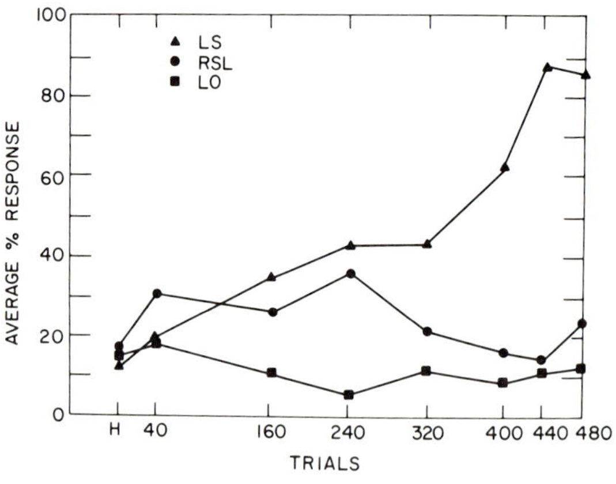

Figure 4 Percentage response to light in trained and control groups (Corning and Freed, 1968). "LS": subjects receiving paired light and shock. "RSL": subjects receiving random light and shock. "LO": subjects receiving light alone. "H" indicates habituation stage where all subjects received light alone.

the experimental treatments and testing them individually in the same compartment. Our findings are presented in Fig. 4 and indicate clear differences between the group receiving paired light-shock treatments (LS) and those groups receiving random light and shock (RSL) and light alone (LO). The RSL group received the same amount of light and shock as the experimental group in any one session.

A further examination of the Halas–McConnell studies reveals that when an attempt was made to replicate exactly the original classical conditioning study, similar results were obtained (Halas *et al.*, 1962) although the general level of light responsivity was higher and a different species of planarian was used. As far back as 1903, Pearl was discussing the differences between *P. dorotocephala* and *P. maculata*, the two species used in the McConnell and Halas studies. The Halas study also showed that shock does not sensitize planarians to light—subjects given shock treatments and tested for light responsivity displayed a decreasing number of "turning" responses and did not change with respect to the number of contractions. The authors chose to interpret their data as showing reflex sensitization rather than learning on the basis that light is a weak US. It is difficult to understand, from the data they present, what is "sensitizing" the animals since one of the experimental groups tends to rule out shock. As the authors stated, "The interpretation of these results represents an interesting challenge" (Halas *et al.*, 1962, p. 971).

To return to the Jensen review, it is curious that this replication of the Thompson and McConnell findings is not mentioned, only the interpretations of the data which are conjectural and not solidly supported by their own data.

Jensen goes on to criticize McConnell's view that in the training of planarians it is frequently necessary to try a number of procedures until the right technique is found. Anyone who has ever attempted to replicate a study realizes that the same conditions do not prevail from laboratory to laboratory. As one example, Rohlich (1968) has recently demonstrated that keeping planarians in absolute darkness for too long a period can bring about structural degeneration in the eyes. Fortunately, in our own laboratories the subjects are exposed to light, but prior to Rohlich's report we never thought to stress our lighting conditions. Factors such as these need to be studied and their relevance to conditioning experiments is quite apparent. The use of identical training procedures would mean nothing if one laboratory used animals with structurally degenerated eyes while another laboratory had normal subjects. Yet, Jensen objects to the improvement of conditions. He writes:

> This advice may provide magic enough. It says to use data only to confirm theory. It says to vary conditions until the data fit the theory that planaria

> learn. It says to change procedures until one is successful and it ignores the possibility that with patience and time one will happen upon an experimental procedure which involves just the right confounding to produce pseudo-learning. Continued manufacture of instances of pseudo-learning seems rather likely if the advice offered . . . is followed [Jensen, 1965, 15].

The above statement is bothersome for two reasons. First, Jensen is falling victim to the very tactic he objects against when he makes the a priori assumption that the demonstration, whatever it is, *will be* pseudo-learning. Science can certainly use someone with a little clairvoyance but the evidence indicates a low degree of perspicacity. Secondly, if we adhere to this sort of reasoning then the first time we attempt a study must also be the last time, because if we try again the probability of success goes up. The reverse situation holds true—one may not change variables until one fails to get planarian learning. Viewed in this way the question of planarian learning is reduced to the earlier work. Fortunately, science has not attempted to progress along the lines suggested by Jensen. A full characterization of variables is a sign of improvement rather than "confounding."

Jensen then deals with other forms of planarian learning and cites Pearl's 1903 monograph as "the major relevant work." This paper is an extensive and well-organized account of planarian reflexes, but it is mostly descriptive and it also does not say what Jensen interprets from it. For example, Jensen notes Pearl's observation that strong stimulation (mechanical) produces negative reactions whereas weak stimulation produces positive reactions. From these observations it is concluded by Jensen that shock ought to increase the light responsivity of the organism. Actually, however, Pearl states the opposite:

> Persistent strong stimulation of the organism rapidly changes the general physiological condition. . . . The animal becomes 'stirred up' generally, moves about with increased rapidity, *its sensitiveness to stimuli becomes diminished* . . . [italics ours; see Pearl, 1903, 582].

For example, in a greatly excited animal Pearl was only able to elicit one response in 23 attempts with weak stimulation, whereas in the normal animal he was able to elicit 22 responses in 35 attempts. From these data one would predict that the general excitation produced by shock ought to decrease the sensitivity to light when they are presented in noncontiguous order—this is in fact, what Halas *et al.* (1962) showed. In the shock control group, the turning responses to light diminished while the number of contractions to light remained unchanged.

Jay Best's maze discrimination studies (Best, 1963; Best and Rubinstein, 1962) are dismissed by Jensen because they had light as a cue. Again Pearl's study is brought in as support for this dismissal, yet, in a statement defining the scope of his research Pearl writes that "No work was done on phototaxis . . ." (1903, 513) and a careful search through the monograph yields nothing which indicates any systematic study of light effects.* We may again quote Jensen:

> According to Pearl (1903), light could produce turning toward or away from the light, depending upon other factors which influence whether the positive or negative reaction is given; this explanation is consistent with the finding by Best and Rubinstein (1962) that periods of rejection occur [Jensen, 1965, 17].

Additionally, if the maze studies are *read* it is readily seen that planarians were trained to go to either a lighted or a darkened alley, that the learning curves for either cue were identical, and that the "periods of rejection" were found to be a function of the maze peculiarities rather than the obscure "other factors" to which Jensen refers. To obtain the results that were reported, according to Jensen's reasoning, Best would have had to be very consistent in doing something to one group that he didn't do to the other, and in order to obtain the "periods of rejection," Best would have had to systematically reverse these factors at a specific point.

Turning to Lee's (1963) operant conditioning experiment, Jensen confuses conjecture with fact in what is his most strained effort to refute this study. Again, one of Pearl's observations is mentioned: activated animals required a period of rest after an activity cycle. The Lee paradigm required planarians to move through a tunnel and break a photoelectric beam which in turn, caused an overhead light to be turned off for 15 minutes. The explanation Jensen offers for Lee's observation of increased response enhancement (worms moved through the tunnel more frequently) goes as follows:

> The experimental procedure in question undoubtedly stimulated the planaria repeatedly and accordingly would be expected to produce a general readiness to come to rest soon after the overhead light was extinguished. The temporal patterns of light stimulation administered to the pairs of experimental

*Pearl's references to his observations of light responses are primarily the following:

> Regarding the effect of different agencies on the rate of the gliding movement no special study has been made, and I can only report a few incidental observations. Such a study should be made by exact quantitative methods, and this I have not had the opportunity to do . . . light of great intensity (that obtained from a projection lantern with an electric arc as its source of illumination) causes a definite increase in the rate of gliding, but this increase has not been measured [1903, 547].
>
> . . . planarians tend to come to rest in regions of comparatively low light intensity. . . . In aquarium dishes placed close to a window, and containing considerable plant material, the worms will be found resting practically always in the half of the dish away from the window. The largest number of individuals will be entangled in the plant material. . . . This behaviour towards light is not, however, an absolutely precise reaction. Many times during experiments I have seen specimens come to rest in the very lightest parts of the dish and remain there . . . [1903, 564–565].

> and control animals were exactly the same, but where the control and experimental animals were when the overhead light was extinguished differed markedly. When the overhead light was extinguished the control animals could be anywhere, but by virtue of the reinforcement contingency the experimental animals were invariably near the light beam. If the planaria would occasionally remain at rest for the 15 minutes that the overhead light remained out, the experimental subjects would be found in the immediate vicinity of the light beam when the overhead light again came on. As a result the experimental subjects would be more likely to interrupt the central light beam than would the control subjects whose presence near the central beam is not so insured [Jensen, 1965, 17].

The certainty of this statement almost makes one believe that there are data, that there have been observations, but the surprising thing is that no data are offered. Best (1965), who was taken the trouble of observing planarians in the operant task, replied to these conjectures by noting that planarians do not stop when the stimulus goes off but may circle the apparatus many times after breaking the detector beam. Identical behavior patterns were observed in both experimental and control groups for the first nine hours. There was no comment by Jensen with respect to Best's observations. Apparently, the conjecture must supercede fact but this is consistent with the strategy of the skeptic.

At this point, one begins to get suspicious. Why this reliance upon a 1903 descriptive monograph to "explain" modern research, why the consistent misinterpretation of studies, why the selective avoidance of studies whose data conflict with the pseudo-learning hypothesis, why the fear of changing variables because positive results might be obtained, why the *carte blanche* acceptance of any negative study no matter how badly done, and why the need to spend so much effort at conjecture to refute any positive study? The answers to all these questions lie in the opening remarks concerning *expectations* and it is clear that in this case the expectations are overcoming scientific objectivity and the available evidence.

Bullock and Quarton (1966), in a Neurosciences Research Program conference, sum it up nicely when they write, "The reviews of McConnell (1965a) and Jacobson (1963; 1965) are very useful, although written by protagonists; critiques by Jensen (1965) and Bennett and Calvin (1964) do not deal with the bulk of the literature and in many respects are more vulnerable than the first named" (1966, 115). It is difficult to see how anyone who reads the literature can conclude otherwise.

SUMMARY OF NEGATIVE COMMENTS

Excluding the comments of Jensen which we already have dealt with in detail, and which in our opinion do not represent the most objective sort of approach, we will summarize at this point the major experimental objections to planarian learning research.

Halas and Associates

The major argument of Halas and his group is that extinction is a valid measure of learning, and during extinction tests the differences between experimental and control animals do not hold up. They also note the extreme variability of planarians' reactivity to light, and their data indicate that naive animals can display a very high response rate. In response to criticism that their extinction studies were never a replication of any one study, the reply is that techniques are changed quite often and it is difficult to know which ones to follow.

Bennett and Calvin

Several different attempts were made to condition planarians including classical conditioning, habituation, maze learning, operant conditioning, and an approach task. While the experiments demonstrated learning in the classical conditioning paradigm and hinted at learning in the approach task, the experimenters were not satisfied with either the variability in planarian behavior nor with the low levels of response achievement in the training situations. Their major point is that the factors responsible for altering planarian behavior ought to be consistently described so that the biochemist can effectively use planarians. They conclude that training methodologies are so far inadequate to produce reliable training.

Brown and Associates

H. M. Brown's laboratory at Utah has recently reported on an extensive series of experiments which have in general, failed to repeat some of the more basic planarian findings (Brown, 1964, 1967a; Brown *et al.*, 1966a, 1966b). It was found that:

1. Short worms were more sensitive to light than long worms.
2. Head regenerates were more sensitive than middle or tail regenerates.
3. Head and tail regenerates displayed equal sensitivity to light at high light intensities but the heads were more sensitive than tails at low light intensities.
4. No evidence for habituation was found.
5. Worms were more active at night but more sensitive to light during the day.

6. No evidence of conditioning was found using light and shock. Spaced light and shock produced the same results as paired light and shock.

Additional studies attempted to test the thesis that the nucleic acids are involved in retention of the conditioned response in planarians; but since the basic phenomenon, learning, was not demonstrated by Brown, there is no need to review these investigations at this point.

Miscellaneous Studies

VanDeventer and Ratner (1964) attempted a replication of the retention after regeneration results and found no differences between their experimental and control groups. The tail regenerates were also much more reactive than the head regenerates. Subsequent studies attempted to define some of the variables that might be affecting results. Temperature, shape of the trough, animal size, dark adaptation, etc. were all found to be possible confounding factors in planarian research.

Reynierse *et al.* (1966) report on the effects of shock on adhesiveness in planarians and find that the tendency of worms to stop moving is a function of shock. These data are used to question the conditioned discrimination results of Griffard and Peirce (1964) on the basis that a less active subject would have a higher probability of displaying the responses Griffard and Peirce were recording.

There are other more specific findings that pertain to selected aspects of various investigations. The reader is referred to the tables in the following sections for a documentation of all studies and for comparisons between specific studies.

RECENT DEMONSTRATIONS PERTAINING TO NEGATIVE COMMENTS

Classical Conditioning

Jacobson has recently paid considerable attention to some of the more serious criticisms raised by Halas and others and has also furthered the definition of critical variables in the classical conditioning situation (Jacobson, 1967; Jacobson *et al.*, 1967). In this way he is violating "Jensen's law" which, if you will recall, states that we may not look for a set of variables that give us proper results, but he is satisfying Bennett's comment concerning the need for the specification of conditions that produce reliable response modification.

Two problems, pseudoconditioning and sensitization, require specific attention. Jacobson has outlined the generally acceptable definitions of pseudoconditioning and sensitization and the procedures which would differentiate between these phenomena and true learning:

> Pseudoconditioning is controlled if groups given backward, simultaneous, or randomly interspersed presentations of CS and US show significantly less response enhancement to the CS than a group given CS and US in paired forward sequence. Sensitization is controlled if differential responsiveness can be established with respect to two stimuli, one paired with the US and the other unpaired [Jacobson *et al.*, 1967, 73].

The initial study compared the effects of presenting planaria with paired and unpaired light and shock trials. As in the Halas studies, mass training was used, but unlike these studies, Jacobson took the trouble to look at the performance of individual subjects during training—worms were randomly selected from the groups and given individual trials. The results of this study are presented in Fig. 5 and show clearly that paired light-shock treatments (CC) produce a different response modification than unpaired light and shock presentations (PC). This finding confirms the observations of Freed's group at Brookhaven. A major difference is that Jacobson finds that the contraction response is the only reliable response measure, whereas in the Brookhaven studies, both contractions and head-turning responses provided a differentiation between groups.

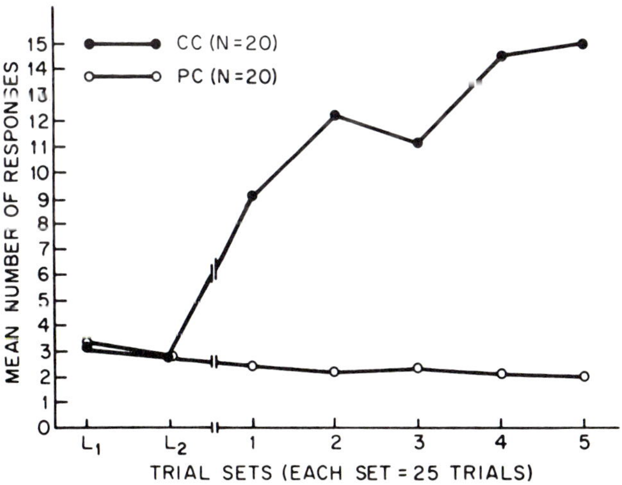

Figure 5 Mean number of responses per set of 25 trials for groups CC and PC. On the abscissa, L_1 and L_2 are sets in which light alone was presented, and 1–5 represent training sets. Dark circle: CC (classical conditioning), N = 20; PC (pseudoconditioned), N = 20 Jacobson, Fried, and Horowitz, 1966).

In subsequent research, more stringent controls were incorporated: a backward conditioning group which received shock followed immediately by light, and a simultaneous conditioning group which received the light and shock stimuli at the same time. These treatments are schematized in Fig. 6a. In Fig. 6b are presented the data obtained during test trials (light alone) administered to the subjects of all groups after mass training.

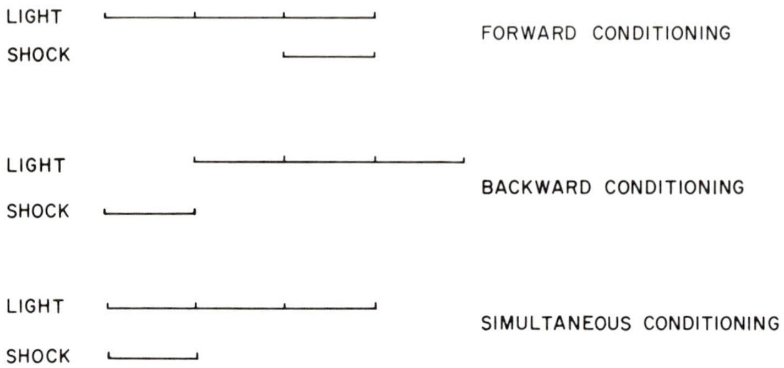

Figure 6a Light-shock relationships in trained and control groups.

It was also found that high-intensity vibration could be effectively used as a CS; having determined this, the way was cleared for answering the sensitization criticism. Animals could be trained under differential conditioning procedures where both light and vibration could be presented to each subject but where the shock would only be paired with one of these stimuli. The results of two separate studies indicated that differential conditioning was possible. In the first, subjects were trained individually, with shock being paired with either light or vibration. Differential responding was attained; subjects displayed a significantly higher response rate to the stimulus paired with the shock than to the stimulus that was not associated with it. Conditions were then reversed and the response rate to the formerly positive CS diminished, demonstrating extinction, while the response rate to the formerly negative CS increased. In the second study, subjects were presented mass training conditions with either light or vibration as the positive CS. At the termination of training treatments, subjects were individually presented a random series of light and vibration test trials. In the test sessions, the subjects were coded and the experimenters did not know whether light or vibration had been the positive cue. The results of this study are presented in Table II and again indicate the differential responsivity of the groups.

Thus, Jacobson has shown that under his experimental conditions, forward conditioning produces a different kind of response modification than either

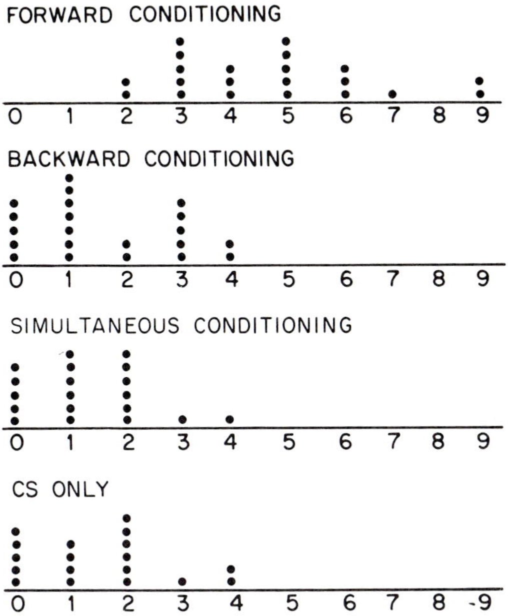

Figure 6b Distribution of numbers of contractions for individual Ss on test series of 25 light presentations (Jacobson, A. L., Fried, C. and Horowitz, S. D. "Classical conditioning, pseudoconditioning, or sensitization in the planarian." *J. comp. physiol. Psychol.*, **64**, 1967, 73. Copyright 1967 by the American Psychological Association, and reproduced by permission.)

backward or simultaneous conditioning procedures, that these response modifications can be differentiated during both acquisition and extinction, that differential conditioning is possible, and that contractions which are seldom elicited by weak light intensities are the more reliable indicator of behavioral plasticity. These investigations representing the most systematic analyses to date do not, as Jensen would have it, represent ". . . an experimental procedure which involves just the right confounding to produce pseudo-learning." Indeed they represent the application of procedures accepted throughout the behavioral sciences as being necessary to demonstrate learning. Of course if it is assumed that the optimum variables may not be delineated if the very first investigation has failed to do so, then these data have no significance and the emphasis upon 1903 science becomes quite understandable.

The Permanency of Memory

We normally think of memory as being relatively persistent in higher animals. Data are available that suggest relative permanency in planarians

TABLE II

FREQUENCIES OF RESPONSE TO LIGHT AND TO VIBRATION DURING EXTINCTION TEST SERIES FOR LIGHT-POSITIVE GROUP AND FOR VIBRATION-POSITIVE GROUP[a]

S	Vibration (+)	Light (−)	S	Light (+)	Vibration (−)
V1	8	0	L1	7	2
V2	8	1	L2	6	2
V3	8	2	L3	6	4
V4	5	0	L4	5	3
V5	4	0	L5	4	2
V6	4	0	L6	3	1
V7	4	1	L7	5	4
V8	3	0	L8	4	3
V9	5	3	L9	4	3
V10	4	2	L10	3	2
V11	3	1	L11	4	4
V12	6	5	L12	1	1
V13	2	1	L13	5	6
V14	3	3	L14	5	6

[a](Jacobson, Fried, and Horowitz, 1967).

(Freed, 1966). Fig. 7 shows the response curves of six *D. dorotocephala* trained by Miss Claudia Ferguson in one phase of the research at Brookhaven. The first 100 trials consisted of light treatments only and as can be seen there is general habituation. Training produced marked response enhancement. After 380 training trials, a 39-day time interval was allowed to lapse; yet, when training was resumed the animals showed response levels that were higher than habituated levels. Subject d is of interest in view of the behavioral variability that so many investigators complain about. This animal ceased responding to the CS and this cessation occurred during a period in which it fissioned. The immediate effect of the altered physiological state is quite apparent at trial 320. After the 39-day rest period, however, the administration of training trials showed that the memory had not been affected even though the responsivity to the CS had been temporarily reduced. In other words, the "read-out" of memory can be affected by a physiological state—in this case, fissioning. An analogous situation might be an attempt to give a woman an I.Q. test while she is giving birth; the "read-out" would certainly differ from the normal state.

It should also be mentioned that if one is arguing that learning in planarians is nothing more than sensitization or pseudo-learning, then it is clear that we also need laws to account for the *acquisition* and long-term *retention*

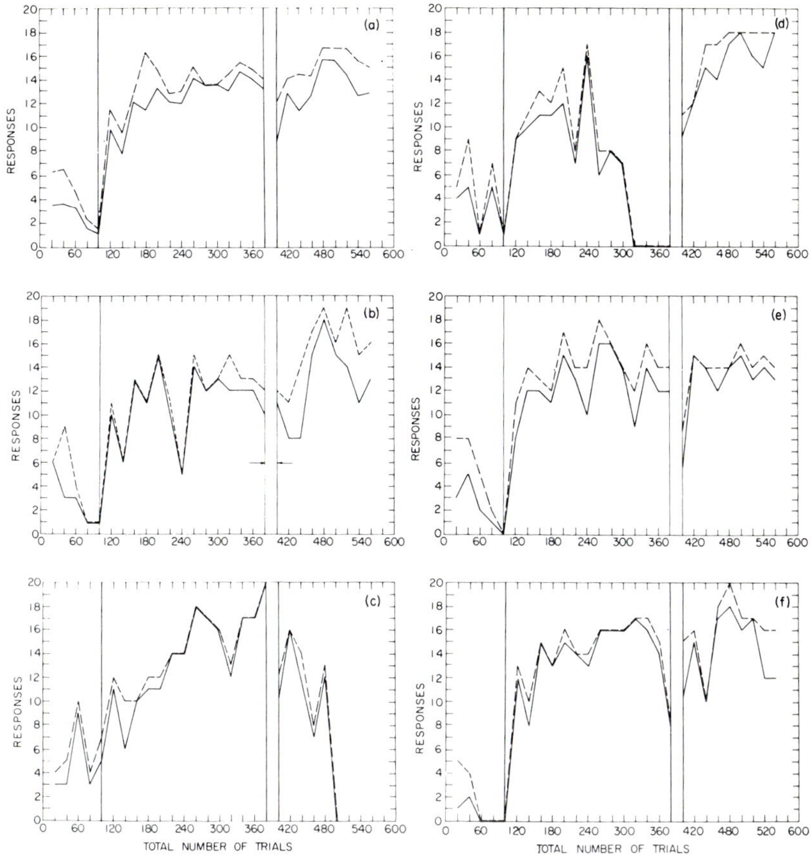

Figure 7 Total (broken lines) and valid (solid lines) responses of six worms during training. The vertical line at 100 trials marks the end of the period of habituation to light. The interval between the vertical lines at 380 and 400 trials was 39 days (Freed, 1966).

of sensitization and pseudo-learning. One can imagine future conferences dealing with the topic of "Molecular Approaches to Sensitization and Pseudo-learning"; in addition to searching for the engram we will have to look for a "pseudo-gram" and a "sensi-gram."

Specific Factors

The frequent variability of planarian behavior and the inability of many laboratories to repeat specific experiments has stimulated efforts towards a more precise definition of factors relevant to the types of behavior that are studied. As we have seen from the data cited above, the response to incoming information can be dependent upon physiological states, and some of the variability that is observed may be due to conditions that are changing the physiology but not the physiology of memory storage. It is becoming apparent that we are going to have to be much more systematic about reporting experimental conditions so that an attempted replication is truly a procedural replication. In some cases the differences between laboratories may depend upon the housing conditions of the animal colonies. In other cases the differences are less obscure, for example, a physicist who wanted to be so scientifically pure that he kept his animals in distilled water. Needless to say, the behavior of his animals changed radically—they died.

In studies such as Mack Brown's, which differ so much from the results reported by others, it is apparent that there are basic differences which do not reside solely in reported experimental procedure (although there are serious deviations in techniques—see footnote 2). His animals were highly variable and sensitive to light. This degree of subject "noise" makes it improbable that there will be any differentiation between experimental treatments and it would seem logical that before one embarks upon any involved iconoclastic research adventure, it ought to be established that the subjects are physiologically the same animals. A good beginning would be to use the same species. In Fig. 8 are data obtained in a study designed to measure the response levels elicited by two different light intensities and to

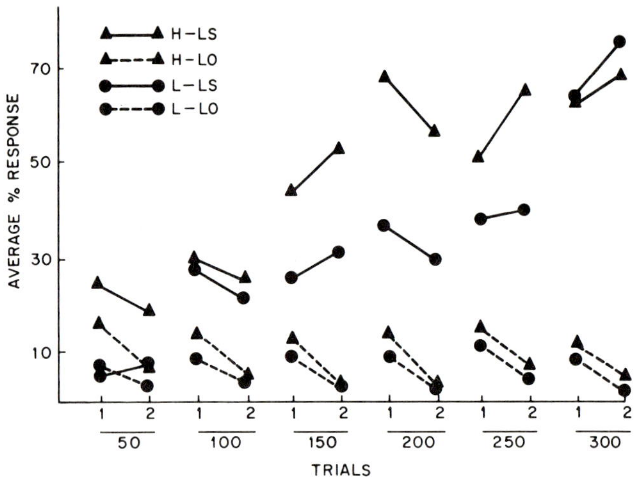

Figure 8 Light responses of trained and control groups under high and low light levels. "H" indicates high light intensity (15 watt bulb 8″ above trough); "L" indicates low intensity ($7\frac{1}{2}$ watt bulb 8″ above trough). "LS": trained groups receiving paired light and shock; "LO": groups receiving light alone on each trial. Each daily session of 50 trials is broken into the first and second half average scores (Corning, Bluemfeld, and John, unpublished observations).

look at the degree of stability in these response levels over time. It can be seen that in those groups receiving the light trials (LO), the response characteristics were replicated from day to day. The data also show a different response curve over time for those animals receiving light-shock pairings. These studies are mentioned because considerable effort was expended in stabilizing our subject conditions prior to asking any questions concerning learning and memory, and the effects of RNase upon these phenomena. It seems paradoxical that when we run rats, we handle our subjects, we specify what breeding line the stock is from, we train then in soundproof boxes, and we specify a large number of factors which when put together give us an output which we call learning. Planarians, on the other hand, are popped into a trough, given a CS and US and are expected to perform like a learning rat, and the concomitant exploration of those factors that can affect the behavior brings the criticism that we are producing ". . . an experimental procedure which involves just the right confounding to produce pseudo-learning."

McConnell (1967b) has recently reviewed the specific factors that are known or suspected to have an effect upon planarian behavior. Included in this survey are such considerations as species, temperature, diurnal variation, seasonal variations, water content, food, lighting conditions, shock parameters, intertrial intervals, slime, and others. A brief discussion of some of these factors will be provided here.

SLIME

In unfamiliar surroundings animals will frequently exhibit a great deal of activity or exploratory behavior and this behavior can interfere with the treatments imposed upon them. Accordingly, subjects are usually given time to adapt to the situation. In a rat this adaptation involves a number of sensory modes such as vision, olfaction, and touch; in the planarian (which possesses more primitive sensory systems) a critical form of "recognition" is by the slime trail deposited by the animal as it crawls about. There may be little visual "recognition" of the situation as in the case of a higher animal. Thus, presenting the planarian with a sterilized learning apparatus every day may result in more behavioral variability.

McConnell and Mpitsos have obtained evidence (McConnell, 1967b; McConnell and Mpitsos, 1965) that planarians learn better when they are in slimed mazes, they will select the slimed portions of the apparatus when given a choice, and that their reactivity to light is enhanced by the absence of slime.

Research in our own laboratories has also established the relevance of slime to the behavior we are interested in (Riccio and Corning, 1969). The first question we asked was whether there was any polarity to a slime trail, *i.e.*, is there a directional aspect to the trail? The results were negative. A second question concerned the preference of planarians for slime in a free-choice situation and the answer was again negative but we tested our animals under "nonstressful" conditions—the room was dark, etc. We have found that the presence of UV light, an altered pH, or other deviant states can change the planarian's preference for slime. Finally, we asked whether the animal's responsivity to light would differ according to slime conditions and found that it did. Our results are summarized in Table III. Animals tested under slimed conditions were half as reactive to light as those in nonslimed conditions. Furthermore, there also appears to be a difference between large and small animals although this was not statistically significant.

This study indicates that in our laboratory, this factor is relevant for certain behavior and we can now take it into consideration in subsequent research. This strategy of defining and controlling for certain factors needs to be adopted for planarian research; it has, after all, been useful in rat studies.

TABLE III

THE EFFECTS OF SLIME ON LIGHT REACTIVITY IN PLANARIANS[a]

	Slime		Nonslime		
Size	Mean response	Trials	Mean response	Trials	N
Large	7.8	60	16.3	60	8
Small	13.5	60	26.8	60	8

Statistical analyses:
- Slime vs. nonslime: $p < 0.01$ (Cochran Q)
- Large vs. small (slime): N.S.
- Large vs. small (nonslime): N.S.

Conditions:
- Subjects: *D. dorotocephala*
- Water: natural aquarium water
- Stimulus: 500 ft.c.; 3 sec. duration; homogeneous in trough.
- Ambient illumination: 1 ft.c.
- Intertrial interval: 60–90 secs.
- Trials: 20/day for three days.
- Size: large Ss greater than 20 mm in length; small Ss less than 11 mm.

[a] Riccio and Corning (1969).

HOUSING

While there are fewer investigators who attempt to keep planarians in distilled water, there is still little consistency from laboratory to laboratory with respect to the medium in which animals are kept. Investigators involved in both positive and negative studies recognize the medium as a critical variable, but little attempt has been made to standardize this important variable for planarian behavior.

An artificial medium has been detailed in McConnell's handbook on planarian behavior (McConnell, 1967a) and we have found this to be adequate in our own colonies. Recently, we have chosen to keep our animals in water removed from a natural aquarium in which fish and snails are kept. A general observation has been an overall reduction in light sensitivity in the subjects although they seem as active as ever. There is certainly a great need for the specification of medium constituents. A good example of some of the consideration that has gone into this problem is provided in Freed's (1966) report of the Brookhaven studies. Not only was the type of food worried about, but the effect of this food on the pH of the water was also examined. In addition, the salt concentrations of the water sources were assayed and when at one point the calcium concentration was too low, adjustments could be made which were reflected in improved behavior. If we assume that such factors are not important we might as well assume behavior takes place in a nonphysical system.

Planarians may also be communicating with each other, a social process that can lead to changes in the physiology and behavior. Reynierse (1967) has shown that species are able to recognize members of their own kind; when two different species are put together in a bowl, there is a high probability that aggregates of like species will be formed. Best (discussion in Corning and Ratner, 1967, 274–275) finds that the probability of fissioning in *D. dorotocephala* is a function of the number of animals in the bowl. Keeping an animal isolated results in an 80–85% probability that it will fission; two per bowl results in a 65% probability; five per bowl results in a 50% probability; and so on until with 20 per bowl the fissioning is reduced to 18–20%. Best has ruled out slime, O_2, and CO_2, and other substances, concluding that contact may be the critical factor. The observation that fissioning can alter the planarian's light sensitivity may make it necessary to control the problem of "loneliness" in flatworms in order to gain greater control over their behavior.

Laboratory lighting conditions have recently taken on some additional significance with the previously mentioned finding of Rohlich (1968) concerning eye degeneration in darkened conditions. In animals (*D. tigrina*) kept in total darkness for seven days, the basal portions of the microvilli were broken up into anastomosing tubules and vesicles. After 21 days of total darkness, the microvillus system was massively disintegrated. These consequences must have some effect on the planarian's processing of visual information. This factor might very well explain some of the discrepancies between studies, particularly the regeneration studies.

DIURNAL FACTORS

Brown (1964) has found an interesting relationship between the time of day and the planarian's motility and light reactivity. Motility was greatest at nighttime whereas light reactivity was greatest during the daylight periods. If this observation holds true for other laboratory conditions it is going to make it imperative that training times be the same if we are to compare data in a fully meaningful way.

SHOCK PARAMETERS

The choice of the US is obviously an important consideration. For example, Barnes and Katzung (1963) have shown that it is possible to condition planarians when the heads of animals were oriented towards the cathode but not when the head orientation was anodal. This study has been difficult to replicate. Hydén *et al.* (in press), in a study of RNA base ratio changes in planarian nerve cells point to initial failures to condition animals using the Barnes and Katzung procedures. Success was achieved when they replicated their techniques *exactly*. The results of this study, additional confirmation of classical conditioning in the planarian, are presented in Fig. 9.

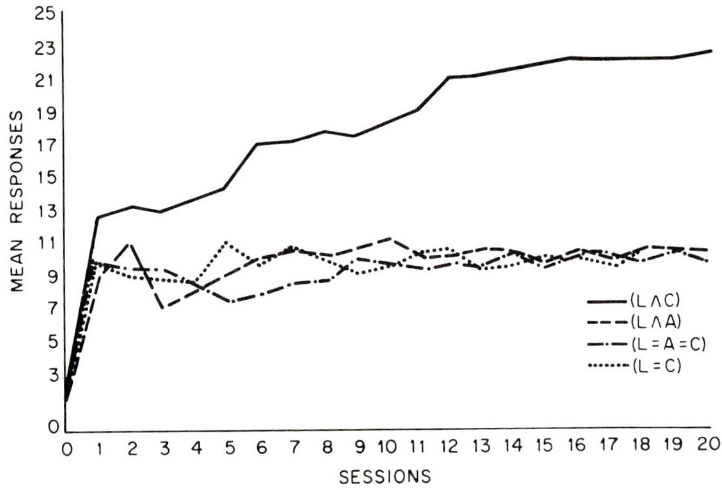

Figure 9 Mean responses to CS as a function of training sessions for the four experimental groups. The four curves shown above represent the mean responses displayed by each of four experimental groups in the training sessions. Group 1 (L ∧ C) consists of animals which receive light paired with shock delivered so that the head of the animal was toward the cathode (N = 13). Group II (L ∧ A) consists of animals which received light paired with anodal head shock (N = 13). Group III (L = A = C) consists of animals which received light which was randomly associated to shock which was sometimes anodal and sometimes cathodal (N = 13). Group IV (L = C) consists of animals which received light randomly associated with cathodal head shock (N = 10) (Hydén *et al.*, in press).

From the growing list of critical variables, it can be seen that the planarian is very much like other organisms—it is a complex beast that cannot be given a "learning machine" treatment. In the following section, specific comparisons between contradictory studies will be made so that a more accurate assessment of the degree of procedural replication may be made.

A QUESTION OF REPLICATION

Specific Comparisons

The areas of controversy center around the attempted replications by Halas and associates of the classical conditioning results of McConnell, the

TABLE IV

CLASSICAL CONDITIONING: COMPARISON OF McCONNELL–HALAS STUDIES

Variables	Thompson and McConnell (1955)	Halas, James, and Stone (1961)	Halas, James, and Knutson (1962)
Species	*D. dorotocephala*	*D. tigrina*	*D. tigrina*
Housing	20–24°C; bkgnd light "minimum"	20–24°C; "complete darkness"	20–24°C; "complete darkness"
Training mode and measures	Individual conditioning; acquisition scores	Mass training; tested 16–22 hr. later under extinction conditions	Individual conditioning; acquisition scores
Trough	Semicircular: 3 × 1.5 × 30 cm.	Rectangular (train): 3 × 1 3/4 × 2 in.; Semicircular (test): 16 × 0.5 × 0.5 in.	V-shape: 14 × 1 × 0.5 in.
Shock (US)	DC: 28 ma via inductorium	DC: 6v via inductorium	DC: 6v at 0.30 ma via inductorium
Light (CS)	2–100W bulbs 1 ft. above trough	2–100W bulbs 1 ft. above trough	2–60W bulbs 1 ft. above trough
Intertrial interval	20 sec.	30 sec.	45 sec.
Trial sequence	3 blocks of 50 trials ea. with 5 min. between blocks	Same	Same

TABLE V

COMPARISON OF STUDIES CONCERNED WITH "RETENTION AFTER REGENERATION"

Variables	McConnell, Jacobson, and Kimble (1959)	VanDeventer and Ratner (1964)	Corning and John (1961)	Brown (1964); Brown, Dustman and Beck (1966a; 1966b)
Species	*D. dorotocephala* (10–24 mm)	*D. tigrina* (8–12 mm)	*D. dorotocephala*	*D. tigrina*
Housing	Aquarium water; minimum light	Melted snow[a]; total darkness	Pond water; indirect room lighting	Spring water; dim light
Trough	Semicircular: 12 × 0.5 in.	"Apparatus similar to McConnell"	Semicircular: 8 × 1 in.	Semicircular: in side X-sectional area of 0.5 cm^2
Light (CS)	2–100W bulbs 6 in. above trough		1–7.5W bulb 8 in. above trough with 0.25 in. plastic shield between bulb and trough	2–11 in. tungsten ribbon lamps operated at color temp. of 2870°K
Shock (US)	DC via inductorium	AC (30 ma)	DC via inductorium	Pulse generator; unipolar shock given (see footnote 2)
Intertrial interval	84 sec.	60 sec.	60 sec.	30 sec.
Trials to criterion	23/25	23/25	34/40	10/15
Trial sequence	50/day	50/day	40/day	30/day
Regeneration time	28 days (bisected Ss)	23 days (bisected Ss)	14–18 days (bisected Ss)	14 days–1 month (most Ss trisected, some bisected)

[a] In planarian research conducted by W. C. Corning at Michigan, it was found that melted snow was an inadequate medium for culturing planarians; a large percentage of subjects died within a week after being placed in the melted snow.

failure of Brown and Ratner to find retention after regeneration, and the failure of Brown to obtain evidence of learning. A comparative analysis of the procedures of these various studies is presented in Tables IV and V. From these considerations it is clear that there has not yet been a 100% procedural replication of any one study.

Some Inconsistencies

Bennett (discussion in Corning and Ratner, 1967, 272) has emphasized the inconsistencies in various positive reports which make it difficult for the beginning investigator to select a set of procedures. This criticism is a valid one and until each investigator can specify accurately all the variables in his studies that are relevant to the behavior under investigation, it is probably not wise to "pick and choose" procedures from several different approaches for a replication attempt. A summary of some of the variables that have differed across a number of positive studies is found in Table VI. It is hoped that as biophysical studies such as Jay Best's (1967a, 1967b), the electrophysiological investigations (for example, Brown, 1967b), and other biological and behavioral studies progress, the reasons for the particular adoption of a certain set of procedures will become more understandable.

TABLE VI

Some Procedural Differences Between Positive Studies

Study	Light (CS)	Shock (US)	Intertrial Interval	Trial Sequence	Responses
Thompson and McConnell (1955)	2–100W bulbs 1 ft. above	DC via inductorium	20 sec.	Individual trials; 150/day in 3 blocks of 50 trials ea.	Head turns + contractions
McConnell Jacobson and Kimble (1959)	2–100W bulbs 6 in. above	DC via inductorium	84 sec.	Individual trials; 50/day. Criterion: 23/25	Head turns + contractions
McConnell, Cornwell, and Clay (1960)		DC via inductorium most effective. AC is more deleterious	30–60 s.		

TABLE VI (*continued*)

Study	Light (CS)	Shock (US)	Intertrial Interval	Trial Sequence	Responses
Jacobson, Fried, and Horowitz (1967)	1–75W bulb 6 in. above (225 ftc.)	AC	20 sec.	Mass training with 25 trials/set; 2 sets on day 1, 3 sets on day 2, 3 sets on day 3, 1 set on day 4. Ss given test trials individually under extinction conditions	Head turns unreliable; contractions are a reliable measure
Corning and John (1961)	1–7.5W bulb 8 in. above (12–15 ft.c.)	DC via inductorium	60 sec.	Individual trials; 40/day. Criterion:34/40	Head turns + contractions

SUMMARY OF LEARNING STUDIES

In Table VII we have attempted to document all the positive and negative learning studies. An annotated bibliography is published regularly in the *Journal of Biological Psychology* and has also appeared in the *Manual of Psychological Experimentation of the Planarian* (McConnell, 1967a).

TABLE VII

Positive and Negative Learning Studies in Planaria[a]

A. Habituation

Positive	Negative
Freed (1966)	Bennett and Calvin (1964)
Westerman (1963)	Brown (1964)

B. Classical Conditioning: Acquisition Measures

Positive	Negative
Barnes and Katzung (1963)	Bennett and Calvin (1964)[b]
Baxter and Kimmel (1963)	Brown (1964)
Bennett and Calvin (1964)[b]	Brown (1967A)
Chapouthier (1967)	Brown, Dustman, and Beck (1966a)
Cherkashin and Sheiman (1967)	Brown, Dustman, and Beck (1966b)

TABLE VII (*continued*)

Positive	Negative
Cherkashin, Sheiman, and Bogorovskaya (1967)	Cummings and Moreland (1959)
Corning and John (1961)	Halas, James, and Knutson (1962)[b]
Corning and Freed (1968)	
Cornwell (1961)	
Cornwell, Cornwell, and Clay (1961)	
Crawford (1967)	
Crawford and King (1966)	
Crawford, King and Seibert (1966)	
Crawford, Livingston, and King (1966)	
Fantl and Nevin (1965)	
Freed (1966)	
Griffard (1963)[b]	
Halas, James, and Knutson (1962)[b]	
Hartrey, Lee, and Morton (1964)	
Hovey (1929)	
Hydén, Egyházi, John, and Bartlett (in press)	
Jacobson (1967)	
Jacobson, Fried, and Horowitz (1966)	
Jacobson, Fried, and Horowitz (1967)	
John (1964)	
Kimmel and Yaremko (1966)	
King, Crawford, and Klingman (1965)	
McConnell (1967)	
McConnell and Mpitsos (1965)	
McConnell, Jacobson, and Kimble (1959)	
McConnell, Jacobson, and Maynard (1960)	
Thompson and McConnell (1955)	
Walker (1966)	

C. Classical Conditioning: Extinction Measures

Positive	Negative
Corning and John (1961)	Baxter and Kimmel (1963)
Crawford (1967)	Brown (1964)
Crawford and King (1966)	Brown (1967)
Crawford, King, and Siebert (1965)	Brown and Beck (1964)
Crawford, Livingston, and King (1966)	Cornwell (1960)
Hullett and Homzie (1966)[b]	Crawford (1967)
Jacobson (1967)	Halas, James, and Stone (1961)
Jacobson, Fried, and Horowitz (1967)	James and Halas (1964)
Kimmel and Yaremko (1966)	
Vattano and Hullett (1964)	

TABLE VII (*continued*)

D. Differential Conditioning

Positive	Negative
Block and McConnell (1967)	Kimmel and Harrell (1964)[b]
Fantl and Nevin (1965)	Kimmel and Harrell (1966)[b]
Griffard and Peirce (1964)	
Jacobson (1967)	
Jacobson, Fried, and Horowitz (1967)	
Kimmel and Harrell (1964)[b]	
Kimmel and Harrell (1966)[b]	

E. Instrumental Paradigms

Positive	Negative
(Instrumental Avoidance)	
Ragland and Ragland (1965)	
(Mazes)	
Best (1963)	Bennett and Calvin (1964)
Best (1965)	
Best and Rubinstein (1962)	
Chapouthier (1968)	
Corning (1964)	
Corning (1966)	
Ernhart and Sherrick (1959)	
Haynes, Jennings, and Wells (1965)	
Humphries and McConnell (1964)	
Jacobson and Jacobson (1963)	
McConnell (1966)	
Pickett, Jennings, and Wells (1964)	
Roe (1963)	
Tushmalova (1967)	
Van Oye (1920)	
Wells (1967)	
Wells, Jennings, and Davis (1966)	
(Operant)	
Best (1965)	Bennett and Calvin (1964)
Crawford and Skeen (1967)	
Lee (1963)	

[a] The columnar arrangement of positive and negative studies is not intended to argue for learning in planarians by providing a "numbers" comparison. It is provided to give the reader a general guide to the existing literature.

[b] *Equivocal findings or interpretations.*

Bennett and Calvin (1964, 4): "Although it would be presumptuous to conclude that planarians cannot be trained, we do conclude that at the present time, the relevant factors necessary for reproducible and reliable training of planarians have not been described with sufficient precision to permit their use in studies of biochemical bases of learning."

Griffard (1963, 600): ". . . in spite of the significance obtained between Group E and all others in both sets of data, the appearance of possible sensitization effects in Group SC renders a conclusion in favor of classical conditioning at least equivocal."

Halas, James, and Knutson (1962, 971): "The results agree in general with Thompson and McConnell's, but it is suggested that reflex sensitization is a more tenable interpretation of the data." ". . . light is a weak US. . . . It would seem that what we may have here is a precursor or primitive type of conditioning."

Hullett and Homzie (1966, 230): "Though the findings support the position of Halas . . . the present experiment does not preclude the possibility that worms can be classically conditioned. The fact that Group B-Lis after 31 days of adaptation responded at a level significantly above that of the control indicates that there may be a classical conditioning component in this paradigm."

Kimmel and Harrell (1964, 228): "Thus, while the differential responding that was found adds some support to the argument that conditioning *does* occur in the planarian, the possibility that sensitization and/or generalization were significantly involved cannot be ruled out."

Kimmel and Harrell (1966, 286): "It does not seem appropriate at present to conclude that the planarian shows this type (differential) of conditioning, although it is possible that a longer training session might permit differentiation to occur."

CURRENT RESEARCH PERTINENT TO MEMORY MECHANISMS

Biochemical Correlates

There has been little work to date concerning the measurement of biochemical events correlative with response modifications in planarians. The small size of the nervous system, the difficulty in identifying neural elements, and the lack of any clear functional differentiation in the nervous system would all tend to make such studies difficult both technically and in terms of interpretation.

Our research at Brookhaven (Corning and Freed, 1968; Freed, 1966) was an attempt to begin a systematic analysis of nucleic acid quantity and turnover in planarians receiving different experimental treatments. These studies were terminated in the preliminary stages but the data have been suggestive of a number of things. Our first goal was to characterize the nucleic acids of whole worms being subjected to classical conditioning. As indicated in a previous section, it was possible to mass condition planarians to a rather high level of response, and this level was significantly different from that of a group receiving random light and shock. Samples of worms were removed at various stages in these treatments, frozen in liquid nitrogen, and analyzed for nucleic acid changes. The administration of P^{32} to the medium in which the animals were kept was found to result in significant incorporation in the RNA and DNA fractions so we could also look at turnover.

The data in Fig. 10 show the response modifications observed in trained and untrained groups in the mass conditioning apparatus, and the RNA/DNA specific activity ratios measured in another set of animals being subjected to identical training treatments. It can be seen that at the beginning and at the termination of treatments these ratios are the same for both groups, but during the middle stages the groups departed significantly. These findings highlight difficulties in searching for the biochemical correlates of behavior —the dramatic events signifying a change in state may precede our behavioral output measures, and these events may also be evanescent in nature.

Crawford (1967; Crawford *et al.*, 1965) has found some evidence of amino acid changes in planarians being trained, but these studies are also in a preliminary stage and difficult to evaluate at the present time.

Single cell analyses of trained and pseudo-trained animals have recently been completed by H. Hydén *et al.* (in press) in a joint investigation. Utilizing the Barnes and Katzung (1963) conditioning procedure, they established that conditioning could only be achieved when animals were presented light paired

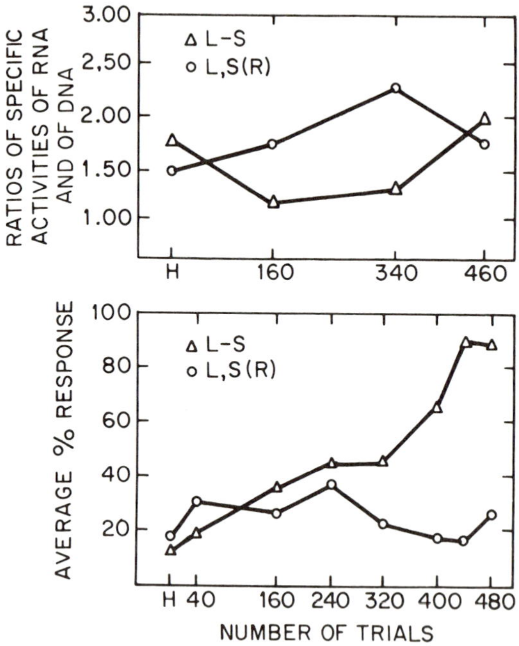

Figure 10 Changes in the responses and in the RNA/DNA specific activity ratio during training of planarians. Subjects receiving paired light and shock (L – S) are compared with controls receiving random light and shock (L,S). "H" marks the end of the period of habituation to light (Corning and Freed, 1968).

with cathodal shock delivered to the anterior part of the worm. Other treatments consisted of light paired with anodal shock to the head, light paired with random anodal and cathodal shock, and light randomly associated with cathodal shock. Nerve cells were dissected from the head ganglia and from the two longitudinal nerve strands. Base ratio analyses indicated no significant differences in the RNA of worms displaying the conditioning, *i.e.*, those that received light paired with cathodal shock to the head.

Transfer of Memory

Jacobson's group has demonstrated that the transfer of training effect is specific to a group receiving paired light and shock and does not occur in animals receiving unpaired treatments (Jacobson *et al.*, 1965). RNA was extracted from animals that had paired light and shock treatment, random light and shock treatment, and from subjects that were naive, and injected into 75 planarians that had been habituated to light. One day after the injection all subjects were given test trials consisting of light only. The results depicted in Fig. 11 show quite convincingly that the performance of animals receiving the RNA from trained animals was different from those receiving the RNA from the pseudoconditioned group and from the naive group. If the RNA were degraded and then injected, the differences did not appear. If the RNA were extracted from trained animals that had also undergone extinction, the transfer effect was as high as in the case of animals receiving RNA from trained but unextinguished subjects.

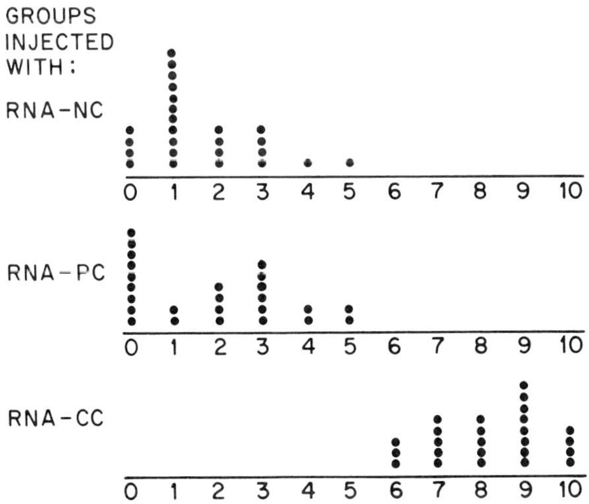

Figure 11 Distribution of scores (responses per 25 trials) for injected planarians (Jacobson *et al.*, 1966).

While Jacobson in his light-shock classical conditioning studies, has been able to observe specific transfer effects, Wells and associates have failed to do so using the Van Oye maze and a T-maze (Haynes *et al.*, 1965; Pickett *et al.*, 1964; Wells, discussion in Corning and Ratner, 1967, 324). Exogenous yeast RNA was found to improve performance in the T-maze. These findings do not necessarily invalidate the Jacobson data—the training situations of Wells were quite different. McConnell (1966b) has reported finding a cannibalistic transfer effect for a maze discrimination. For reviews of planarian transfer experiments, see Jacobson *et al.* (1966), John (1964), and McConnell (1962, 1965, 1966a, 1967c, 1968).

Memory Disruption

The earlier work of Corning and John (1961) demonstrated the possibility of interfering with retention in conditioned planarians by placing them in a solution of ribonuclease (RNAase) during regeneration. These studies have been extended by others (Cherkashin and Sheiman, 1967; Cherkashin *et al.*, 1967) to include central nervous system lesioning as well as biochemical lesioning. Removal of the head ganglion in already conditioned animals (*Ijimia tenuis*) caused the disappearance of conditioned reflexes, but memory was restored when the ganglion regenerated. Removal of the longitudinal nerve strands had no effect on the conditioned responses but when both the head ganglion and longitudinal strands were lesioned (leaving only the lateral trunks and the subcutaneous nerve net) there was no recovery. These studies essentially confirmed what McConnell's group had found in their regeneration studies; that memory in the planarian involves the entire central nervous system and that the cephalic ganglia are necessary for its "read-out."

When trained animals were sectioned transversely and allowed to regenerate, they displayed equal retention of the conditioned response at eight to ten days after the division. This retention was also observed in animals that had undergone two divisions and were presumably composed of entirely new tissue (Cherkashin *et al.*, 1967).

Studies on the effects of RNAase showed that the enzyme was capable of inhibiting formation of the conditioned reflex if planarians were placed in it prior to a training session or during training treatments (Cherkashin and Sheiman, 1966; 1967). Exposure to RNAase after training only inhibited the expression of the conditioned response; putting the animals back in a normal solution was sufficient to produce memory recovery. DNAase, 8-azaguanine, bromine uracil, and aurantine had no effects on conditioning. Investigation of RNAase effects in regenerating planarians indicated that unlike the Corning and John (1961) findings where the enzyme had a differential effect on retention in head and tail regenerates, both halves were affected. Furthermore, the memory deficit was not induced if the animals were placed in the enzyme *after* sectioning but only when they were in it *prior* to sectioning. The discrepancies between these studies and the earlier research of Corning and John could be due to a number of factors such as species differences, type of stimuli, enzyme concentrations, etc.

Tushmalova (1967) has also found that RNAase can reversibly inhibit an avoidance discrimination in *Podoplana olivacea* but also found that the enzyme reduced motor activity. It is concluded that RNA participates in the regulation of cell excitability, but may not be directly involved in learning and memory.

A different approach to assessing the role of nucleic acids in planarian behavior was developed by Brown (1964; 1967a). Planarians that had been subjected to light-shock treatments were exposed to ultraviolet (UV) irradiation at a wavelength (260 mM) which would damage nucleic acids. Responses to light were diminished as a consequence of the UV exposure but photorestorative treatment (which partially reverses the UV effects) could bring about an increase in light reactivity. However, the use of these techniques to further delineate the relationship between nucleic acid states and learning was not successful due to a failure to achieve a clear demonstration of learning.[2]

CAN A FLATWORM FIND HAPPINESS IN MEMORY RESEARCH?

The planarian was one of the first "lower" preparations to be extensively studied in the search for memory mechanisms; it has elicited criticism and instigated great controversies. We may begin to speculate at this point as to

[2] Best (1967a) has presented an argument for the use of the "dirty" waveform of the Harvard Inductorium in the conditioning of planarians. The procedures of Brown *et al.* (1966b) in shocking planarians might explain in part their failure to achieve conditioning. According to their description, "Animals received an equal number of head directed anodal and cathodal shocks" (306). Barnes and Katzung (1963) and Hydén *et al.* (in press) have shown that light paired with cathodal shock delivered to the head could produce conditioning whereas light paired with anodal shock, would not. By intermixing anodal and cathodal trials, Brown *et al.* were in effect, giving their subjects only half as many valid trials, *i.e.*, trials with cathodal orientation. It is also possible that the anodal stimulation was reversing any conditioning that was occurring with the light-cathodal shock pairings—Barnes and Katzung did find this to be the case and it is puzzling that Brown *et al.* would adopt procedures which others had shown would not lead to conditioning. Perhaps with a group receiving consistent light—cathodal shock pairings, they would have achieved a higher rate of response.

Shock parameters appear to be critical. Generally, the Harvard Inductorium or head-directed cathodal shocks appear to be best although Jacobson (Jacobson *et al.*, 1967) has had success with ac shock.

whether the difficulties this beast has encountered are a special problem of the planarian or due to more general problems of definition, technology, and conception. Rumblings can be heard in research on the Aplysia, the roach, and other preparations as the physiologists and the psychologists are getting together and discovering there is no easy rapport between the two disciplines. This is certainly not the time for investigators to retreat into definitions and wage battles over semantics, rather, the energies should be applied to an ever increasing exposure of adaptive mechanisms in as wide a range of species as possible. To assume we now have *the* definition of learning and then use this definition to structure what we will research in

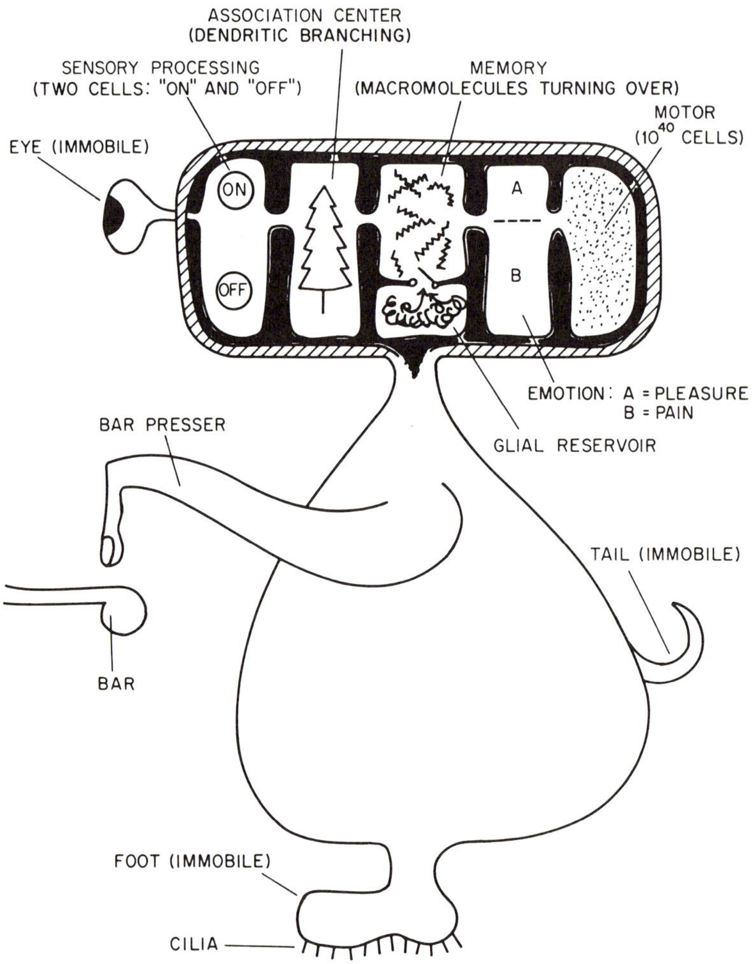

Figure 12 A proposed "organism" for memory research.

the way of plasticity seems to be a reverse process of the way things should be going. As discussed in a recent *Neurosciences Research Program* conference ". . . we are attempting to define a phenomenon as yet inadequately described and then trying to use that definition both to extend the description and to provide an explanation for it" (Bullock and Quarton, 1966, 165).

In planarian research there have been debates over the proper ways to measure learning, the best controls that rule out sensitization and pseudolearning (assuming these are not a part of learning), and what constitutes a replication. There are those who would discount the findings on the basis of a priori assumptions about what level of animal can learn, there are some who are unhappy because the worm does not act like a rat, and there are others who think that research on anything other than a rat is a waste of time. In short, we are just beginning to realize what a complexity the living organism is and how difficult it is to define those set of factors that determine its moment to moment behavior. This complexity frightens some who would retreat to the rat and attempt to answer all questions with an organism who just happens to comply with our assumptions about the learning process.

Perhaps the best solution to the planarian's problems (and other "breakthrough" preparations, past, present and future) would be for us to stop and build an organism that will fulfill all *expectations* of those bent on discovering the engram prior to understanding the organism. An example of such an organism is portrayed in Fig. 12. Having provided a "learning machine" to those who need one, perhaps the rest of us can go about the business of studying the transactions between organism and environment and the biology that relates to these transactions independently of definitions that tell us what to look for.

There are going to be more debates over learning as research reaches out into more organisms and studies more organismic processes. We cannot blame the flatworm, the roach, or any other animal for the debate over what we mean by "learning"—after all, they didn't invent the word.

ACKNOWLEDGMENTS

The authors wish to thank S. Freed, A. Jacobson, E. R. John, and J. McConnell for their assistance in the preparation of this paper.

REFERENCES

Barnes, C. C. and Katzung, B. G. Stimulus polarity and conditioning in planaria. *Science*, 1963, **141**, 729.

Baxter, R. and Kimmel, H. D. Conditioning and extinction in the planarian. *Amer. J. Psychol.*, 1963, **76**, 665.

Bennett, E. and Calvin, M. Failure to train planarians reliably. *Neurosciences Research Program Bulletin*, 1964, **2**, 3.

Best, J. B. Protopsychology. *Scient. Amer.*, 1963, **208**, 55.

Best, J. B. Behaviour of planaria in instrumental learning paradigms. *Animal Behaviour*, suppl. 1, 1965, **13**, 69.

Best, J. B. Major factors in classical conditioning of planarians: Stimulus waveform and neural geometry. In, W. Corning and S. Ratner (Eds.), *Chemistry of Learning: Invertebrate Research*. New York, Plenum, 1967a, P. 255.

Best, J. B. The neuroanatomy of the planarian brain and some functional implications. In, W. Corning and S. Ratner (Eds.), *Chemistry of Learning: Invertebrate Research*. New York, Plenum, 1967b, P. 144.

Best, J. B. and Rubinstein, I. Maze learning and associated behavior in planaria. *J. comp. physiol. Psychol.*, 1962, **55**, 560.

Block, R. A. and McConnell, J. V. Classically conditioned discrimination in the planarian, *Dugesia dorotocephala*. *Nature*, 1967, **215**, 1465.

Brown, H. M. Experimental procedures and state of nucleic acids as factors contributing to "learning" phenomena in planaria. Unpublished doctoral dissertation, University of Utah, 1964.

Brown, H. M. Effects of ultraviolet and photorestorative light on the phototaxic behavior of planaria. In, W. Corning and S. Ratner (Eds.), *Chemistry of Learning: Invertebrate Research*. New York, Plenum, 1967a, P. 295.

Brown, H. M. Some characteristics of the light-evoked electrical response of the planarian eyecup. In, W. Corning and S. Ratner (Eds)., *Chemistry of Learning: Invertebrate Research*. New York, Plenum, 1967b, P. 164.

Brown, H. M. and Beck, E. C. Does learning in planaria survive regeneration? *Fed. Proceed.*, 1964, **23**, 254.

Brown, H. M., Dustman, R. E. and Beck, E. C. Experimental procedures that modify light response frequency of regenerated planaria. *Physiol. and Behav.*, 1966a, **1**, 245.

Brown, H. M., Dustman, R. E. and Beck, E. C. Sensitization in planaria. *Physiol. and Behav.*, 1966b, **1**, 305.

Bullock, T. H. and Quarton, G. C. Simple systems for the study of learning mechanisms. *Neurosciences Research Program Bulletin*, 1966, **4**, 105.

Chapouthier, G. Conditioning in the European planarian, *Dendrocoelum lacteum:* The effects of prolonged conditioning. *J. Biol. Psychol.*, 1967, **9**, 23.

Chapouthier, G. Relations entre deux reactions des Planaires face à une discrimination droite-gauche. *C.R. Acad. Sci.*, Paris, 1968, **266**, 905.

Cherkashin, A. N. and Sheiman, I. M. The use of simple biological models in memory mechanisms. Paper presented at International Psychology Congress, Moscow, 1966.

Cherkashin, A. N. and Sheiman, I. M. Conditioning in planarians and RNA content. *J. Biol. Psychol.*, 1967, **9**, 5.

Cherkashin, A. N., Sheiman, I. M. and Bogorovskaya, G. I. Conditioned reflexes in planarians and regeneration experiments. *Neuroscience Translations*, 1967, **1**, 12.

Corning, W. C. Evidence of right-left discrimination in planarians. *J. Psychol.*, 1964, **58**, 131.

Corning, W. C. Retention of a position discrimination after regeneration in planarians. *Psychon. Sci.*, 1966, **5**, 17.

Corning, W. C. and Freed, S. Planarian behavior and biochemistry. *Nature*, 1968, 219, 1227.

Corning, W. C. and John, E. R. Effect of ribonuclease on retention of conditioned response in regenerated planarians. *Science*, 1961, **134**, 1363.

Corning, W. C. and Ratner, S. C. *Chemistry of Learning: Invertebrate Research*. New York: Plenum, 1967.

Cornwell, P. Classical conditioning with massed trials in the planarian. *Worm Runner's Digest*, 1960, **2**, 34.

Cornwell, P. An attempted replication of studies by Halas *et al.* and by Thompson and McConnell. *Worm Runner's Digest*, 1961, **3**, 91.

Cornwell, P., Cornwell, G. and Clay, M. Retention of a conditioned response following regeneration in the planarian. *Worm Runner's Digest*, 1961, **3**, 34.

Crawford, T. Behavioral modification of planarians. In, W. Corning and S. Ratner (Eds.), *Chemistry of Learning: Invertebrate Research*. New York: Plenum, 1967, P. 234.

Crawford, T. and King, L. Spontaneous recovery of a classically conditioned response in the planarian. *Psychon. Sci.*, 1966, **6**, 427.

Crawford, T., King, F. and Siebert, L. Amino acid analysis of planarians following conditioning. *Psychon. Sci.*, 1965, **2**, 49.

Crawford, T., Livingston, P. and King, F. Distribution of practice in the classical conditioning of planarians. *Psychon. Sci.*, 1966, **4**, 29.

Crawford, F. T. and Skeen, L. C. Operant responding in the planarian: a replication study. *Psychol. Rep.*, 1967, **20**, 1023.

Cummings, S. B. and Moreland, C. C. Sensitization vs. conditioning in planaria: some methodological considerations. *Amer. Psychol.*, 1959, **14**, 410.

Ernhart, E. N. and Sherrick, C. Retention of a maze habit following regeneration in planaria (D. *maculata*). Paper presented at Midwestern Psychological Association, St. Louis, 1959.

Fantl, S. and Nevin, J. A. Classical discrimination in planarians. *Worm Runner's Digest*, 1965, **7**, 32–34.

Freed, S. Endogenous biochemistry of planarians correlated with learning experiments at Brookhaven National Laboratory. BNL Report No. 981 (T–414), 1966. (Available from the Clearinghouse for Federal Scientific and Technical Information, National Bureau of Standards, U.S. Dept. Commerce, Springfield, Virginia.)

Griffard, C. D. Classical conditioning of the planarian *Phagocata gracilis* to water flow. *J. comp. physiol. Psychol.*, 1963, **56**, 597.

Griffard, C. D. and Peirce, J. T. Conditioned discrimination in the planarian. *Science*, 1964, **144**, 1472.

Halas, E. S., James, R. L. and Knutson, C. An attempt at classical conditioning in the planarian. *J. comp. physiol. Psychol.*, 1962, **55**, 969.

Halas, E.S., James, R.L. and Stone, L. A. Types of responses elicited in planaria by light. *J. comp. physiol. Psychol.*, 1961, **54**, 302.

Hartrey, A. L., Keith-Lee, P. and Morton, W. D. Planaria: Memory transfer through cannibalism reexamined. *Science*, 1964, **146**, 274.

Haynes, S. E., Jennings, L. B. and Wells, P. H. Planaria learning: Nontransfer and nonfacilitation in a Van Oye maze. *Amer. Zool.*, 1965, **5**, 713.

Hovey, H. B. Associative hysteresis in marine flatworms. *Physiol. Zool.*, 1929, **2**, 322.

Hullett, J. W. and Homzie, M. J. Sensitization effect in the classical conditioning of *Dugesia dorotocephala*. *J. comp. physiol. Psychol.*, 1966, **62**, 227.

Humpheries, B. and McConnell, J. V. Factors affecting maze learning in planarians. *Worm Runner's Digest*, 1964, **6**, 52.

Hydén, H., Egyházi, E., John, E. R. and Bartlett, F. RNA base ratio changes in planaria during conditioning. *J. Neurochem.*, in press.

Jacobson, A. L. Learning in flatworms and annelids. *Psychol. Bull.*, 1963, **60**, 74.

Jacobson, A. L. Learning in planarians: current status. *Animal Behaviour*, suppl. 1, 1965, **13**, 76.

Jacobson, A. L. Classical conditioning and the planarian. In, W. Corning and S. Ratner (Eds.), *Chemistry of Learning: Invertebrate Research*. New York: Plenum, 1967. P. 195.

Jacobson, A. L., Fried, C. and Horowitz, S. D. Planarians and memory. *Nature*, 1966, **209**, 599.

Jacobson, A. L., Fried, C. and Horowitz, S. D. Classical conditioning, pseudoconditioning, or sensitization in the planarian. *J. comp. physiol. Psychol.*, 1967, **64**, 73.

Jacobson, A. L. and Jacobson, R. Maze learning in planaria—a case history. *Worm Runner's Digest*, 1963, **5**, 69.

James, R. L. and Halas, E. S. No difference in extinction behavior in planaria following various types and amounts of training. *Psych. Record*, 1964, **14**, 1.

Jensen, D. D. Paramecia, planaria, and pseudolearning. Draft of paper presented at symposium "Learning and related phenomena in invertebrates," Cambridge University, 1964.

Jensen, D. D. Paramecia, planaria, and pseudolearning. *Animal Behaviour*, suppl. 1, 1965, **13**, 9.

John, E. R. Studies on learning and retention in planaria. In, M. A. Brazier (Ed.), *Brain Function*, Vol. II. Univ. California Press, 1964, P. 161.

Kimmel, H. D. and Harrell, V. L. Differential conditioning in the planarian. *Psychon. Sci.* 1964, **1**, 227.

Kimmel, H. D. and Harrell, V. L. Further study of differential conditioning in the planarian. *Psychon. Sci.*, 1966, **5**, 285.

Kimmel, H. D. and Yaremko, R. M. Effect of partial reinforcement on acquisition and extinction of classical conditioning in the planarian. *J. comp. physiol.*, *Psychol.*, 1966, **61**, 299.

King, F. J., Crawford, F. T. and Klingman, R. L. A further study of amino acid analysis and conditioning of planarians. *Psychon. Sci.*, 1965, **3**, 189.

Lee, R. M. Conditioning of a free operant response in planaria. *Science*, 1963, **139**, 1048.

McConnell, J. V. Memory transfer through cannibalism in planarians. *J. Neuropsychiat.*, Suppl. 1, 1962, **3**, s42.

McConnell, J. V. Cannibals, chemicals, and contiguity. *Animal Behaviour*, Suppl. 1, 1965, **13**, 61.

McConnell, J. V. Comparative physiology: learning in invertebrates. *Ann. Rev. Physiol.*, 1966a, **28**, 107.

McConnell, J. V. New evidence for the "transfer of training" effect in planarians. Paper presented at International Congress of Psychology, Moscow, 1966b.

McConnell, J. V. (Ed.). *A Manual of Psychological Experimentation on Planarians*, 2nd edition. Ann Arbor: *J. Biol. Psychol.*, 1967a.

McConnell, J. V. Specific factors influencing planarian behavior. In, W. Corning and S. Ratner (Eds.), *Chemistry of Learning: Invertebrate Research.* New York: Plenum, 1967b. P. 217.

McConnell, J. V. The biochemistry of memory. In, W. Corning and S. Ratner (Eds.), *Chemistry of Learning: Invertebrate Research.* New York: Plenum, 1967c. P. 310.

McConnell, J. V. In search of the engram. In, W. Corning and M. Balaban (Eds.), *The Mind: Biological Approaches to its Function.* New York: John Wiley, 1968. P. 49.

McConnell, J. V., Cornwell, P. and Clay, M. An apparatus for conditioning planaria. *Amer. J. Psychol.*, 1960, **73**, 618.

McConnell, J. V., Jacobson, A. L. and Kimble, D. P. The effects of regeneration upon retention of a conditioned response in the planarian. *J. comp. physiol. Psychol.*, 1959, **52**, 1.

McConnell, J. V., Jacobson, R. and Maynard, D. M. Apparent retention of a conditioned response following total regeneration in the planarian. *Amer. Psychologist*, 1958, **14**, 410.

McConnell, J. V. and Mpitsos, G. Effects of the presence or absence of slime on classical conditioning in planarians. *Amer. Zool.*, 1965, **5**, 122.

Pearl, R. The movements and reactions of fresh-water planarians; a study in animal behaviour. *Quart. J. Micr. Sci.*, 1903, **46**, 509.

Pickett, J. B. E., Jennings, L. B. and Wells, P. H. Influence of RNA and victim training on maze learning by cannibal planarians. *Amer. Zool.*, 1964, **4**, 411.

Ragland, Rae S. and Ragland, J. B. Planaria: interspecific transfer of a conditionability factor through cannibalism. *Psychonom. Sci.*, 1965, **3**, 117–118.

Reynierse, J. H. Aggregation Formation in planaria, *Phagocata gracilis* and *Cura Foremanii*: Species differentiation. *Animal Behaviour*, 1967, **15**, 270.

Reynierse, J. H., Jensen, D. D. and Schroeder, O. F. Shock duration and adhesion in the planarian, *Phagocata gracilis*. *Psychon. Sci.*, 1966, **4**, 335.

Riccio, D. and Corning, W. C. Slime and planarian behavior. *Psychol. Rec.*, 1969, **19**, 507.

Roe, K. In search of the locus of learning in planarians. *Worm Runner's Digest*, 1963, **5**, 16.

Rohlich, P. Fine structural changes of photoreceptors induced by light and prolonged darkness. In, *Invertebrate Neurobiology*, J. Salanki (Ed.), New York: Plenum, 1968.

Thompson, R. F. *Foundations of Physiological Psychology*. New York: Harper–Row. 1967.

Thompson, R. and McConnell, J. V. Classical conditioning in the planarian, *Dugesia dorotocephala*. *J. comp. physiol. Psychol.*, 1955, **48**, 65.

Tushmalova, N. A. Conditioned reflexes in the Baikal planarian *Podoplana olivacea* after injection of ribonuclease. *Zhurnal vysshei nervni deyatel 'nosti im. I. P. Pavlova*, 1967, **17**, 359.

VanDeventer, J. M. and Ratner, S. C. Variables affecting the frequency of response of planaria to light. *J. comp. physiol. Psychol.*, 1964, **57**, 407.

Van Oye, P. Over het geheugen bij fr flatwormen en andere biologische waarnemingen bij deze dieren. *Natuurwet. Tijdschr.*, 1920, **2**, 1.

Vattano, F. J. and Hullett, J. H. Learning in planarians as a function of interstimulus interval. *Psychon. Sci.*, 1964, **1**, 331.

Walker, D. R. Memory transfer in planarians: an artifact of the experimental variables. *Psychon. Sci.*, 1966, **5**, 357.

Walker, D. R. and Milton, G. A. Memory transfer versus sensitization in cannibal planarians. *Psychon. Sci.*, 1966, **5**, 293.

Wells, P. H. Training flatworms in a Van Oye maze. In, W. Corning and S. Ratner (Eds.), *Chemistry of Learning: Invertebrate Research*. New York: Plenum, 1967. P. 251.

Wells, P. H., Jennings, L. B. and Davis, M. Conditioning planarian worms in a Van Oye type maze. *Amer. Zool.*, 1966, **6**, 295.

Westerman, R. A. A study of the habituation of responses to light in the planarian *Dugesia dorotocephala*. *Worm Runner's Digest*, 1963, **5**, 6–11.

Zelman, A., Kabat, L., Jacobson, R. and McConnell, J. V. Transfer of training through injection of "conditioned" RNA into untrained planarians. *Worm Runner's Digest*, 1963, **5**, 14.

COMMENTS

Edward L. Bennett

Lawrence Radiation Laboratory,
Berkeley, California.

I would like to commend Drs. Corning and Riccio for their comprehensive summary of the art of planarian training. It is gratifying to me to see that there has been more effort by some investigators to describe more adequately than previously the essential parameters of planarian training. I am happy to see Dr. Corning reemphasize the importance of being "systematic about reporting experimental conditions so that an attempted replication is truly a procedural replication". I find it distressing, however, in spite of ten years or more of research in this area, that even Dr. Corning cannot point to a 100% procedural replication of any one training study. Thus, it is impossible in the absence of further experiments to determine if the relevant factors necessary for reproducible and reliable training of planarians have been described even yet. I do feel, though, that significant steps may have been taken in this direction.

For the biochemist who is interested in planarians as a tool to study learning mechanisms, a key experiment is the transfer of memory as described by Jacobson *et al.* in 1966. Two years have elapsed since this potentially important experiment, and as yet no additional data either confirming or extending these observations have been forthcoming. In this, as in the case of mammalian transfer studies which have been the subject of other papers at this Symposium, it is essential that several laboratories can replicate and extend the primary observations.

Last, but not least, when does Dr. Corning expect to have his "learning machine" (Fig. 12) in quantity production? This will certainly be an aid for all of us interested in mechanisms of memory, and hopefully it will be a superior organism for such studies than either the worm or the rodent.

A COMPARISON OF "TRANSFER" RESULTS OBTAINED WITH TWO DIFFERENT TYPES OF EXTRACTION AND INJECTION PROCEDURES, USING IDENTICAL BEHAVIORAL TECHNIQUES*

E. J. Fjerdingstad† W. L. Byrne

Department of Biochemistry, Duke University

Thomas Nissen‡ H. H. Røigaard-Petersen§

Institute of General Zoology, University of Copenhagen

In the beginning of 1964, Drs. Nissen, Røigaard–Petersen and Fjerdingstad carried out a pilot study intended to show if any characteristics of a conditioned behavior would appear in rats injected with a ribonucleic acid extract derived from the brain of trained donor rats. For this purpose we constructed a completely automatic training and testing set-up, a two-alley runway in which approach to a lit alley was reinforced by a water reward for rats deprived of water for 24 hours. We prepared a cold phenol RNA extract from trained and untrained donors and injected it intracisternally into naive animals. A third group of animals was not injected but was anesthetized, as were the animals of the two other groups.

Though our group sizes were only five and four animals, a difference of

*Supported in part by U.S. Public Health Service Grant No. MH-12871 and by Statens Åndssvageforsorg.

†Currently at Department of Anesthesiology, Baylor College of Medicine, Houston, Texas.

‡Currently at The Danish Institute for Educational Research, Copenhagen, Denmark.

§Currently at Department of Biochemistry, Aarhus University, Aarhus, Denmark.

test performance between recipients of "trained" and of "untrained" extract was obtained in the direction of the "trained extract" recipients making fewer errors during five days of testing. This difference was statistically significant at the $P = 0.05$ level. (Fjerdingstad *et al.*, 1965a).

Very much encouraged by these results, we proceeded to test the specificity of the effect. This was simply done by reinforcing one group of donor animals for running to the dark alley and another for running to the lit alley, as before. From these donors two extracts were prepared, labeled "light trained" and "dark trained" extract, respectively. Both extracts were tested in recipients in two experiments. In the first all recipients were reinforced for running to the dark alley. In this experiment recipients of "light trained" extract performed better ($P = 0.055$). In the second experiment all recipients were reinforced for running to the lit alley, and in this experiment recipients of "dark trained" extract performed better ($P = 0.013$) (Nissen *et al.*, 1965). Thus, already at this stage we met the peculiar "reversed" or negative effects also described by Rosenblatt and Miller (1966), Rucker (1967) and Ungar and Irwin (1967), and which seem to be a characteristic of transfer results under certain circumstances.

At this stage a return was made to the simpler design of the first experiment in order to further confirm these results before proceeding with experiments on specificity. A somewhat more time consuming blind injection procedure was now used in which both extracts were injected alternatingly from coded syringes. No significant differences were found between recipients of "trained" and of "untrained" extracts in these experiments. However, when comparing the three experiments it was consistently found that the animals injected first in the group receiving "trained" extract had a greatly improved performance, but that the later the animals were injected, after dissolving of the extract, the less apparent was any effect of "trained" extract (see Fig. 1). This was highly significant ($P < 0.001$), and seemed to indicate that a chemical change of some kind occurred when the extract was kept dissolved for injection, as it was heated somewhat from the hand holding the syringe. Preliminary studies with the ultracentrifuge seemed to indicate that during this time there was a decrease in the sedimentation velocity of the major peak of the sedimentation pattern (Røigaard–Petersen *et al.*, 1968). It has been reported that breakdown may take place upon slight heating in RNA extracts prepared without any special precautions to eliminate ribonuclease (Kok *et al.*, 1966).

In order to escape the difficulties described above, some experimentation was done with a number of changes in the procedure out of which emerged a technique that seemed to be capable of yielding very uniform results. This will be described in somewhat more detail.

The original extract was found by a biuret method to contain at least

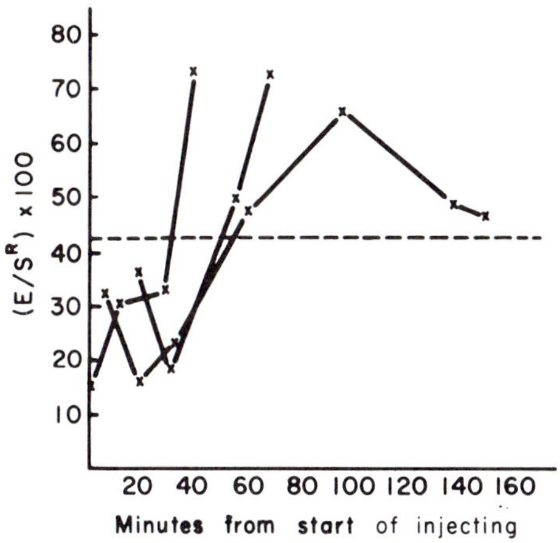

Figure 1 Effect in individual rats of extract from trained donors in the first complete session. Ordinate is errors in per cent of correct choices (errors/correct choice × 100), abscissa indicates the time in minutes expired from injection of the first animal in the group. The dashed line is the mean level of performance of rats injected with extract from untrained donors. The data arise from three separate successive experiments run blindly, the difference in performance of animals injected before and after 40 minutes is highly significant ($P < 0.001$) according to Mann–Whitney U-test.

50% protein (percentage of RNA content) and about 15% DNA. It was thought that the protein content might be responsible for the lability of the extract and partly for this reason, partly to restrict the possibilities of which component was the active in the extract, alternative methods were tried in an attempt to remove this protein contamination (Røigaard–Petersen *et al.*, 1968). Preliminary experiments were carried out with the method of Ralph and Bellamy (1964) in which RNA is precipitated with cetyltrimethylammonium-bromide. This method had to be discontinued when in several experiments it was found that the extracts thus obtained contained a component that killed recipient animals within minutes after intracisternal injection of as little as 0.25 mg of the extract. Intracisternal injections of 1% cetyltrimethyl-ammoniumbromide showed that this was without any effect on the animals, and thus could not be responsible for the lethal effect. Other experiments showed, however, that by centrifuging the water phase at 25,000 *g* after the initial separation and applying the same treatment to the redissolved RNA immediately before dialysis, quite large amounts of highly biuret positive materials could be removed, leaving an RNA extract containing only less than 10% protein and so little DNA that the indole test according to Keck (1956) was negative on one mg of the material. The differences in composition between the two types of extract are also very apparent from the ultraviolet absorption curves (Fig. 2).

When preliminary experiments showed that this extract was still not very stable, it was decided to try to speed up the injection procedure by using an anesthetic that would allow to bring all recipient animals of one group into narcosis at the same time, and making the necessary preparations such as opening the skin of the neck before the extract was dissolved. Thus, ten animals could be injected within ten minutes instead of within an hour.

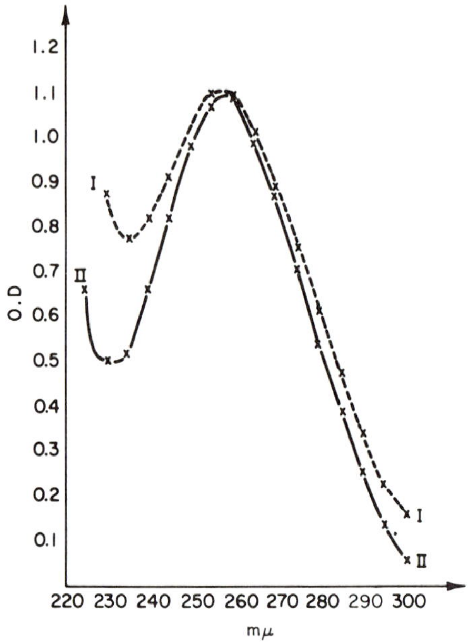

Figure 2 Ultraviolet absorption spectra of the original RNA preparation (I) (Fjerdingstad *et al.*, 1965) and on improved RNA preparation (II) (Røigaard-Petersen *et al.*, 1968).

Chloral hydrate was the agent decided upon, and it was administered by the subcutaneous injection of a 12% solution (0.7 ml injected into a 200 gm rat).

In all these experiments one mg of RNA (determined by U.V. absorption) was injected in both groups (earlier brain equivalents had been injected), and this was as before dissolved in a 0.145 *M* sodium bicarbonate-0.005 *M* potassium bicarbonate solution (Dingman and Sporn, 1961). The volume injected into each animal was 50 μliters, injection was carried out with a tuberculin syringe fitted with a European no. 19 needle (corresponds to American no. 26).

The original training set-up was also modified to some extent. The area around the water cup where reinforcement was given was made triangular to eliminate a tendency in some of the animals to hide in a corner during the earlier phases of training. The stimulus lights were changed from naked bulbs to 10 × 10 cm frosted glass plates lighted from behind by five watt bulbs and located six cm above the response platforms. Fig. 3 gives further details of the training apparatus.

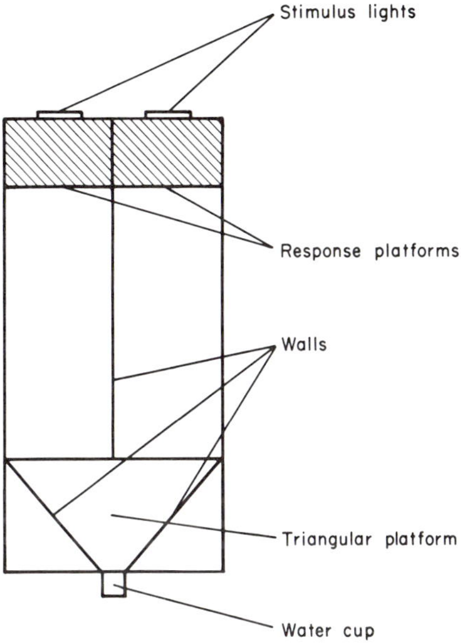

Figure 3 Vertical view of the modified two-alley runway used in experiments SS1–SS4. Scale 1:18. Change of illumination (and correctness) between alleys is carried out by means of a telephone selector, which is only activated after a correct choice. After all responses the stimulus light goes off, and only goes on again when the rat returns to the Triangular platform, which is also the point where the rat is put down at the start of a session, facing the two alleys. A weak houselight (1 watt, shielded by a translucent white plastic sphere) is placed above the water, cup, near the top of the box. The training set-up is kept in a dark, nearly soundproof room.

The donor training and the recipient testing procedures were identical. Before starting training or testing, the rats were put on a water schedule of three days with water *ad lib* for 30 minutes once a day at noon. Thereafter they were deprived of water for 24 hours, and subsequently put into the training set-up for 30 minutes or until 30 trials had been completed. This is called a "short session" and in the curves will be indicated by a I. On the following days the rats were left in the set-up until they had obtained 60 reinforcements of 0.1 ml water each, irrespective of the number of errors made. Animals that did not obtain ten reinforcements in the first hour were taken up, and given water in the home cage so as to get six ml. If an animal was running vigorously after an hour it would generally run to 60 reinforcements without stopping. Scores of the few animals obtaining less than 45 reinforcements were not used. The time to finish a complete session decreased from more than an hour to less than ten minutes during training. Animals were run until they attained a criterion of less than 10% errors (percentage of reinforcements) in two successive days. The average starting performance was about 95% in the short session and 70% in the first complete session.

Using these procedures in a series of four experiments, SS1–SS4, each with approximately ten animals in each group, we have consistently found an initial reverse effect, *i.e.*, animals injected with "trained extract" performed initially less well than animals injected with the "untrained extract" (Fig. 4). When comparing with the groups of donors used, it is even apparent

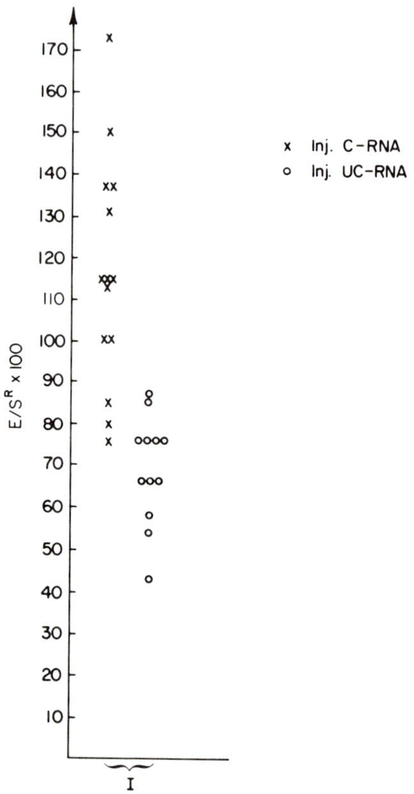

Figure 4 Performance in the first short session of experiments SS1 and SS2 of animals injected extract from trained donors (C–RNA) and from untrained donors (UC–RNA). The initial reverse effect is seen to be highly significant, P<0.001 (Mann–Whitney U-test).

that this is not simply an inhibiting effect of the trained extract. The "untrained extract" recipients actually make fewer errors during the first day than do uninjected animals, that is, they show an increased light preference. In view of the fact that the animals get very few reinforcements during the first day, there is some justification for considering the performances during that session to be guided more by initial preferences than by learning as shown in Fig. 5.

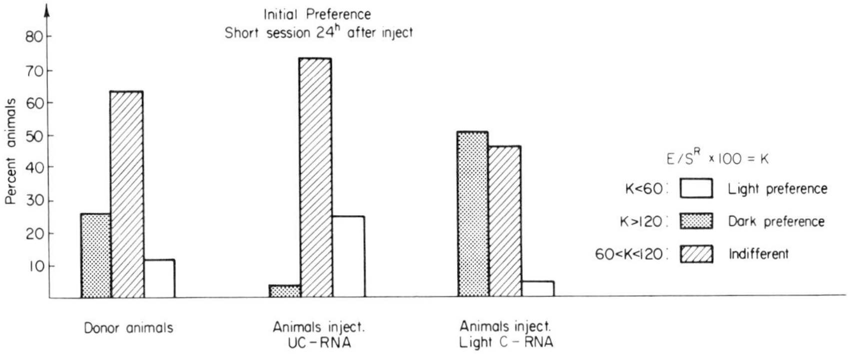

Figure 5 Diagram showing the percentage of animals in the short sessions of experiments SS1, SS2 and SS4 that preferred the lit alley, preferred the dark alley or behaved indifferently. The performance of donor groups used for preparing trained extract is also included. UC–RNA is untrained extract, light–C–RNA is extract from animals trained to run to the lit alley.

On the following days, the complete sessions, the intergroup difference in the reverse direction decreases, to be followed by a difference in the predicted direction which is generally also significant. This is shown in four experiments by the curves in Figs. 6–9. Table I gives individual scores. The overall effect is thus one of the animals injected with "trained extract" making fewer errors during learning (Fig. 10). Table II gives the result of the statistical evaluation of the data.

The reason why this changed procedure should give an initial reversal of the transfer effect is not easy to see. It should be pointed out, however, that by standardizing the amount of RNA-injected at one mg it was actually somewhat decreased in comparison with earlier experiments in which a whole brain equivalent, corresponding to 1.3–1.6 mg was used. Rosenblatt and Miller (1966) have found indications that lower doses of brain extract may sometimes give a reversal. In this connection it is interesting to consider that the effect later shifts in the predicted direction. It is possible that the uptake of the active component is sufficiently slow to account for the initial reversal, and later direct effects are due to a slow increase in concentration during the days of testing. This is, of course, purely hypothetical at present.

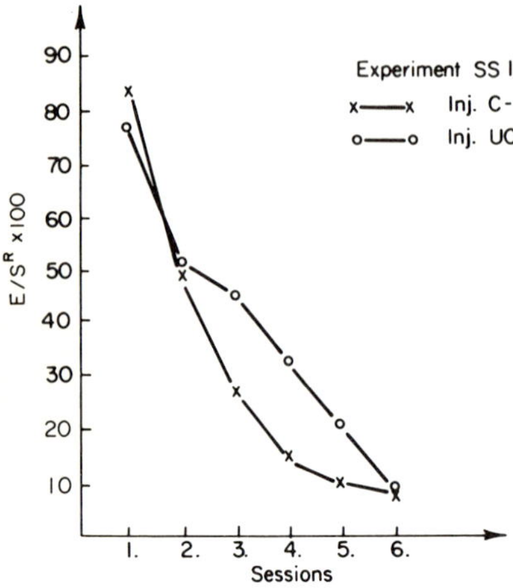

Figure 6

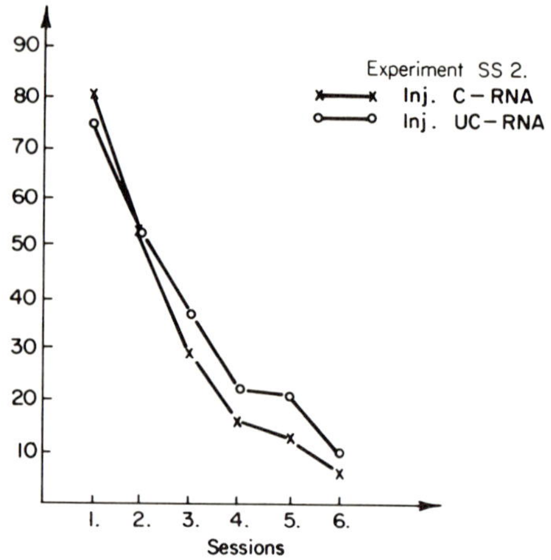

Figure 7

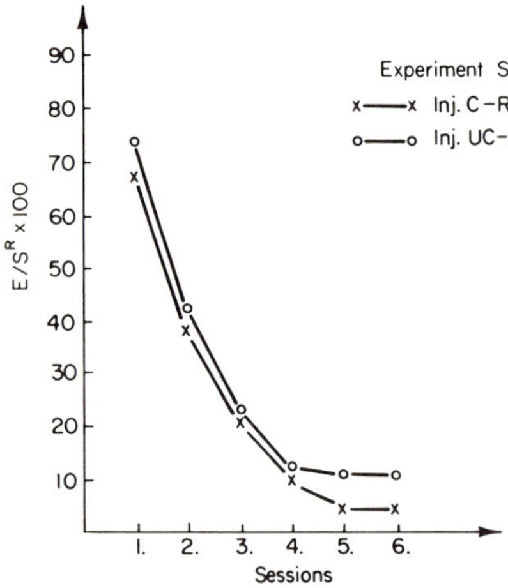

Figure 8

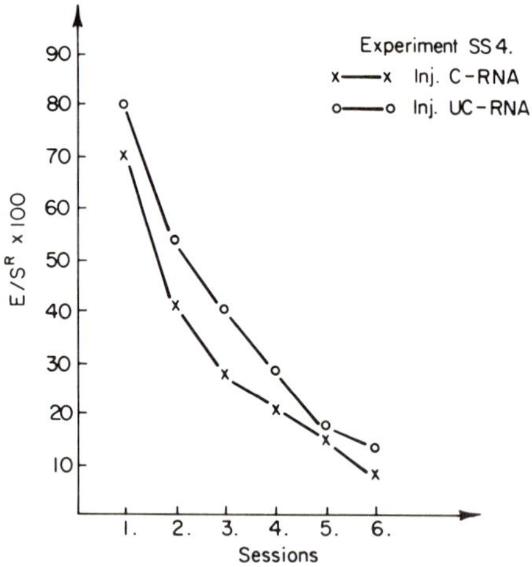

Figure 9

Learning curves representing means of (E/Sr) × 100, *i.e.*, errors as percentage of reinforcements during six complete sessions with 24 hours interval in experiments SS1–SS4. x——x, recipients of trained extract; ○——○, recipients of untrained extract.

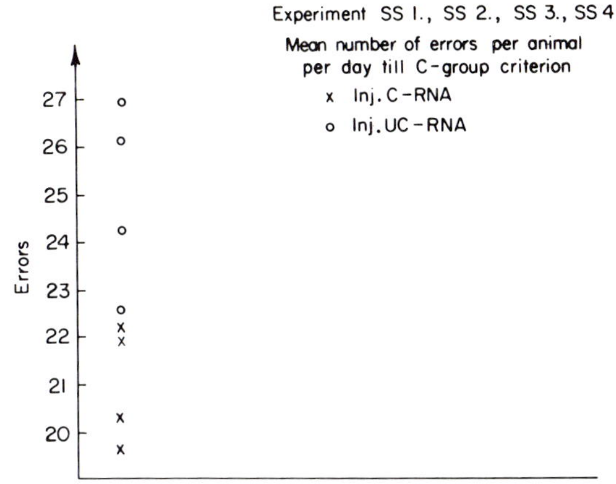

Figure 10 The experiments SS1–SS2 as compared by the calculation of the mean number of errors made per animal per day till the day the trained extract group reached criterion. x, animals injected trained extract; ○ animals injected untrained extract.

TABLE I

Single and Mean Values of Performance in the Experiments, SS1–SS4

	I (short initial sess.)	1st regular sess.	2nd	3rd	4th	5th	6th	7th
SS1 Trained Extract								
Rat								
706	100	60.0	31.7	20.0	18.3	10.0	13.3	—
715	172	86.7	46.7	31.7	11.7	0.0	1.7	—
714	113	152	26.7	10.0	23.3	16.7	13.3	—
710	86	70.0	60.0	31.7	8.3	10.0	10.0	—
708	114	56.7	53.3	40.0	23.3	6.7	10.0	—
707	250	75.0	51.6	25.0	15.0	10.0	6.7	—
705	131	68.3	60.0	31.7	13.3	20.0	11.7	—
704	114	98.5	63.3	31.7	5.0	8.3	3.3	—
Mean	132.5	83.4	49.2	27.7	14.8	10.2	8.8	—
SS1 Untrained Extract								
Rat								
700	58	73.3	46.7	36.7	16.7	26.7	13.3	—
699	55	82.5	66.7	35.0	31.7	15.0	3.3	—
697	86	67.5	28.3	56.7	48.3	26.7	15.0	—
696	77	90.0	78.3	56.7	25.0	10.0	3.3	—
695	77	50.0	53.3	50.0	56.7	30.0	16.7	—
694	122	95.0	43.4	38.3	18.4	16.7	5.0	—
Mean	79.2	76.4	52.8	45.6	32.8	20.9	9.4	—

TABLE I (*continued*)

	I (short initial sess.)	1st regular sess.	2nd	3rd	4th	5th	6th	7th
SS2 Trained Extract								
Rat								
767	not	96.7	40.0	41.7	18.3	25.0	6.7	—
768	recorded	88.4	55.0	31.7	11.7	5.0	1.7	—
772	due to	45.0	30.0	17.7	6.7	3.3	5.0	—
770	apparatus	141.5	78.3	35.0	26.7	11.7	3.3	—
766	failure	48.4	48.4	26.7	10.0	16.6	10.0	—
765		63.3	63.3	23.4	21.7	6.7	8.9	—
Mean		80.6	52.5	29.4	15.9	11.4	5.9	—
SS2 Untrained Extract								
Rat								
787	not	121.0	63.3	38.3	25.0	21.6	6.7	—
786	recorded	72.0	73.3	50.0	28.3	6.7	5.0	—
785	due to	91.8	58.3	60.0	51.1	50.0	30.0	—
783	apparatus	80.5	31.3	8.3	13.3	8.3	1.7	—
782	failure	85.0	33.3	20.0	3.3	13.3	5.0	—
781		78.4	60.0	35.0	23.4	38.3	28.3	—
780		92.3	93.3	66.7	38.4	18.5	5.0	—
779		88.4	36.7	10.0	8.3	26.7	5.0	—
777		53.4	30.0	35.0	11.7	15.0	5.0	—
776		81.7	53.6	51.6	21.7	10.0	10.0	—
Mean		84.5	53.3	37.5	22.5	20.8	10.2	—
SS3 Trained Extract								
Rat								
826	77	81.7	30.0	13.3	13.3	8.3	11.7	—
830	—	—	45.1	23.3	1.7	1.7	1.7	—
832	114	53.2	36.6	15.0	5.0	3.3	0.0	—
834	150	78.4	50.0	28.3	16.7	3.3	5.0	—
840	80	53.2	28.3	28.4	5.0	6.7	8.4	—
Mean	105.3	66.6	38.0	21.7	8.3	4.7	5.3	—
SS3 Untrained Extract								
Rat								
825	67	35.0	41.6	33.3	15.0	8.3	13.3	—
827	67	75.0	55.0	31.6	20.0	3.3	5.0	—
829	67	60.0	40.0	20.0	5.0	5.0	11.7	—
835	88	80.4	18.4	13.3	8.4	15.0	5.0	—
841	77	103.3	31.7	23.3	11.6	18.3	20.0	—
845	222	86.7	40.0	5.0	8.4	15.0	6.6	—
847	43	73.4	65.0	30.0	26.7	11.6	15.0	—
Mean	90.0	73.4	41.7	22.4	13.6	10.9	10.9	—

TABLE I (*continued*)

	I (short initial sess.)	1st regular sess.	2nd	3rd	4th	5th	6th	7th
SS4 Trained Extract								
Rat								
915	140	60.0	18.3	8.3	6.7	13.3	6.7	15.0
917	170	83.3	26.7	30.0	35.0	25.0	13.3	3.3
919	58	66.6	38.3	16.7	11.8	13.3	1.7	3.3
921	57	50.0	71.6	43.3	40.0	16.7	6.7	8.3
923	123	60.0	26.7	20.0	11.7	10.0	21.7	3.3
925	173	125.0	103.0	25.0	18.3	6.7	10.0	8.3
927	125	75.0	46.7	36.6	28.3	6.7	3.3	1.7
929	142	63.3	23.2	15.0	1.7	—	5.0	3.3
931	77	51.6	75.0	51.6	35.0	26.7	5.0	26.7
Mean	118.3	70.5	47.7	27.4	20.9	14.8	8.2	8.1
SS4 Untrained Extract								
Rat								
914	100	155.0	65.0	85.0	56.7	50.0	20.0	1.7
916	50	80.0	23.3	48.3	43.3	16.7	22.4	10.0
920	73	58.3	28.3	11.7	5.0	16.7	6.7	1.7
922	77	46.6	53.3	33.3	11.7	3.3	10.0	1.7
924	150	46.6	43.4	36.7	21.7	3.3	11.6	1.7
926	100	—	76.8	45.0	65.0	51.7	56.7	56.7
928	107	111.5	50.0	36.6	16.7	16.7	1.7	5.0
930	100	100.0	50.0	20.0	8.3	0.0	0.0	5.0
932	71	61.6	38.4	11.7	8.3	11.7	6.7	1.7
934	77	163.0	108.1	78.3	31.7	10.0	5.0	3.3
936	133	50.0	55.0	40.0	38.4	5.0	5.0	10.0
Mean	94.4	79.3	53.8	40.6	27.9	16.8	13.3	9.0

TABLE II

RESULTS OF APPLYING THE MANN–WHITNEY U-TEST (ONE-TAILED, NO CORRECTION MADE FOR TIES) TO THE DATA OF THE EXPERIMENTS, SS1–SS4

Probabilities, P, of the distributions of (E/Sr) × 100 found in the sessions of experiments SS1–SS4.

Sessions	I	1	2	3	4	5	6
Exp. SS1	.01	n.s.	n.s.	.002	.01	.03	n.s.
Exp. SS2	—	n.s.	n.s.	.05–.10	.05–.10	.05–.10	n.s.
Exp. SS3	.11	n.s.	n.s.	n.s.	>0.1	.053	.04–.10
Exp. SS4	n.s.	n.s.	n.s.	n.s.	n.s.	n.s.	n.s.

Probability of distribution found when comparing E/animal/day till C–RNA groups reach criterion level, in all four experiments.

$P = .014$

n.s. = not significant.

Having thus arrived at a relatively reproducible experimental technique, it was considered desirable to approach the problem of the actual nature of the active component. In view of the fact that the different groups reporting "transfer" results had not only been using divergent behavioral techniques but also quite varying extraction and injection methods, it was considered desirable to carry out "hybrid" experiments, in which the extraction and injection procedures reported from one laboratory were tested with the behavioral technique of another laboratory.

As the simplest approach to hybrid experimentation we chose to use our own behavioral technique together with the intraperitoneal injection of a crude Tris-NaCl brain extract described by Byrne and Samuel (1966). In making this type of extract the brains were homogenized immediately in a Waring Blender using 4.0 ml of a 0.01 *M* Tris-0.9% NaCl buffer, pH = 7.5 per brain. After centrifugation for 20 minutes at 8000 *g* the supernatant is poured off and injected intraperitoneally.

Three experiments were made using 1, 1.5, and 1.7 brain equivalents, respectively. While essentially nothing was seen with the two lower doses, quite large intergroup differences appeared in the third (see Fig. 11). These were in exactly the same direction as in the experiments already described, *i.e.*, an initial reversal followed by a later difference in the predicted direction. The group sizes were quite small, several recipients died, and the difference was not significant ($P = 0.11$), but still the similarity to the learning curves of earlier experiments was sufficiently striking to encourage further hybrid experimentation.

At this point I took up a temporary position at Duke University and a second type of hybrid experiment was carried out in collaboration with Dr. Byrne. This of course furnished an opportunity to carry out exactly the opposite kind of hybrid experiment, that of testing the intracisternal injection of RNA-type extract with the behavioral methods of Byrne and Samuel (1966). However, before describing this, it will be more convenient to describe in some detail the training and testing procedure used in these experiments, and also the type of results that were obtained when using the complete procedure of Byrne and Samuel.

The extraction and injection procedure has already been described. Both donor training and testing of recipients were carried out in two bar Skinner boxes, which were completely symmetrical to the smallest detail. Great precautions were taken to this end, because earlier experiments by Byrne and Samuel had seemed to indicate that extract from donors trained to press the left bar was not as potent as extract from donors trained to press the right bar, these results were obtained using asymmetrical boxes with a door in one side.

Before start of training, the donor animals, male Holtzmann rats of about

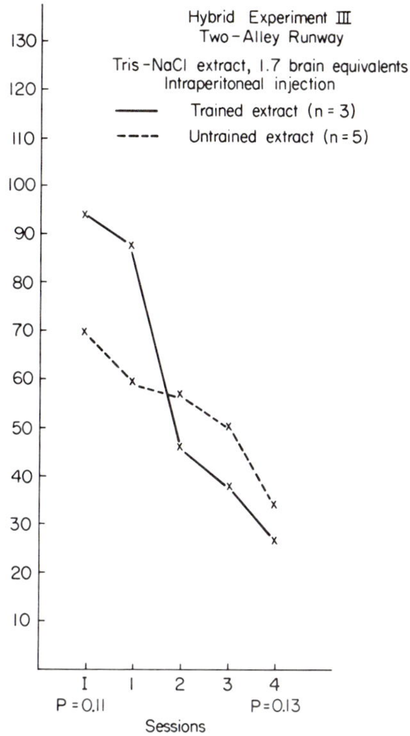

Figure 11 Outcome of a hybrid experiment in which the two-alley runway training and testing was combined with the use of Tris–NaCl brain extract injected intraperitoneally, 1.7 brain equivalents per recipient. ———, recipients of trained extract; ……, recipients of untrained extract.

200 gm weight, were kept on "environmental complexity." That is, they were housed in very large cages 20 or 30 together, and once a day for 20 minutes given access to a large box in which were kept a large variety of small cardboard boxes and similar things with which they could satisfy their explorative tendencies.

After a week on environmental complexity, the rats were deprived of food with an aim to reducing their body weight to 85% of the intial weight, and keeping this constant during training. At the same time they were rehoused in separate cages. Training was started with the rats being put three at a time into the Skinner box, for 20 minutes without reinforcement. At the start of this and all following sessions, there were always five food pellets in the foodcup.

Some hours after the "group session" the rats were put individually into the Skinner boxes with both bars reinforcing, remaining there until a total of 50 presses had been made, or until 20 minutes had passed. Thereafter, they were trained for 20 minutes, one session a day, with only one bar reinforcing, thus yielding two donor groups of right- and left-trained donors on a fixed ratio schedule starting at 1:1, the next day being 3:1 and increasing each day until 9:1 on the last day. After the final session the animals were left in their cages for two to four hours and then sacrificed by means of a guillotine, the brains were removed within 40 seconds and placed in crushed dry ice. Extraction was carried out as already described. When using the crude Tris-NaCl extract, the brains were homogenized individually or two at a time, depending on whether one or two brain equivalents were to be injected into the recipients. For RNA extraction the brains of a donor group were pooled.

Naive donors were also kept at 85% of initial weight but were not handled in any other way.

Recipient animals were exposed to the same program of environmental complexity. Before being injected they were put through the group session and a ten-minute unreinforced individual session in which their presses on right and left bar were recorded, this measure was used to select uniform recipient groups. Some hours after this they were injected, generally in the evening. The next day, about 15 hours after injection, they were put in the Skinner boxes for an unreinforced session of ten minutes (UN-1), which was repeated after six hours (UN-2). On the following days this was followed by 20-minute sessions once a day, with both bars reinforcing for all recipients, on a 1:1 schedule. Thus, because the rat was rewarded for pressing either bar, reinforcement was not able to influence the choice of bar, but did serve to keep the rat working in the set-up. This represents a very nice compromise between testing under a learning schedule and the unreinforced testing schedules. The latter have the drawback that they are actually identical with extinction procedures, *i.e.*, they should tend to make the desired response disappear.

Two experiments using this behavioral procedure have been run. In the first, only crude extract was used, in the second, some recipients were injected intracisternally with RNA-type extract.

In the first experiment four different groups of animals were run: one group received extract from donors trained to press the left bar (L-recipients); a second group received extract from donors trained to run to the right bar (R-recipients); a third group received extract of naive donors (N-recipients); and the fourth group (Tris-NaCl-recipients), receiving only buffer injected in a corresponding volume. Furthermore, the animals receiving brain extracts were injected either one brain equivalent or two brain equivalents in each of the three brain extract groups.

In the earlier experiments using this technique, an enhancement of learning had been found, recipients of the trained extracts learning to press the bars faster than recipients of naive extract (Byrne and Samuel, 1966), and this effect was most pronounced in recipients of R-extract. However, only insignificant differences in bar preference between R-recipients and L-recipients, *i.e.*, in terms of specificity, had been found.

In the present experiment this situation was reversed, there was no significant difference in learning performance of any brain extract groups, or between them and the tris-NaCl group. In contrast there was very clear evidence of specificity of the effect of R-extracts and L-extracts. It was found that due to the variation in individual initial preference of the animals this could most strikingly be expressed as the $\Delta\%$R, *i.e.*, the change from one session to the next in the percentage of presses on the right bar. When considering this parameter from the first to second reinforced session (Tables III and VI) it was found that R-recipients had an increasing right preference and L-recipients had a decreasing right preference. When comparing all R and all L recipients this was significant ($0.05>P>0.025$), and when comparing only R and L recipients receiving two brain equivalents it was highly significant ($P=0.002$). Furthermore, in both groups the increase in effect found when comparing recipients of two brain equivalents to recipients of one brain equivalent in the same group was also striking ($P=0.026$ in R recipients). All L animals were significantly different from all N animals ($P<0.02$), this was increased if considering only two brain recipients ($P=0.004$). Differences between R and N animals were not significant though approaching this when considering only two brain animals ($P=0.12$). When considering the Tris-NaCl recipients the reverse was true, there was significant differences between these and all R's ($P=0.025$) and still more compared to only two brain R's ($P=0.011$). This was in reality the effect of a difference between the N and Tris-NaCl animals of which the former went more to the right and the latter more to the left. This difference also was significant ($P=0.05$ when considering all N animals, 0.026 if considering only two brain N's). In all cases where significant intergroup differences were seen these were thus increased by considering only two brain recipients, therefore there was a very clear dose effect.

When comparing these results to those obtained earlier with a similar procedure, the question arises why in this case no enhancement of learning was found, but instead a specificity of effect as far as bar preferences go. There may be several causes, one of which could possibly be the fact that in the Skinner boxes used earlier, the bars had to be pressed quite hard to be activated, 25 gm, while in the present boxes they reacted to a pressure of one or two grams. Thus they would be much more likely to be activated by the rat incidentally touching them, and this might conceivably mask intergroup differences in terms of bar pressings scored against time.

TABLE III

RANKING OF Δ%R Re-1 TO Re-2 IN ALL RECIPIENTS OF EXPERIMENT D-I, AND P VALUES CALCULATED ACCORDING TO THE MANN–WHITNEY U-TEST. INJECTION WAS INTRAPERITONEAL TRIS-NACL EXTRACT VALUES IN PARENTHESIS ARE SCORES OF ONE BRAIN RECIPIENTS

Rank	L	R	N	Tris	Rank	L	R	N	Tris
1		+59.2			21	(−4.4)			
2		+51.2			22				−5.0
3				+50.0	23	(−6.6)			
4		+30.2			24			−8.5	
5			+29.3		25		−10.0		
6			(+26.5)		26	(−10.1)			
7		(+24.9)			27		(−11.3)		
8		+21.0			28		(−11.9)		
9			(+20.3)		29			(−14.3)	
10			(+16.2)		30				−16.1
11			+16.1		31				−16.2
12			+15.8		32		(−17.4)		
13		+13.0			33	−17.5			
14			+11.7		34	−19.4			
15	(+6.1)				35				−20.0
16	+2.7				36			(−20.6)	
17	−1.0				37				−25.2
18			−2.4		38				−40.1
19	(−3.3)				39	−40.4			
20		(−3.4)			40	−50.0			

All R versus all L: U = 31, .05 > P > .025.
All R versus all N: U = 55, P > .1.
All L versus all N: U = 25, .025 > P > .010.
All R versus Tris: U = 16, P = .025.
All L versus Tris: U = 31, P > .1.
All N versus Tris: U = 16, P = .025.
2 brain R versus 1 brain R: U = 4, P = .026.
2 brain L versus 1 brain L: U = 8, P = .12.
2 brain N versus 1 brain N: U = 15, P = .53.
1 brain R versus 1 brain L: U = 8, P = .21 in reverse direction.

As far as the obtaining of "symmetrical" intergroup specificity in this experiment goes, it may be the result of using a completely symmetrical box. If this is true, it certainly shows an amazingly large effect of a quite trivial alteration of the set-up, and serves to point out how well the experimental situation should be controlled in transfer experiments.

In the next experiment, using an identical behavioral technique, it was decided, in addition to running an R and L group, which had been injected two brain equivalents as above, to also conduct a hybrid experiment in which intracisternal RNA injection, one mg per recipient was substituted for the intraperitoneal crude extract injection, and the experiment carried out as described earlier.

The result of this experiment is seen in Table V. While this time there was no significant intergroup differences in the crude extract groups, RNA extract recipients yielded much the same results as in the earlier experiment, Δ %R from Re-1 to Re-2 was significantly ($P = 0.05$) different in these groups. Some of the animals in these groups only pressed very little. If among the 32 animals tested in this experiment only those RNA recipients that ranked among the 20 highest as far as bar pressing goes (TTL15, *i.e.*, the time taken to learn to press the bars 15 times was used as measure) considered, then the RNA recipients left after this screening by an independent measure are highly significantly different in the predicted direction ($P = 0.002$) when considering the Δ %R Re-1 to Re-3.

These results, if further confirmed, would seem to indicate that the same degree of specificity of transfer effects can be obtained when using the intracisternal RNA injections as with the intraperitoneal injection of a crude extract. An important aspect is that with the crude extract two brain equivalents must be used in order to get any effect, while with the RNA extract effect was obtained with one mg per recipient, an amount that only corresponds to about 0.6 brain equivalents of RNA. Furthermore, the crude

TABLE IV

RANKING AND P VALUES IN SKINNER BOX EXPERIMENT D-I WHEN CONSIDERING ONLY RECIPIENTS OF TWO BRAIN EQUIVALENTS
POSITION PREFERENCE—LEFT VS. RIGHT
Δ %R Test Re-2 vs. Test Re-1

	Type of Injection		
	Left	Right	
Rank	Δ %R	Δ %R	
1		+59.2	Experiment D-1
2		+51.2	Tris-NaCl Extract
3		+30.2	2 brain equivalents
4		+21.0	Intraperitoneal Injection
5		+13.0	
6	+ 2.7		
7	− 1.0		
8		−10.0	$U = 2$
9	−17.5		$P = .002$
10	−19.4		Mann–Whitney U Test, one-tailed
11	−40.4		
12	−50.1		

TABLE V

RANKING AND P VALUES, ALL RNA RECIPIENTS OF SKINNER BOX EXPERIMENT D-II. INJECTION WAS INTRACISTERNAL RNA EXTRACT POSITION PREFERENCE—LEFT VS. RIGHT Δ%R Test Re-2 vs. Test Re-1

	Type of Injection		
	Left	Right	
Rank	Δ %R	Δ %R	
1	+58		Experiment D-2
2		+53	"RNA", Phenol Extraction and Dialysis
3		+26	1 mg (2/3 Brain equivalent)
4		+24	Intracisternal injection
5		+ 7	
6	− 1		
7		− 2	
8	− 9		
9	−12		
10		−12.2	
11		−13	U = 18
12		−13	P = 0.05
13	−20		Mann–Whitney U Test, one-tailed
14	−21		
15	−28		
16	−36		
17	−49		

extract (two brain equivalents) contained about 60 mg protein while the protein content of the RNA (less than 10%) would amount to only 0.1 mg per recipient. Thus there is a reduction in protein content of about 600 times, without any concurrent change in activity. This would at least seem to show that the likelihood of the active component being a protein is not too large. Also the long dialysis (18 hours against three shifts of about 100 volumes dialysis fluid) would seem to rule out components of lower molecular weight. Of course nothing can really be stated about what is responsible for the activity until enzymatic digestion has been tried out on the RNA extract.

As a further indication of the effectiveness of the intracisternal RNA injection method I should like to mention that Dr. Joan Martin at Duke University has succeeded in transferring aversion to saccharin in rats by this technique. When rats, which generally prefer saccharin solution to water, are irradiated (200 roentgens were used) at the same time or shortly after being exposed to a 0.1% saccharin solution, they will avoid this for many days, drinking only plain water if offered the choice. Rats that were thus trained for three trials were trained donors, control donors were rats exposed to exactly the same treatment except for having a lead shield in the X-ray machine, so that no irradiation resulted. Trained extract recipients consumed significantly less saccharin than controls during eight days of testing (means of amounts (sacch./water + sacch.) × 100 were 48.5 and 70.3 respectively, there were seven rats in both groups and P was 0.04 according to the U-test).

Another possible interpretation of the striking similarity of effect obtained with these two widely different extraction and injection methods and of the phenomenon that indeed transfer results have been obtained with a still wider spectrum of extraction methods, might of course be that this is caused by the active component being identically the same in all the different extracts. This would seem to indicate that the active component had to be extremely potent, but cannot be ruled out under present circumstances, as cannot either the possibility that there may be several active components, for instance RNA and protein.

In conclusion, it may be stated that there seems to have been much information gained by the use of these hybrid methods, which would perhaps not so easily have been obtained by carrying out straightforward replications of the original work. One striking observation made is that much the same results may be obtained in a given test situation even when testing quite different types of extract and injection procedures, or in other words the effect obtained (initial reverse or not, etc.) seems to be more dependent on the behavioral technique than on the biochemical and injection procedures, that are more or less equivalent at least as far as qualitative aspects goes. As mentioned before this may indicate a similarity or indentity of the active components of the different extracts. We believe these to be the most important conclusions to be drawn from the results presented.

REFERENCES

Byrne, W. L. and Samuel, D., 1966. Behavioral modification by injection of brain extract prepared from trained donors. *Science*, **154**, 418.

Dingman, W. and Sporn, M. B., 1961. The incorporation of 8-azaguanine into rat brain RNA and its effect on maze learning by the rat: an inquiring into the chemical basis of memory. *J. psychiat. Res.*, **1**, 1–11.

Fjerdingstad, E. J., Nissen, Th. and Røigaard-Petersen, H. H., 1965. Effect of ribonucleic acid (RNA) from the brain of trained animals on learning in rats. *Scand. J. Psychol.*, **6**, 1–6.

Keck, K., 1956. An ultramicro technique for the determination of deoxypentose nucleic acid. *Arch. Biochem. Biophys.* **63**, 446–451.

Kok, I. P., Skuratovskaya, I. N., Bilai, T. I. and Gulko, M. A., 1966. Use of bentonite in isolation of native high-polymer ribonucleic acid from animal tissues. *Ukr. Biokhim. Zh.*, **38**, 206–211.

Nissen, Th., Røigaard–Petersen, H. H. and Fjerdingstad, E. J., 1965. Effect of ribonucleic acid (RNA) extracted from the brain of trained animals on learning in rats.—II. Dependence of the RNA-effect on training conditions prior to RNA-extraction. *Scand. J. Psychol.*, **6**, 265–272.

Ralph, R. K. and Bellamy, A. R., 1964. Isolation and purification of undegraded nucleic acids. *Biochim. biophys. Acta*, **87**, 9–16.

Røigaard–Petersen, H. H., Nissen, Th. and Fjerdingstad, E. J., 1968: Effect of ribonucleic acid extracted from the brain of trained animals on learning in rats.—III. Results obtained with an improved procedure. *Scand. J. Psychol.*, **9**, 1–16.

Rosenblatt, F. and Miller, R. G., 1966. Behavioral assay procedures for transfer of learned behavior by brain extracts, II. *Proc. Nat. Acad. Sci. U.S.A.*, **56**, 1683–1688.

Rucker, W. B., 1967. Memory: Antagonistic transfer effects by RNA extracts and relation to donor training. Abstr. Ph.D. diss., Univ. Chicago.

Ungar, G., 1966. Chemical transfer of learning. Its stimulus specificity. *Fed. Proc.*, **25**, 207.

Ungar, G. and Irwin, L. N., 1967. Transfer of Acquired Information by Brain Extracts. *Nature*, **24**, 453–455.

A SIMPLIFIED BEHAVIOR TEST OF BRAIN EXTRACTATE TRANSFER EFFECTS IN RATS

Richard Gay

Michigan State University
East Lansing, Michigan

Alfred C. Raphelson

University of Michigan, Flint College
Flint, Michigan

One of the problems appearing in the early work on transfer effects with brain extract was that recipient subjects (Ss) were not always tested for the identical responses as those for which the donor Ss had been trained. In addition, the observed response was often too molar to allow the measurement of any subtle transfer effect that might be present.

It appeared reasonable, therefore, that there would be some advantage in research on this problem, to use an experimental situation which produces in untrained Ss a clear-cut, uniformly observed response. Then, one group of donor Ss could be trained to respond otherwise in this situation and tests could be carried out on naive recipient Ss which would determine whether they demonstrate the natural or trained response.

Advantage was taken of the easily demonstrated fact that when given a choice, rats prefer a dark, inclosed black space to a well-lit white space. A linear maze was constructed in which three boxes (11 in. × 12 in.) were connected by two runways (24 in. × 6 in.) [see Fig. 1]. The boxes were arranged so that there was one in the middle of the runway which served as a Start Box (SB). A second box, painted white (WB) was connected to the Start Box by the left runway (LRW) and a third box connected to the right

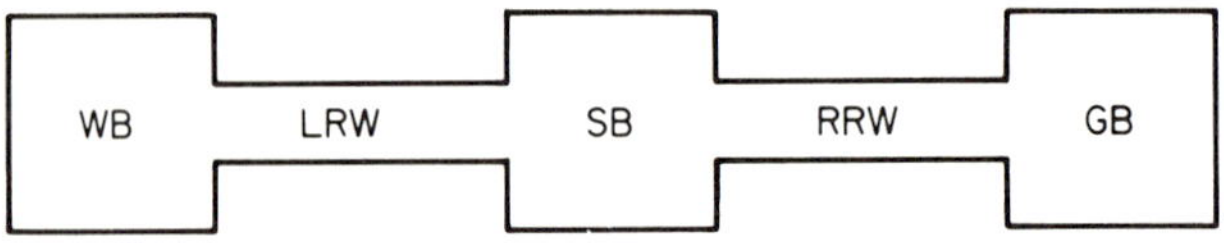

Figure 1

runway (RRW) served as the Grid Box. The whole apparatus, with the exception of the Grid Box was covered with clear plexiglass. Twenty-five-watt lights were placed directly over the White and Start Boxes. The Grid Box was covered with a wooden top with a small dark blue plexiglass window which transmitted a minimum of light. The floor of the Grid Box was covered with shock grid and a guillotine door controlled the entrance and exit from it.

Preliminary work with this apparatus verified the fact that rats, being negatively phototropic when allowed free exploration of the runway, would show a definite preference for the Grid Box, *i.e.*, the darkest enclosed area. A pilot group of six Ss were placed in the Start Box and allowed a three-minute exploration period. All Ss performed in the same manner. They remained in the Start Box anywhere from ten to thirty seconds, and then after looking down both runways, proceeded to the Grid Box where they remained for at least two-thirds of the 180-second test period. There was little exploration of the left runway. No animal ever entered the White Box nor proceeded farther than half way down the left runway. It was found useful to obtain three definite measures of the Ss behavior: latency to reach the Grid Box from the Start Box, time spent in the Grid Box, and time spent in the White Box.

The experimental learning task utilized the demonstrated preference for the darker Grid Box of two Donor groups of S. Then utilizing a passive avoidance paradigm, one of the Donor groups was trained to avoid the Grid Box through use of shock-escape training. After this training, the Donors were sacrificed and brain extract taken. After receiving the respective extract, two naive groups (Recipients) were tested in the same apparatus for the same response, *i.e.*, the amount of time spent in the Grid Box and White Box as well as the latency to the Grid Box.

SUBJECTS

Two experimental tests were run:

Experiment One: Ss (19 male Long–Evans rats, approximately 65 days of age) were divided into four groups: Learned-Donor (N = 7); Control-Donor (N = 4); Learned-Recipients (N = 4); and Control-Recipients (N = 4).

Experiment Two: Ss (22 female Long–Evans rats, approximately 100 days of age) were divided into four groups: Learned-Donor (N = 7); Control-Donor (N = 7); Learned-Recipient (N = 4); and Control-Recipient (N = 4). The test trials were carried out by an assistant who did not know the designation of individual Ss nor the purpose of the experiment. In all other respects the two experiments were similar.

PROCEDURE

All Ss in the Learned-Donor and Control-Donor groups were individually placed in the Start Box and allowed free exploration of the entire apparatus for 180 seconds. The amount of time each S spent in the dark Grid Box and the bright White Box was recorded. As expected there was little exploration with all Ss showing a definite preference for the dark Grid Box. The average time spent there for all groups was about 120 seconds. Only one of the 25 Ss entered the White Box during this period, spending less than five seconds there and almost all the remaining time in the Grid Box.

After all Ss had demonstrated their preference for the Grid Box, the Ss in the Learned-Donor groups were confined in the Grid Box and administered a 27-volt shock for five seconds after which the gate was raised and the Ss were allowed to escape the shock by retreating up the runway. Four shock trials were given each day for four days. All Ss in this group demonstrated an inhibition of their original preference for the Grid Box along with the usual signs of emotionality (e.g., urination, defecation, respiratory increase, freezing). Two hours after the last training session, all Ss were sacrificed and their brains removed.

Within each group the brains were pooled and processed together according to the following extraction procedure: The brains were homogenized for five minutes in a ten ml mixture of equal parts of 92% phenol and isotonic saline, then centrifuged at 15,500 rpm for 45 minutes. The aqueous phase was drawn off and the mixture centrifuged again at 6,000 rpm for five minutes. The mixture was then brought up to a concentration of 0.1 *M* $MgCl_2$ and two volumes of cold 95% ethanol were added to precipitate the remaining extractate. The precipitation was enhanced by a final centrifugation at 6,000 rpm for 15 minutes. The remaining ethanol was evaporated, and the extractate dissolved in a 1.25 ml of isotonic saline.

Within one hour after the extraction was completed, the Recipient Ss received 0.25cc intraperitoneal injections of the appropriate extractate. In each of the two experiments, four Ss received extractate from the Control-Donor group and four Ss received extractate from the Learned-Donor group.

Two hours after the injection, all recipients were individually placed in the Start Box and allowed to explore the entire apparatus for three min. The amount of time each S spent in Grid Box and White Box was recorded, as well as the time it took each S to reach the Grid Box after being placed in Start Box. Ss were tested in this manner five times at intervals of two, four, six, eight, and sixteen hours after the injection.

TABLE I

Average Time (in sec.) Spent in Grid Box and White Box by Learned-Recipients and Control-Recipients During First Transfer Test

Experiment	Goal Box		White Box	
	Control Recipients	Learned Recipients	Control Recipients	Learned Recipients
One	123	0	0	62
	153	58	0	0
	126	54	0	0
	120	0	0	35
Two	83	0	0	45
	150	0	0	2
	117	0	0	0
	84	12	0	28
Mean	119.5	15.5	0.0	21.5

TABLE II

Average Time (in sec.) Spent in Grid Box and White Box by Learned-Recipients and Control-Recipients Over Five Transfer Tests

Experiment	Goal Box		White Box	
	Control Recipients	Learned Recipients	Control Recipients	Learned Recipients
One	127	2	0	17
	144	62	0	2
	135	56	0	21
	13	0	0	14
Two	76	40	17	16
	107	0	5	2
	108	80	1	4
	145	13	0	23
Mean	106.9	31.6	2.9	12.4

RESULTS

The results of both experiments are presented in Tables I and II. Table I shows the amount of time each recipient S spent in the dark Grid Box and the bright White Box in the first transfer test given two hours after the extractate injection. It should be remembered that all Donor groups showed a definite initial preference for the Grid Box, but the Learned-Donor groups acquired an avoidance reaction to it. Five out of eight Ss receiving brain extract from the Learned-Donor groups did not enter into the dark Grid Box during the entire three-minute free exploration period. All of the Ss receiving brain extract from the Control-Donor groups not only entered the Grid Box, but, on the average, spent close to two-thirds of the period there.

In addition, none of the Control-Recipients entered the bright White Box whereas five of the eight Learned-Recipients entered this box.

The significance of these data are obvious from an inspection of the individual scores in Table I. However, tests of significance were run on the mean differences between the two groups in time spent in the Grid Box and the White Box and reached at least the 0.05 level of significance in both instances.

These quantitative data do provide clear evidence of some kind of differential effect being present in the two groups of recipients. They do not, however, do justice to the dramatic character of the observed behavior. In hope of conveying a more complete picture of what was observed, the qualitative observations of the test behavior of the four Learned-Recipient Ss in Experiment One on the first transfer test are presented in Table III.

Table II presents the same kind of data as previously considered for all five of the transfer tests. The trends which appeared on the first test (Table I) continued throughout the 16-hour test period.

DISCUSSION

It appears from the quantitative and qualitative data presented that some kind of "transfer" was obtained in these two experiments. In that regard, these results are replications of the several studies that have been reported earlier. However, in face of the large number of negative findings that have been reported one can only be cautious in generalizing from these results.

It is hoped that this study may contribute to the continuing discussion of "transfer-effects" by suggesting the use of a type of situation that elicits a well-defined, uniform preinjection response that can be easily assessed in donor and recipient groups alike. The response, though simple and reliable, lends itself to an easily quantified description. However, as revealed in the

TABLE III

Descriptions of Behavior and Actions of the Experimental–Recipient Rats During the First Transfer Test

SS	Hours Tested after Injection	Latency to G.B. (sec.)	Time in W.B. (sec.)	Time in G.B. (sec.)	Observations
1	2	180	62	0	SS in S.B. sniffed and groomed then ran down LRW into W.B. after brief exploration of the LRW. SS slowly approached G.B. from RRW (after 90 sec. in maze) in a flattened crawling position. It touched its nose to the inside of the G.B., jerked back quickly and ran back to the S.B. where it froze and groomed for remaining time.
2	2	71	0	58	SS spend 65 sec. in S.B., grooming and breathing heavily. Then ran down RRW, stopped before G.B., poked its head in and out of box twice, went in for three sec., ran out and back to S.B. and groomed briefly. Then reapproached G.B. and went in for 30 sec. before vacillating between G.B. and LRW for rest of time.
3	2	26	0	54	Froze when put in S.B. groomed briefly, then went down RRW slowly, stopped at the G.B., jerked back then flattened out and entered slowly, backed out, reentered slowly again and stayed 42 sec. then came out and explored all of maze with exception of W.B. Groomed and vacillated in and out of G.B. to various parts of the maze.
4	2	180	one-half body in for 35 sec.	0	From S.B., SS 'looked' both ways stopped and groomed. Did not enter RRW at any time during the three min. but crouched and froze when it poked its head into RRW. Withdrew to S.B., sat, heavy breathing then turned and investigated LRW fully and projected one-half its body into W.B. then spent rest of time between LRW and S.B. (with grooming).

data presented above, the results can be so clear that they hardly need statistical analysis.

Further research is currently underway to determine if the transferred avoidance response is situationally specific. A series of test situations will be set up in which the recipient Ss will be given exploration choices in several situations which vary on a similarity continuum with the original training situation. Observations will be made on differences in preferred areas by the recipient groups as the situations decrease in their similarity from the original training situation.

REFERENCE

Gay, Richard and Raphelson, Alfred C. " 'Transfer of learning' by injection of brain RNA: a replication." *Psychon. Sci.*, 1967, **8**, 369–370.

CHEMICAL TRANSFER OF LEARNED INFORMATION*

G. Ungar

Baylor University College of Medicine
Houston, Texas

Over the past three years we have reported several series of experiments in which extracts of brain taken from animals trained for certain tasks, injected into naive recipients, could induce behavioral changes related to the training of the donors. We became interested in this area when we were able to induce morphine tolerance by injecting material extracted from the brain of tolerant animals (Ungar, 1965; Ungar and Cohen, 1966). These results encouraged us to explore the possibility that tolerance to physical stimuli, *i.e.*, habituation, could be similarly induced by brain extracts from habituated donor animals. We found that animals injected with brain extracts from animals habituated to a sound showed a significant decrease in their startle responses to the same stimulus (Ungar and Oceguera–Navarro, 1965). This transfer was subsequently shown to be stimulus-specific (Ungar, 1966, 1967a, 1967b).

In later experiments, other situations were examined. In one group of experiments, donors were trained to avoid electric shock with the sound of a buzzer as the conditioned stimulus. After they had reached criterion, their brains were administered to naive recipients which were trained identically with the donors. It was observed that these animals learned the avoidance task significantly faster than the recipients of untrained brain extracts. Similar observations were made with escape training in a Y-maze, using light as a clue (Ungar, 1967a, 1967b).

However, it appeared desirable to use other criteria of transfer than rate

*This work was supported by USPHS grant No. MH 13361.

of learning. In a further series of experiments, therefore, the recipients received no reinforcement whatsoever, so that any change of behavior following the administration of extract could be attributed only to the information contained in the latter (Ungar and Irwin, 1967). In one type of experiment, the donors were trained to avoid electric shock by running into the left or right arm of a Y-maze. Extracts of their brains were injected into recipients whose left or right turns were recorded for several days before and after injection of the extracts. In several experiments, administration of the extract did change the behavior of the recipients in the direction predicted by the training of the donors (Fig. 1). However, the results were not consistently reproducible and were influenced by a number of factors of which only a few could be analyzed (Ungar, 1967a).

More consistent results were obtained in experiments in which the donors were trained to escape into the lighted arm of a Y-maze. The difference between pre- and postinjection behavior or between experimental and control groups were in most cases statistically significant but small. Table I summarizes some of the results (Ungar, 1967a; Ungar and Irwin, 1967).

All the experiments just mentioned have strongly suggested that the

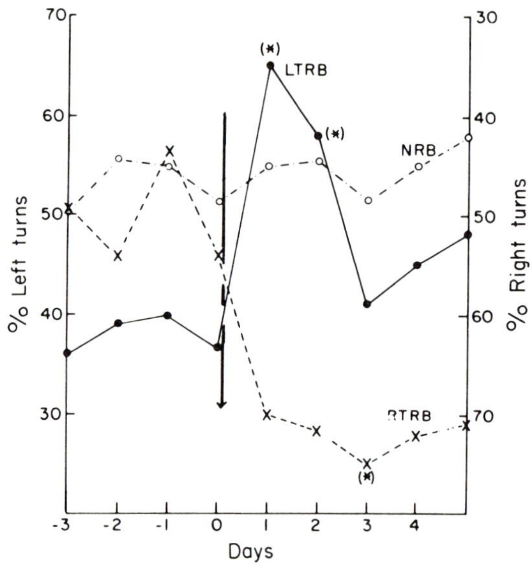

Figure 1 *Left and right turns of mice injected with brain extracts from donors trained to escape to the left (LTRB), to the right (TRRB) and from untrained rats (NRB).* Abscissa: days before and after injection (indicated by arrow) of 0.75 g per mouse; ordinate: percentage of left and right turns. Asterisks indicate significant difference from preinjection mean (U-test). Number of animals: LTRB and RTRB, six; NRB, ten. (Ungar 1967a, with permission).

TABLE I

ACTION OF BRAIN EXTRACTS FROM DONORS TRAINED TO ESCAPE TO LIGHT (1 TO 5) AND FROM UNTRAINED RATS (6, 7) ON BEHAVIOR OF UNREINFORCED RECIPIENTS. (UNGAR, 1967 WITH PERMISSION)

	% Escape to Light						
Days	(1)	(2)	(3)	(4)	(5)	(6)	(7)
−1	62.5	57	52	54	56	57	59
0	57.5	63	59	59	59.5	59	55
Preinjection mean	60	60	55.5	56.5	58	58	57
1	59	63	61	67(1)	76.5(1)	63	57.5
2	65	64	65(2)	68(2)	77.5(1)	64	57.5
3	60	62	69(3)	70(3)	73.5(3)	61	58
4	59	67.5(1)	69.5(3)	69.5(3)	71(3)	58	63
5	59	71(1)	71(3)	70(3)	64(2)	60	58
Postinjection mean	61	65.5(1)	67(2)	69(3)	73(3)	61	59
Difference	1	5.5	11.5	12.5	15	3	2
Brain (g/mouse)	.5	.75	1.0	1.25	1.25	1.25	1.0
Days to criterion	10	10	5	10	5	—	—
N	4	6	20	20	10	16	12

Significance of difference with preinjection mean (U-test) (1) P = .05–.01; (2) P = .01–.001; (3) P < .001.

behavior of the recipients was modified by injection of donor brain and the modification was related to the training received by the donors. In Table II we tried to assess the statistical probability of such a modification by dividing all the recipients into two groups: (1) those whose behavior was changed in the direction expected by the training of the donors, and (2) those whose behavior remained unchanged or changed in the opposite direction. Among the recipients of untrained brain the same classification was made. Table II shows that among 350 recipients of trained brain 73.2% changed in the expected direction while only 8.8% of the controls showed this change. The probability of these modifications being due to chance, calculated by the χ^2 method, is shown in Table II for each type of experiment and for the total of 475 recipients.

TABLE II

BEHAVIORAL CHANGES IN ANIMALS INJECTED WITH EXTRACTS OF BRAIN TAKEN FROM TRAINED (TRB) OR UNTRAINED (NRB) DONOR RATS

	Number of Recipients				P^b
	TRB		NRB[a]		
	+	0 or −	+	0 or −	
Habituation	41	1	0	22	<.001
Conditioned Avoidance	10	5	2	11	<.01
Left or right escape	110	67	2	28	<.001
Escape to light					
Reinforced	15	9	5	15	<.02
Unreinforced	80	12	2	38	<.001
	256	94	11	114	<.001
Percentage	73.2	26.8	8.8	91.2	
N	350		125		

+ = Change in the direction expected from the training of donors.

0 = No change; − = Change in opposite direction.

[a] In control animals, the direction of the change was counted as in the corresponding experimental animals.

[b] χ^2 method.

In spite of the high probability of transfer shown by the pooled results, individual experiments remained unpredictable, indicating the need for a paradigm which would yield clear-cut and readily reproducible evidence that extracts of trained brain can influence the behavior of the recipients. The impressive results obtained by Gay and Raphelson (1967) using passive avoidance promised to offer this possibility. These experiments are based on the well-known dark preference of rats and mice and its reversal when the animals are submitted to electric shock when they seek refuge in the dark. The following is an account of our experiments confirming the original results, specifying some of the conditions of success and starting to define the chemical properties of the active material.

Donor animals (Sprague–Dawley rats, male, 200 to 250 g) were trained in two boxes (the starting box, *S*, made of white plastic with a transparent lid, 30 × 20 × 15 cm; and the dark box, *D*, of black plastic on five sides and an electrified grid floor, 25 × 10 × 5 cm) connected by a passage five cm long with cross section of 8 × 8 cm provided with a gate. The rat placed in *S* usually ran into *D* after a few seconds. At this moment, the gate closed and the animal received two ma electric shocks for five sec. through the grid floor. Afterwards the gate was lifted and animal ran back to the starting box or was returned to it. The procedure was repeated five times: the five shock periods of five sec. were separated by intervals of 10 to 15 sec. One such session was given to the donors daily for varying periods as specified below.

Approximately two hours after the last training session the brains were rapidly removed, frozen in dry ice and stored at −20°C until use. Brains from groups of animals similarly treated were pooled, weighed, and homogenized in water with a motor-driven Teflon pestle. The homogenate was stirred for five to six hours in an ice bath, kept in the refrigerator for 14 to 16 hours and centrifuged at 80,000 × g for one hour. The volume of the supernatant was adjusted so that one ml was equivalent to two g of brain. In recent experiments the homogenate was poured into cellophane bags and dialyzed against 20 volumes of distilled water for 20 to 24 hours. The dialyzate was lyophilized and the residue dissolved in distilled water. The extracts were kept frozen at −20°C until use.

The recipients were male Swiss mice, 20 to 25 g. They were tested in the same boxes as those used for the training of the donors but a third compartment was attached to the starting box. This white box (*W*), made of the same material as *S*, was connected to it by a passage identical with the one linking *S* to *D*.

Before injection, mice were screened to eliminate those that did not spend at least 90 out of a total of 180 sec. in the dark box. In 400 mice, the mean time in the dark box was 140 ± 17 (S.D.) sec. After screening, the mice were injected intraperitoneally with the brain extracts and tested at the intervals specified below. The animal was placed in *S* with both gates open. By means of stop watches the time spent in all three compartments was recorded for 180 sec. The results are tabulated below in terms of the mean number of seconds spent in *D* (TDB).

In the experiment summarized in Table III, the recipients were divided into two groups: group one received an extract of brain taken from donors trained for six days (amounts equivalent to one g of brain were injected into each mouse); group two was injected with the same dose of similarly prepared extract of untrained rat brain. The results show that the extract from untrained animals had no significant effect on the behavior of the recipients, but the trained extract spectacularly reduced TDB. The significance between the means, computed by the *t* test, indicates $p < 0.001$ until the fourth day when the majority of animals reverted to the preinjection pattern of behavior.

In the next series of experiments, the effect of varying amounts of brain and varying length of training was studied. As shown in Fig. 2, the effect increased with the dose of brain material injected and with the length of

TABLE III

EFFECT OF BRAIN EXTRACTS ON TIME SPENT IN DARK BOX (TDB)

	Experimental[a] TDB	± S.D.	Control[b] TDB	± S.D.	P
Pre-injection	139.5	21	139	23	—
20 h	20	19	146.5	27	<.001
42 h	15	12	139	27	<.001
66 h	43	18	144	25	<.001
88 h	95	53	135	20	>.05
N	12		12		

[a] Injected with dialyzate of brain extract from rats trained for six days (one g per mouse).
[b] Dialyzate of untrained rat brain extract, one g per mouse.

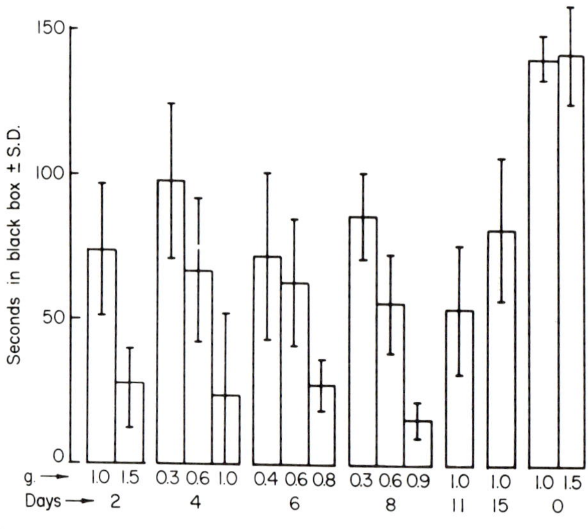

Figure 2 *Influence of donor training and dose of brain injected on recipients.* Avoidance in terms of time spent in dark box (sec. ± S.D.) during 180 sec. trials. Training of donors indicated in days during which they were given electric shock in dark box for five periods of five sec. each. Dose of brain injected in g per mouse. Each column represents the mean values obtained in six mice, tested three times during each of the three postinjection days.

training, up to eight days. For longer periods of training, the effect declined sharply. The decline was less marked with donors trained for eight days and given five or ten days rest before removal of the brain.

Based on the quantitative relationship between dose and effect and the slope of the dose-response curve, a unit of activity was defined as the amount of material that reduces TDB to 90 sec., and the content of the brain in transfer factor could thus be expressed in units per gram (Fig. 3).

In all the experiments just summarized, the donors were given five periods of five sec. of shock every day. In other experiments, varying amounts of shock were given daily. It is seen in Fig. 3 that one five sec. period per day for six days (total of 30 sec.) gave results that might have been obtained in one day with five shock periods. When shock was given for 50 sec. daily, the results resembled those of 11 days training with 25 sec. per day. It seemed likely, therefore, that the total duration of exposure to shock might have some relation to the amount of transfer factor present in the brain. Experiments were set up to observe whether giving 30 periods of five sec. in one day would give results similar to five periods of training for six days. However, in these experiments, the results depended also on the interval between the last trial and the removal of the brain. Within the first six hours and after 24 hours, the results were unsatisfactory. The best yield was obtained after 12 to 24 hours.

The chemical properties of the transfer factor are shown in Fig. 4. It is

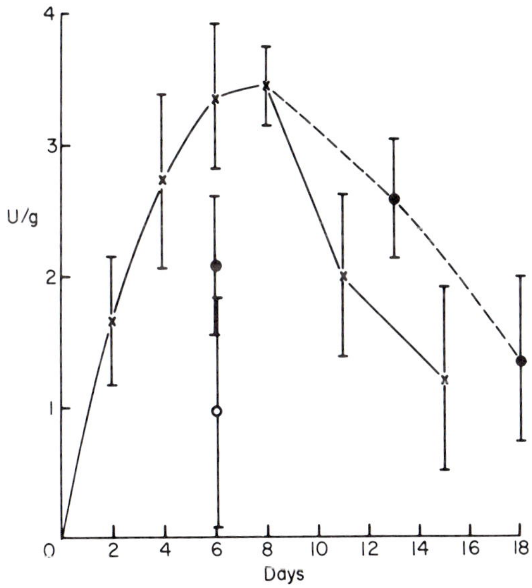

Figure 3 *Effect of training on the content of donor brain in transfer factor.* Abscissa: days from beginning of training to removal of brain; ordinate: units of transfer factor per g of donor brain (calculated from empirically established equation U = (125.5 – TDB)/35.5). Line x—, donors received shock for five periods of five sec. daily for number of days indicated on abscissa; line ●——, donors trained for eight days and killed five or ten days after last session. Point ○ on sixth day indicates donors trained with one period of five sec.; point ⓐ, with ten periods of five sec.

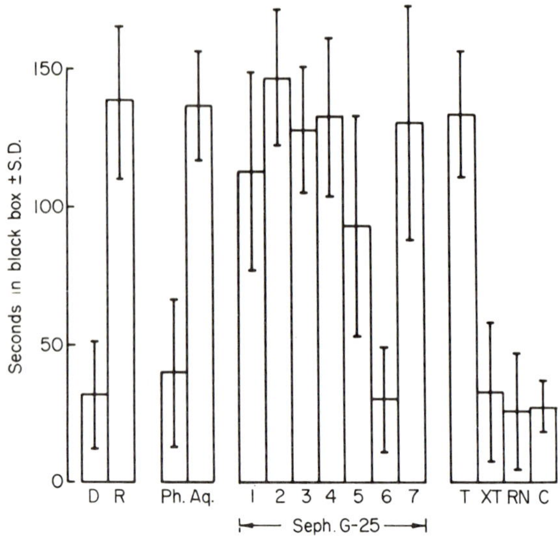

Figure 4 *Chemical properties of transfer factor*. Effect of purification and characterization procedures of donor brain extract (trained for six days) on passive avoidance in recipients (in TDB ± S.D.). D, dialyzate; R, retentate after dialysis of homogenate against 20 volumes of distilled water for 20 hours at 2°C; Ph., phenol phase; Aq., aqueous phase after partition of the dialyzate in equal volumes of 80% phenol and water. Separation on Sephadex G-25 column (each tube contained three ml): fr. 1, tubes 25–35; fr 2, 36–44, fr 3, 45–51, fr 4, 52–60, fr 6, 61–70; fr 7, 91–110. Incubation with enzymes: T, crystalline trypsin, 0.5 mg/ml; XT, crystalline chymotrypsin 0.5 mg/ml; RN, pancreatic RNase 0.4 mg/ml; C, control, no enzyme. Conditions of testing as in Fig. 2.

seen that the substance responsible for the transfer is dialyzable. When the dialyzate was submitted to phenol partition the activity was found in the phenol phase. After incubation of the dialyzate for one hour with trypsin, chymotrypsin and pancreatic RNase at the optimum pH of these enzymes, it was found that trypsin destroyed the activity while the other enzymes had no effect. The transfer factor has been purified by precipitation with acetone, passage on Sephadex G-25 and Dowex 50. At the present state of purity, one unit of activity corresponds to approximately five μg of material.

The experiments just summarized suggest that passive avoidance can be transferred to naive animals by means of brain extracts. Reproducibility of the experiments depends on the following conditions: (1) length of training of the donors; (2) interval between last test and removal of the brain; (3) amount of brain material injected; and (4) conditions of testing of the recipients. On this last point, screening of the recipients to eliminate those with weak dark preference and low activity may be critical. It is recommended to test the recipients three to five times each day and use the mean value.

The main obstacle to a wider agreement on the possibility of transferring acquired information by chemical means has been the lack of a conceptual framework. Superficially, chemical transfer seems irreconcilable with the generally accepted idea that the nervous system operates by impulses travelling along specific anatomical pathways. We have, however, proposed a hypothesis in which the molecular mechanism intervenes in labelling these pathways and establishing the new connections necessary for the learning process. A peptide code can readily account for marking the number of neural pathways present in the most complex central nervous system. One can assume that intensive training results in increased synthesis of the code molecules involved in the behavior pattern being learned by the animal. When an extract of brain taken from such a trained animal is administered to a naive recipient the code molecules injected attach themselves to the neurons of the same pathways in which they were located in the donor's brain. This can establish in the recipient the neural connections corresponding to the behavior learned by the donor. This tentative explanation is part of a broader hypothesis on the molecular mechanism of learning presented in detail elsewhere (Ungar, 1968).

REFERENCES

Gay, R. and Raphelson, A. (1967). "Transfer of learning" by injection of brain RNA: a replication. *Psychon. Sci.*, **8**, 369.

Ungar, G. (1965). Transfer of morphine tolerance by material extracted from brain. *Fed. Proc.*, **24**, 548.

Ungar, G. (1966). Chemical transfer of learning, its stimulus specificity. *Fed. Proc.*, **25**, 207.

Ungar, G. (1967a). Transfer of learned behavior by brain extracts. *J. Biol. Psychol.*, **9**, 12.

Ungar, G. (1967b). Chemical transfer of acquired information. *CINP Symposium*, Washington, 1966. In *Excerpta Medica*, ICS **129**, 169.

Ungar, G. (1968). Molecular mechanisms in learning. *Persp. Biol. Med.*, **11**, 217.

Ungar, G. and Cohen, M. (1966). Induction of morphine tolerance by material extracted from brain of tolerant animals. *Int. J. Neuropharmacol.*, **5**, 183.

Ungar, G. and Irwin, L. N. (1967). Transfer of acquired information by brain extracts. *Nature*, **214**, 453.

Ungar, G. and Oceguera-Navarro, C. (1965). Transfer of habituation by material extracted from brain. *Nature*, **207**, 301.

SOME APPLICATIONS OF "MEMORY TRANSFER" IN THE STUDY OF LEARNING

S. Reiniš and D. R. Mobbs*

Departments of Physiology and Biochemistry
Ghana Medical School
Accra, Ghana

One of the ways to study memory mechanisms is to look at the effects of brain extracts from trained animals. In this paper we present a working hypothesis as to the role of active extracts from trained brains.

The first group of experiments used rats in which the alimentary conditioned reflex was formed. Food was placed behind a semitransparent door whose opening was registered. The conditioned responses of the animals were recorded, *i.e.*, the opening of the door and the taking of food using light as the conditioning stimulus. The intertrial reactions were also recorded, *i.e.*, when the rats opened the door and tried to find the food in the period between conditioning stimuli. At the end of training the rats were sacrificed and their brains homogenized and injected into rat recipients intraperitoneally. The recipient rats were then starved for 48 hours and subsequently tested in the same experimental situation.

The number of conditioned reactions was significantly increased in animals injected with homogenates from trained brains (Table I). This increase of positive responses was not found in rats injected with homogenates from experimentally "naive" brain or liver. The increased activity was only found in animals injected with trained brain homogenates when the animals were starved for 48 hours in the period between injection and the first testing

*D. R. Mobbs is presently at the Department of Biochemistry, University of Zambia, Lusaka, Zambia.

TABLE I

NUMBER OF CORRECT CONDITIONED RESPONSES IN RATS AFTER THE INJECTION OF BRAIN HOMOGENATES

Experimental group	Percentage of correct responses
Animals starved for 48 hours:	
Control rats	14.7
Injection of trained brain	63.3
Injection of untrained brain	17.1
Animals starved for 16 hours:	
Control rats	19.0
Injection of trained brain	17.1
Injection of untrained brain	11.4

session. The rats which were fed after injection and subsequently starved for 16 hours before the first testing session showed no such increase of activity. Injection of homogenates from "naive" brains produced a lower number of conditioned responses than the control rats when they were fed after injection.

There was a significant increase (Table II) in the number of intertrial reactions of rats injected with brain homogenates, such an increase being found with rats injected with homogenates from either trained or "naive" brains. If the recipient rats, however, were starved for 16 hours after an initial feeding, then the number of intertrial reactions was lower.

This showed that trained brain homogenate injections increased not only

TABLE II

NUMBER OF INTERTRIAL REACTIONS PER MINUTE OF INTERTRIAL TIME IN RATS AFTER THE INJECTION OF BRAIN HOMOGENATES

Experimental group	Number
Animals starved for 48 hours:	
Control rats	0.78
Injection of trained brain	2.12
Injection of untrained brain	2.27
Animals starved for 16 hours:	
Control rats	1.72
Injection of trained brain	0.56
Injection of untrained brain	0.94

the number of conditioned responses of "naive" recipients in the first session, but also increases the number of intertrial reactions. This suggests that the animals' activity increases both specifically with regard to the conditioning stimulus and nonspecifically in the intertrial period. This is confirmed by the increase in the general activity but not of the conditioning stimulus reaction in animals injected with untrained brains.

An attempt was made to affect the instinctive behavior of animals: Mouse killing by rats served as the model. About 30% of rats consistently kill mice, the rest never kill them. This behavior pattern is inborn and instinctive. An attempt was made to transfer this instinctive killing by means of injection of brain homogenates from killer rats into nonkiller rats. The conditions of transfer were the same as the previous experiment. Homogenates of whole brain in saline were injected intraperitoneally.

When either brain homogenates from individual killer rats were injected into nonkiller rats, or homogenates from nonkiller rats injected into killer rats, no change of behavior was observed after 48 hour starvation period. Transference of instinctive behavior was not achieved under conditions similar to successful learned transference.

There is probably an analogy between instinctive and trained behavior. The instinctive behavior is coded in the DNA of the nerve cells and the production of the corresponding proteins change the activity of the nerve cells. This protein production is controlled by derepressors which are independent of the normal functional activity of the cell. Formation of environmentally-dependent permanent memory trace requires a mechanism which reacts on the activity of the nerve cell so as to induce the formation of new RNA. The resultant new protein synthesized from the RNA template alters the functional activity of the cell.

The difference between instinctive and learned behavior patterns is in the derepressing mechanism, although it may also be partly due to differences in the section of the chromosomal DNA being derepressed.

Comparison of this model with experimental data has led to the proposal of the following hypothesis as to the role of "memory transfer" substances.

When the nerve cell is repeatedly stimulated, then the synapses, postsynaptic membranes, lysosomes, or some other part of the cell produce derepressors. These derepressors influence the corresponding parts of the chromosomal DNA and lead to the formation of the corresponding messenger RNA. The protein formed from the new RNA by the ribosomes increases the synaptic transmission, probably by fixation on the postsynaptic membrane or by some other change of synaptic transmission. This facilitated synaptic transmission would then cause a more frequent firing of the cell with subsequently more extensive formation of derepressor.

The derepressor may not only influence the formation of new RNA in the nerve cell, but might also diffuse out into the surrounding glial cells, where it would derepress the chromosomal DNA present there.

The derepressor cannot be firmly bound to any of the cellular structures as it is released during extraction procedures and transferred into the brain of another animal where it induces similar processes to those in the brain of the donor.

The agent of memory transfer is not, therefore, the memory trace substance itself, but rather the factor which induces the formation of the memory trace. This agent could be the low molecular weight polypeptide isolated by Ungar and Rosenblatt. Such an agent would be specific for a given region of the brain and also for the complexity of the given task.

A similar mechanism can be derived by analogy with recent hypothesis of antibody formation. This supposes that for certain genes there is a randomizing element which enables the particular gene to produce a potentially large number of different messenger RNA molecules. Stimulation by antigen or other activating substance causes a "fixing" of the DNA so that in the future the messenger RNA produced is the template for a particular protein. We can consider learned behavior as a case whereby the stimulating agent causes a fixing of the random DNA so as to produce the messenger RNA corresponding to the memory trace protein. If extracts from the brain are injected into another animal this same stimulating agent will induce the transformation of the random DNA of new nerve cells to the same fixed DNA, thus "transferring memory." In instinctive behavior the DNA is fixed from birth so there is no stimulating agent in this case to be transferred.

This hypothesis supposes that memory transfer need not be too specific. The formation of a stable memory trace depends on the fixation of a new protein which may be different in different synapses or postsynaptic membranes. This hypothesis is only one of several possible models illustrating the role of memory transfer substances, but it explains several features of "memory transference." The hypothesis may explain both the generalized and specific effects of extracts, the blocking effects of some extracts on learning, and unlike previous hypotheses explains the difference between the transfer of trained and inborn behavior.

A second possible explanation of unsuccessful transference of instinctive behavior is simply that the quantity of active substances is very small. This suggests that the highest quantity of active factors is present during training, and decreases after training has finished. The following experiment was performed to test this possibility.

Mice were trained to open a transparent door placed in front of food. The conditioning stimulus accompanying the food was a light placed above the door. In each session, ten trials were carried out. The opening of the transparent door was registered electronically. The number of the conditioned reactions and the number of intertrial reactions were recorded. The brains of mice-donors were homogenized with distilled water, left in refrigerator overnight and centrifuged. Clear supernatants were injected intraperitoneally into the mice recipients. The donors were divided into three groups: the first group consisted of experimentally naive, untrained mice. The second group was of trained mice who reached ten conditioned responses in one session, and the number of intertrial reactions was very high—between 60 and 80 in one session. The number of training sessions was 12–14. The third group of donors, killed after 22–26 sessions, had ten conditioned responses in one session and less than ten intertrial reactions.

Injections of brain extracts from incompletely trained mice of the second group produced no subsequent increase in the activity of the recipient animals (Table III), whilst injections of brain extracts from completely trained mice of the third group produced significant increases in activity.

TABLE III

NUMBER OF CORRECT CONDITIONED RESPONSES IN MICE AFTER THE INJECTION OF BRAIN EXTRACT FROM INCOMPLETELY AND COMPLETELY TRAINED MICE

Donors	Average number of positive responses
Untrained	2.70
Uncompletely trained	2.50
Completely trained	6.25

A fourth group of donor mice was fully trained to the criterion (in 22–26 sessions), then left fifteen days without any training. After fifteen days they were once again tested in the conditioning chamber and killed. The brain extracts were then injected again to the naive mice-recipients. The donors after two weeks of rest showed almost complete retention (average number of conditioned responses was 9.53), but their brain extracts were fully ineffective, the performance of recipients in the first session being 2.12. This experiment therefore shows that "memory transfer" is a phenomenon connected with the process of training, not with the ready memory trace, and supports our derepressor hypothesis.

INDUCTION OF DISCRIMINATORY BEHAVIOR BY MEANS OF BRAIN EXTRACTS*

Frank Rosenblatt

Section of Neurobiology and Behavior
Cornell University
Ithaca, New York

INTRODUCTION AND REVIEW OF PREVIOUS WORK

The progress of science is necessarily cautious, conservative, and skeptical. Reports of basic phenomena which appear to be at odds with basic expectations require far more evidence, more replication, and closer scrutiny to be accepted, than is the case with results which merely confirm or extend existing knowledge. In the new and puzzling field of "transfer experiments," with which this report is concerned, this necessity for extra-careful experimental design and analysis has been amplified by the frequently conflicting character of results from different laboratories, or even from the same laboratory at different times. It has sometimes required a considerable measure of faith not to shrug off all results as a freakish whimsicality of nature, to be attributed to undiscovered sources of bias which somehow vary in the most confusing fashion from experiment to experiment. The fact that the observed "biases" occur so frequently and strongly has led us on through two years of intensive experimentation in an attempt to make them occur regularly and controllably. In our most recent experiments which are reported here, we feel that we have begun to accomplish this. Undoubtedly, however, it will require many additional repetitions, many variants of the

*The work described in this report has been supported by NIH Research Grant MH 13138–01 and ONR Contract NONR 401(59).

experimental design, and above all an acceptable integration of these phenomena with the rest of biology, before they will be generally accepted. These results are presented, therefore, not as proof but as evidence that a "transfer phenomenon" exists, that specific kinds of behavior may be induced by brain extracts which depend for their effect on the training or experience of donor animals, and that the effects obtained depend in a complicated way upon the dosage of extract used over an extremely large dynamic range.

It is only recently that experimental evidence suggesting a biochemical basis for memory and learning has begun to accumulate. Most of this evidence falls into four main classes:

1. Experiments on changes in brain RNA due to training or experience;
2. Experiments on the relation of protein synthesis to retention of learning;
3. Invertebrate transfer experiments;
4. Vertebrate transfer experiments.

Recent reviews of this subject have now appeared in books by Gaito (1966), John (1967), and Grossman (1967). The most influential experiments have been those of Hydén and Egyházi (1962) on RNA changes following learning experiences in rats, and the experiments of Flexner *et al.* (1963) and Agranoff (1964) on the abolition of memory for a learned task due to the administration of drugs which interfere with protein synthesis. Inspired largely by the work of Hydén, the earliest transfer experiments were undertaken with planarians in an attempt to determine whether the most rudimentary types of "learning" in primitive organisms might be transferable by chemical means. Some of the best known of these experiments are those of Corning and John (1961) and McConnell (1962), who demonstrated that the retention of conditioned responses in regenerated tail sections of severed planaria could be abolished by RNase, and that "transfer" of simple conditioned responses might be transferred by cannibalism of trained worms. Subsequent attempts at replication and clarification of these results, however, left considerable doubt as to the specificity of the information transferred, and seemed to indicate that a general facilitation effect, enhancing the responsiveness of the recipients, might be all that had occurred (Hartree *et al.*, 1964).

The first vertebrate transfer experiments were carried out independently at four different laboratories. In the three studies by Fjerdingstad *et al.* (1965), Reiniš (1965), and Babich *et al.* (1965), the working hypothesis was that RNA might somehow directly record the experience of the animal, and that an extract containing RNA might therefore transfer the learned behavior to a recipient. In the fourth study, Georges Ungar (1965) assumed by analogy to previous work on transfer of morphine tolerance, that a polypeptide might be capable of transferring behavioral habituation effects.

Ungar's extracts, unlike the others, were therefore designed to obtain a protein factor from the brain of the donor animals.

It was characteristic of all the initial experiments that the behavioral measures employed were sensitive chiefly to rate or speed of performance, or speed of learning of a given task. In no case was a discrimination measure used to determine whether the facilitation induced by the brain extracts was specific to a particular learned task or was a general enhancement of activity, curiosity, sensitivity, or learning ability. The previous demonstration that the planarian transfer effect might be a general facilitation of the latter sort made it imperative that the specificity of the "transfer phenomenon" be checked as unequivocally as possible. Our program at Cornell began with an attempt to replicate the Babich–Jacobson design (Rosenblatt *et al.*, 1966), but following a partial success in early experiments, shifted emphasis to devising a better task for the assay of specific information transfer (Rosenblatt *et al.*, 1966; Rosenblatt and Miller, 1966). Differences between rats injected with extracts from experimental and control donors were found in a large number of test situations. The tasks included the following: bar-pressing in a Skinner box for a food reward, shock-avoidance in a shuttle box, selecting a bar with a light or dark bulb over it, learning to open the right or left door of a pair for a food reward, working a right or left bar, going to a right or left food cup, or making right or left turns in a Y-maze. Recently, differences have also been obtained with a visual discrimination of white vs. black cards in a Lashley jumping stand. Of all these tasks, the right-left discrimination involving a choice of one of two bars in a modified Skinner box has so far seemed most suitable for a systematic assay procedure involving a discriminatory response. This procedure is readily automated; involves no handling of the rats during an experiment; is efficient for the testing of large numbers of animals; and is relatively bias-free, since the right and left choice can be made perfectly symmetric by a careful balancing of levers and maintaining strict identity between the two sides of the test chamber.

As data on discrimination experiments accumulated, several new phenomena began to appear. The most puzzling of these was the "inversion phenomenon." Recipient rats, instead of selecting the response for which the corresponding donors were trained, seemed to do just the opposite (*e.g.*, making a left turn or pushing a left bar if the donors were trained to make a right turn or push a right bar, or vice versa). Similar effects had been noted by Nissen *et al.* (1965), but were initially met with considerable skepticism. A preliminary interpretation suggested that at low doses a vague or incomplete transfer of the donors' learning might not be sufficient to induce the rats to mimic the behavior of the donors. This incomplete transfer, however, might be sufficient to induce a satiation effect for the donors' task, causing them to explore alternatives which would be novel and unsatiated for the donors. Subsequent work, however, has demonstrated that there may be a considerable number of maxima and minima in the dose response curve. The above interpretation has accordingly been replaced by the idea that different active components, differing in their behavioral effects, may come into play with different dosage thresholds. As doses are increased, one therefore observes a complicated spectral profile with a succession of different effects produced as each component comes into play. These results have led us to prefer the term "behavioral induction" in place of the previous term "behavioral transfer." It is apparent that there may be no simple correspondence between the behavior of a recipient animal and that of the donor, even though the induction effects are specific for the type of training received by the donor.

While the work described above was in progress, many other laboratories were attempting parallel perlication programs. These studies were based in most cases on the Babich–Jacobson design which seemed to produce the most dramatic results and had received the greatest amount of public attention. Most of the earlier results were disappointing, and the difficulty of replication rapidly became evident. Many experimenters concluded that the original claims were in error, and abandoned further work. Most notable among the negative reports was a statement by 23 signatories from eight different institutions, summarizing their failures to achieve transfer (Byrne *et al.*, 1966). In each of these cases, however, RNA extracts were employed following the prescriptions in earlier reports, and later studies by some of the same investigators began to show positive results when the method of extraction and testing was modified (Byrne and Samuel, 1966). The demonstration of inversion effects also offered a possible explanation for some of the replication failures, since there are points on the dose response curve, between the various maxima and minima, where all effects appear to vanish, and the expected difference between experimental and control groups is zero. A more difficult problem, however, is the extreme variability of individual rats, some of whom may show positive effects at the same dose that induces negative effects in other individuals. We have demonstrated that rats which differ in preinjection activity measures may behave differently when injected with the same extract; at present we suspect that these differences are simply a difference in sensitivity, but the effect may be to shift the dose-response curve so that a maximum for one subgroup becomes a minimum for another. Precise control over the activity level, the degree of starvation, age or weight of rats, and cage conditions seems to be essential. There is also evidence that strain differences may be important. The following sections will report on the substantial amount of work that has been done on the effects of concomitant variables.

At the present time, we believe that despite the complexities of interpretation, we have obtained a usable (though cumbersome) assay procedure, and the emphasis in our work has shifted towards the purification of the active component in our extracts. A fractional precipitation procedure has been developed which provides a first step in purification, and further separation on the basis of molecular weight is now being attempted. Preliminary results of these studies, as well as details of the procedures, will be found in the following sections.

GENERAL METHODS AND PROCEDURES

Choice and Maintenance of Animals

With the exception of one pilot experiment performed with cats, all experiments in our laboratory have been done with albino rats, generally of the Holtzman strain. Comparisons with rats obtained from other suppliers have not been systematic enough to permit a definitive statement as to relative advantages or disadvantages of different strains. Of three stocks of rats which we have tried at different times, the Holtzman has been best suited to our purposes. It is docile, submits readily to handling, and has a high level of spontaneous exploratory activity in each type of apparatus which we have used. It learns well, with little or no shaping in the positive reinforcement tasks which were used in the great majority of our experiments to date. However, recent experiments have begun to reexamine the use of negative reinforcement. This has the advantage that the rats need not be starved, and consequently remain in better and more uniform condition throughout the experiment. There are some indications, however, that the Holtzman strain may not be an optimum choice for avoidance tasks. A different strain will be tested shortly.

Rats are delivered to our laboratory in weekly shipments, and are placed in large stock cages (each holding 12 to 20 rats) for at least a few days before use in an experiment. Our rats are shipped at an age of 60 days, and are generally used within ten days of arrival. Older rats (up to 90 days) have occasionally been used as donors, but younger rats appear to be preferable as recipients. The older rats show a marked reduction of activity levels, learn more slowly, and yield poorer quality data, often giving no responses in the course of a typical test session. Still younger rats might be used, but (as with rats from newly arrived shipments) their resistance to anesthetics and to arduous experiments is less, and it is more difficult to establish a satisfactory weight level for them in a starvation regime. At the start of an experiment our females typically weigh 180 to 200 gm and males are about 50 gm heavier.

Environmental conditions with respect to light and humidity have not been closely regulated, but temperature is controlled at about 75°F. The stock cages and cages for experimental rats are kept in the same room or in adjoining rooms. The test equipment is located within a few feet of the experimental cages, minimizing transportation and handling of rats during transfer from home cage to test apparatus. The diet, consisting of standard grain pellets (Big Red Laboratory Animal Food), is available in the stock cages at all times. In some of our earlier experiments terramycin was kept in the drinking water to combat enteritis, but this has proven unnecessary in the present environment and has been discontinued. There have been no outbreaks of either respiratory or gastro-intestinal problems in our colony during the last year.

The rats used for an experiment are placed in individual wire cages, seven × seven × ten in., and the cages are stacked six to a row in large wagons. Water is available at all times, but food is withheld until their weight is reduced to the desired level. The rats are weighed when they are first caged, and preconditioning procedures are not started until the following day. A card on each cage is used to record the weight and food given to each rat. On subsequent days, weighing is done late in the day after the rats have been run. On the day of injections, however, rats are weighed while still under anesthesia, and this weight is used for computations of weight loss at time of injection. We have aimed for a reduction of 20% from the starting weight (unless otherwise noted for particular experiments). For overweight rats (over 200 gm initial weight for females) this is generally increased to about 25%. It seems undesirable to let female rats of this age group fall much below 150 gm, or for males to fall below 180 gm, as the incidence of deaths from anesthesia or experimental procedures will markedly increase. With female rats over 150 gm, deaths in the course of an experiment can generally be held to one or two percent or less, most of these occurring during anesthesia or surgery. When a rat has reached its desired weight (which typically takes three or four days) it is fed one large pellet (about five gm) for each five gm or fraction thereof that it falls below the weight limit. This is given after the rat has completed all runs for the day. Food used for reinforcement in our experimental apparatus consists of 45 mg Noyes grain pellets. Rats may obtain anywhere from zero to over a hundred such pellets during a day's work.

In several of our earlier experiments we have attempted to maintain two rats to a cage to conserve space, or to equate cage factors between experimental and control rats. Under these circumstances, it is essential that the rats be separated during feeding; otherwise, a dominance relationship generally develops in each cage, with one rat acquiring all of the food and the other starving. This necessitates an awkward and time-consuming procedure for feeding large numbers of rats, and individual caging has been the rule in all of our recent experiments.

It is important to assign rats to experimental groups according to a system which is independent of the order in which they are removed from stock cages. There are generally significant activity differences between the first rats to be withdrawn and the last rats in the cage, reflecting differences in timidity, ability to escape, curiosity, or other such factors. Differences due to a "capture bias" may be large enough to substantially affect the results of an experiment. Assignment of rats to experimental groups has been done randomly in a number of experiments (using numbered poker chips to assign rats to different treatment groups), but a more controlled procedure is now used which guarantees that each test box (of which there are six) will receive, if possible, an equal number of rats in each treatment group. This leads naturally to a factorial design which can be used for an analysis of variance if desired. In spite of these precautions, we occasionally find significant differences between preinjection characteristics of experimental and control groups. These may include differences in weight, activity, or pretest bias. When these differences appear to be likely to affect the outcome of an experiment, the only alternatives are to control them statistically (by selecting matched subgroups of rats for analysis or by regression techniques) or to repeat the experiment. Each of these methods has been tried, with no important changes in qualitative conclusions resulting in any case. Consequently, it seems unlikely that there is a selection or assignment bias running systematically through our experiments.

When first caged for an experiment, there is appreciable variability among the rats with respect to levels of activity, nervousness, and response to handling; but the preconditioning procedures described in the following chapter appear to produce a much more uniform group within three to four days. Signs of nervousness tend to disappear, and activity levels (measured by the number of traverses between the operant bar and food cup in our preconditioning boxes) are generally quite uniform after five days of preconditioning. A sixth day, included in the design of several experiments, produces still better uniformity. The sixth day is used in most of our experiments for a pretesting procedure which involves a box change, and again tends to bring out individual differences between the rats. We have not experimented with the more elaborate preconditioning procedures employed by Byrne and Bennett, or by Wolthuis (personal communications), but there seems to be general agreement among laboratories using such procedures, that they improve the uniformity of the rats and consequently reduce the variance due to individual factors.

A single experiment has been run with male littermates, in a balanced factorial design, to ascertain the extent of genetically-dependent variability relative to other individual factors. Forty-eight litters of four rats each were used. Unfortunately, the rats were almost a month older than usual at the time of the experiment. This may have affected the results, and data rates were rather low, as is typical for older rats. An analysis of variance showed no significant differences between litters in bar-pressing activity on either the first or third day of preconditioning, after controlling for box differences. Differences among the six training boxes, however, were significant.

A comparison of 90 males and 90 females of about the same age (the females were a week older) was carried out in one experiment. The activity rate of the females (number of bar presses in the third preconditioning session) was 16.3 ± 1.8, and of the males was 12.3 ± 1.8. The measures used in our tests (percentage of left responses) are designed to be as free of activity effects as possible. It may be, however, that this difference in activity level could have an appreciable influence on the outcome of experiments with male or female rats due to the correlation which has appeared between rate of activity and sensitivity to the extracts. Since somewhat clearer results are obtained from the less active rats in a given population, it appears that the males, with the lower activity rates, are to be preferred over the females. In recent experiments we have begun using male rats exclusively to avoid possible complications from the oestrus cycle.

Preparation and Storage of Extracts

In all our experiments, rats are killed without anesthetization by decapitation. Brains are generally removed by a team of three workers. One kills the rats, and two others remove the brains. The whole brain is used (unless otherwise noted), including the olfactory lobes when these can be removed easily, and all of the brain stem severed at the upper end of the spinal cord. An average wet weight of about 1.6 gm per rat brain is obtained. Brains are removed within 30 seconds and placed in a chilled homogenizer tube. In earlier experiments, homogenization was done with a Potter–Elvehjem homogenizer, followed in some cases by sonification. In all of the experiments done during the last year, however, we used a Sorvall omnimixer, with a speed setting of four for one minute. Brains are homogenized in groups of 30 or less at one time. This method appears to give much more reproducible results and permits the mixture to be refrigerated more effectively throughout homogenization. While we have no evidence to suggest that extreme speed or rapid chilling of the brains is essential in this phase of the preparation, we still attempted to reduce the material to 0–4°C as quickly as possible, and to perform all further operations up to the time of injection at that temperature.

Four different extraction media have been used in our published experiments: phenol, Triton X-100, TCA, and water. The phenol method was used where we attempted to replicate procedures from other laboratories. Triton has been used in experiments in which we hoped to solubilize as much as possible of the brain protein, but it is extremely difficult to remove subsequently, and can be quite toxic when injected intravenously. The TCA procedure was used in an attempt to separate TCA-soluble substances (which we believed might include our active substance) from the bulk of TCA-insoluble proteins. After several attempts, however, it was decided that this procedure had no advantages over the milder and more readily reproducible water extracts. It has been our experience that any reagent which tends to form a precipitate in the early stages of extraction should be avoided, since it is likely to trap activity which would otherwise appear in the soluble phase. When precipitating agents are added later (as in our acetone fractionation procedure) it is imperative that they be added slowly and under closely controlled conditions, to avoid precipitating active material. Extraction with glass distilled water (three times the wet weight of the brain) is the method presently preferred over the addition of any reagents. Comparisons with the Triton procedure have shown that three successive water extractions of the homogenate leave very little additional detergent-soluble protein, provided sufficient time is allowed for each successive extraction. In our procedure, the initial homogenate is shaken for one hour (in the cold) on a Burrell wrist action shaker. It is then centrifuged for one hour using a Spinco L-2 preparative ultracentrifuge with either a Type 21 or Type 19 rotor and a speed setting of 15,000 rpm, at 0–4°C. The supernatant is stored, and the pellet rehomogenized in another three weights (3W) of water and shaken for two hours. This is again centrifuged for one hour, the pellet rehomogenized in the same amount of water, and shaken 12 to 15 hours. It is then centrifuged a third time and the three supernatants are pooled, giving a final volume of about 15 ml per brain.

A sucrose extract has been used as the initial step in subcellular fractionation experiments, and homogenization in this case is done with the Potter–Elvehjem homogenizer in order not to disrupt subcellular constituents. Where this procedure is used as an initial step, a modified water extraction of the fraction of interest should then be carried out, but in the one experiment which we have so far completed, the various subcellular fractions were further extracted with Triton.

The use of saline for initial extraction is to be discouraged, since it may lead to very high concentrations of salt in the later stages of the procedure and in the final extract. This can not only lead to osmotic shock effects on injection, but makes the precipitation of the extract uncertain and hard to control. We considered it advisable, however, to control the pH of our media, by buffering the aqueous supernatant at pH7.4, with 0.03*M* sodium phosphate buffer. When large dosages are used, the residual phosphate can present a problem, as it leads to a high ionic strength and may induce shock reactions when injected intravenously. In the dosages used in most of our recent experiments, however, the residual phosphate has been very slight. Our procedure has been to measure its final concentration in the extract, adjusting all doses to contain the same total phosphate, and adjusting the final ionic strength to an approximately isotonic level.

Having obtained an extract which presumably contains most of the desired activity in soluble form, the next step of the procedure in each case is to precipitate the active material by the addition of suitable solvents. Organic solvents, particularly ethanol and acetone, have proven most successful for this, although TCA does seem to yield an active precipitate as well. In experiments where Triton or phenol is used as the extractant, ethanol was used for precipitation, usually washing the precipitate several times with ethanol to remove as much residual Triton as possible. The ethanol precipitates obtained in this way are hard to resolubilize and are generally less tractable than the acetone precipitates. The present procedure is to add acetone very slowly (dripping it into the solution at a rate of 10–15 ml/min. with constant stirring), and to allow about 24 hours for precipitation, at 0–4°C. The precipitate is then collected by centrifuging for 30 minutes in a 19 or 21 rotor at 15,000 rpm. The pellet is then homogenized in 1.5 volumes of H_2O (based on original wet weight of brain), and lyophilized to dryness to remove any residual acetone. The resulting acetone fractions which come down at low concentrations of acetone (up to 50%) are difficult to resolubilize, and are extracted by rehomogenizing in 1.5 volumes of water, shaking 24 hours, and recentrifuging for one hour, under the same conditions as above. This extraction of the pellet is repeated three times, the three supernatants combined, and lyophilized to the final injection volume.

Fortunately, the acetone precipitates which come down at concentrations over 50% acetone are those which seem to contain most of the activity. These fractions are readily resoluble, and do not require the prolonged re-extraction procedure just described. In all cases, the final solution is centrifuged at 3,000 rpm for five minutes to remove suspended material.

Acetone fractions which have been studied include a fraction from 0–10%, 10–40, 40–60, 60–90, 0–50, 50–75, and 75–95% acetone. The bulk of material is precipitated in the early fractions (under 50% acetone), leaving only a small portion in the 50–75% fraction, and trace amounts in the fraction over 75%. A comparison of activity in these fractions will be presented in the following section. Fig. 1 shows the main steps in the present procedure and the distribution of material in the different fractions. The 50–75% fraction contains five to ten mg of protein, and about 0.2 mg of RNA.

The acetone fraction from 60–90% which was obtained from our first few experiments (in which acetone was added relatively rapidly) caused immediate

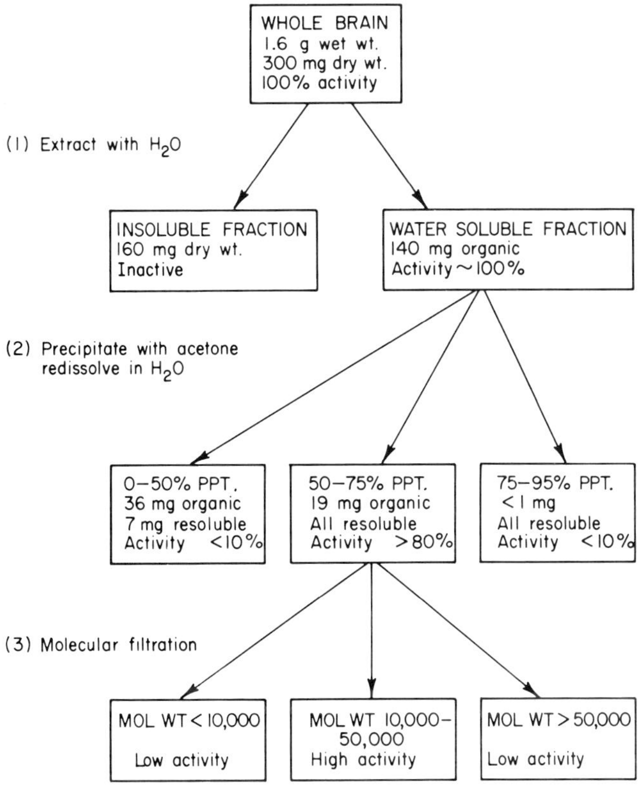

Figure 1 Acetone fractionation procedure.

respiratory arrest when injected intravenously, even at low doses (0.5 brains per rat or less). A rat showing this syndrome could generally be saved by immediately suspending the injection and applying artificial respiration. An adaptation effect appeared to follow, and the remainder of the dose could generally be administered safely. Very slow injection was generally successful in preventing the occurrence of this syndrome. When the slow acetone addition procedure was adopted, this toxic effect was not found in any subsequent extracts. Some of the low acetone precipitates have occasionally shown toxic effects, usually producing death over a period of minutes or hours, rather than the acute effects found with the 60–90% fraction. These effects have not occurred regularly, and may have been due in some cases to partial degradation or bacterial contamination of these fractions. Adverse physiological effects were also noted in a pilot experiment with cats. Intravenous injection produced an acute syndrome resembling a decortication syndrome in a number of animals, while other animals seemed unaffected by the same dosage. The reaction typically lasted only a few minutes in its acute phase, followed by complete recovery within an hour. Similar effects have not been seen in rats, although occasionally mild convulsive effects appear briefly after injection. In all of the most recent experiments where low doses were used, no adverse physiological effects have been noted.

In some of our most recent experiments, an attempt has been made to go on to a further stage in purification of the extract by a fractionation based on molecular weight. Our early experiments with Sephadex chromatography (Rosenblatt *et al.*, 1966) suggested that the active substance should have a molecular weight of about 5,000. An attempt has been made to obtain a first division of the extract by filtering the 50–75% acetone fraction with membrane filters (Carl Schleicher and Schuell Co.) which are supposed to cut at molecular weights of about 50,000 and 10,000. While the filtrate obtained from the 50,000 filter has been highly active, the actual cutting points are probably quite different from those advertised. Present studies have not yet established the actual molecular size range of this filtrate.

Injection Methods

In different experiments we have used intraperitoneal (IP), intracranial, and intravenous (IV) injection techniques. Unless otherwise noted, all of the injections for the experiments described in the following chapters were given intravenously, via the external jugular vein. Early experiments (Rosenblatt and Miller, 1966) suggested that this method gave a more potent effect than IP injections, but a quantitative comparison of the magnitude of the effect would require a full set of dose-response curves obtained with each method in a parallel design; this has not been carried out.

Experiments through C25B used ether as an anesthetic in carrying out the IV injections; beginning with Expt. C25C a 50:50 mixture of O_2 and CO_2 was used. With the latter procedure injections can be carried out faster, and the percentage of rats lost due to anesthetic is virtually zero. The rat is fastened to a board with padded aligator clips held by rubber bands. A wad of cotton moistened with ether is used to maintain anesthesia, after the rat has been lightly anesthetized in a glass jar. After swabbing the area over the right jugular vein with ethanol, an incision is made with surgical scissors, completely exposing the vein for several millimeters, and clearing it of covering membranes and residual fat. The injection is made with the needle entering the vein through the margin of the overlying muscle, which stabilizes the vein and needle. Disposable syringes are used, with a number 26 needle for IV injections. The incision is dusted lightly with sulfathiazole, and closed with a single wound clip. If a vein is damaged during the operation, the opposite vein may be used, but with a practised operator this should rarely happen. Over 95% of the injections made in this manner can be verified to place the entire dose directly in the vein, with at most a few percent of cases in which a portion of the dose is discharged under surrounding membranes due to slippage or improper positioning of the needle. Bleeding is normally negligible, and particularly with the O_2–CO_2 anesthetic rats recover almost immediately and could easily be tested within an hour after the injection.

We have speculated about the possibility that a slow and sustained release of material to the rat's system might produce a better effect than the sudden massive dose of the intravenous method. Intramuscular injections and pellet implantations may be tried in future experiments. The IP method seems to be a compromise of sorts, but seems poorly suited for a phenomenon which is as dosage-sensitive as we have found our effects to be. The rate of absorption and transport would vary greatly from one rat to the other even if we were certain of similar placement of the injection. With the IV method, as will be shown in the following chapter, effects can be obtained with minute dosages far below anything that we have attempted with intraperitoneal injections. The intracranial method (Rosenblatt and Miller, 1966) should also be explored further in later experiments. One should probably be cautious, however, in employing a method which might lead directly to injury of the brain in experiments of this sort, since this could greatly complicate the interpretation of results. Intracisternal injections which are probably closely comparable to our intracranial method have been used regularly by Fjerdingstad *et al* (1965) with positive results.

When very large numbers of rats have been injected with ether anesthesia, it has sometimes required many hours (over 15 in several cases) to complete the injections. In the earlier experiments, no special care was taken to maintain the extracts at constant temperature throughout this period, but after our attention was called (by Fjerdingstad) to the possibility of extract deterioration during the injection period, we have kept injection bottles chilled throughout the procedure.

Experiments on the thermal stability of our extract have not yet been carried out, but it appears to be highly stable when stored in the cold. A batch of material stored in dry form (the final stage before making up the solution for injection) has retained its activity for three months without any clear sign of deterioration. Since a round of injections has taken upwards of 15 hours in some cases we have initiated a practice of keeping the extracts in an ice bath during injection. At present, injections of up to 180 rats rarely take over five hours to complete, and unless otherwise noted are done in the afternoon of the day preceding the start of testing. A split-half comparison of the first half of the injected rats against the second half has shown significant differences in some experiments, but it is uncertain how much of the difference is due to extract deterioration, and how much may be due to the fact that the second half of the rats came from a different shipment than the first half (sometimes differing by a week or more in age and duration of time spent in our laboratory after shipment). In Expt. C23A, for example (reported in more detail in the following section), the first 60 rats given a dose of 0.5 brains showed significant effects at a five percent level in three out of six tests, and a probability of 0.01 for the mean performance over the six tests. Rats injected later in the series (which included a total of 300 rats injected over 16 hours) showed biases in the same direction, but no significant test sessions. In Expt. C24A, on the other hand (which employed the same fraction—60% acetone—as C23A), the reverse seemed to be true. The second half of the rats showed significant effects which were not found in the first half. In C23C (40% acetone) a significant inversion effect was found with a dose of one-half brain among the first 90 rats, while the second 90 rats showed a rather strong (though nonsignificant) bias in the positive direction at the same dosage. A similar reversal of bias was found in the same experiment with the ten percent acetone extract at a dosage of two brains, although at 0.5 brain there was merely a diminution of a significant inversion effect in the second half of the rats. In general, C23C showed much more significant results in the first half than the second half of the rats. (P values for the first 90 rats for the six tests were 0.9730, 0.9397, 0.9710, 0.9829, 0.9977, and 0.9977, while with the second 90 rats they were 0.2733, 0.2207, 0.1932, 0.3175, 0.2950, and 0.1616 respectively, with a Mann–Whitney U test.) In several other cases examined (C23B, C24B, and C24C), there were no clear differences in significance between first and second halves of the rats, although only in C23B was there a really high correlation between the dose-response curves obtained for the two halves. Although the interpretation of these results is uncertain (as to whether they reflect changes in the extract, differences in timing from injection to testing, or differences in rat shipments), the discrepancies between halves have been substantial enough to make us much more careful about uniformity of recipient populations in later experiments, and to suggest the refrigeration of extracts during injection as a regular procedure.

A further refinement of injection procedure in all recent experiments has been the control of dosage by concentration rather than volume. In earlier experiments volumes ranging from a few tenths of a ml to two ml were used, but since we found that large volumes of buffer or distilled water could produce modifications in the left/right bias of the rats, it seemed most desirable to keep the total volume of the injection medium constant. At present, injection volumes of 0.8 ml are standard in all experiments. Dosages are obtained by successive dilutions of the most concentrated dose to be used in a given experiment, with adjustment of phosphate levels if the amount present is appreciable. We suspect that the effects of large volume injections, particularly with distilled water as a medium, are due to osmotic shock. This makes the rats less responsive to cues or strange objects which they otherwise tend to investigate, or leads for some other reason to an abolition (and sometimes a reversal) of behavioral biases which have been measured on pretests. This drastic reduction of pre- and postinjection correlations is rarely found with actual extracts, particularly when the volume is held down. Nonetheless, the possibility of an artifact due to differences in the injection medium or volume is a serious one where experimental and control extracts are not carefully matched in all particulars.

Standardization of Equipment and Recording of Data

Most of the experiments to be described in this report have made use of modified commercial Skinner boxes produced by Scientific Prototype, Inc. (Fig. 2). Preconditioning boxes have a single lever at one end and a foodcup in the center of the opposite end, with a floor treadle to record when the rat has traversed the box from the lever to the foodcup to obtain a pellet. Automatic counters record the number of times such a cycle is completed. Multiple bar-presses are counted as a single cycle, until the rat returns to the foodcup. The boxes are installed in slide drawers in standard soundproof chambers. Training of donors and testing of recipients is carried out in two-bar boxes similar in design, except that the rat has a choice of a left or right lever in the wall opposite to the foodcup. A single response (left or right) is recorded automatically for each cycle between the foodcup and the levers; in case of multiple responses, only the first response of the cycle is counted. This reduces the weight of bursts of responses by counting only initial choices, and prevents the rat from obtaining more than one pellet for each trip from bars to foodcup. If a donor rat pushes the wrong bar during training, the response is recorded, but the rat is not rewarded. Recipients receive a pellet for pushing either bar, to prevent the systematic training of a side bias, and to keep them from extinguishing their bar-pressing behavior after injection.

Recording of data is done by automatic counters which register the number of times a rat has pushed each bar in the course of a session. Training sessions for donors and preconditioning sessions for recipients (in which they are trained to bar-press in the one-bar boxes) last 15 minutes; test sessions, including pretests prior to injection, last five minutes. Six boxes are available, the one-bar and two-bar boxes being plugged into the same soundproof cases. The rat will always be run in the same box through all tests, and the box number is recorded. Although an effort has been made to keep all adjustments (lever weight, timing, etc.) as nearly identical as

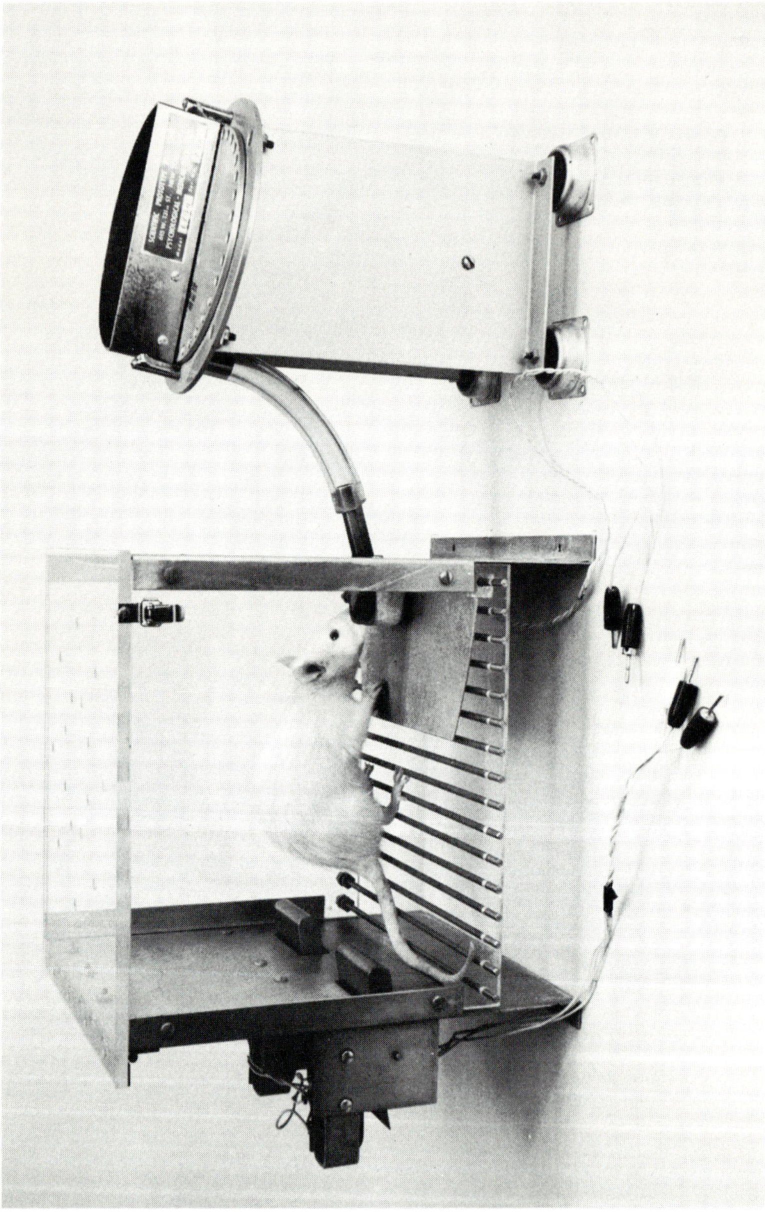

Figure 2 Photograph of 2-bar box.

possible between boxes, systematic differences do occur, as can be seen through an analysis of variance which takes account of box variation in a factorial design. Our present practice is, therefore, to compute separate means for each box in balanced designs, and to use these to correct our data for box-biases, which sometimes favor the pressing of the left or the right bar.

We have found evidence that these box biases can be affected by the condition of the rats. A breakdown and occasional reversal of an otherwise systematic bias occurs in rats of some of our control groups who have been injected with large volumes of distilled water. It is important to be aware of this phenomenon as a possible artifact influencing the interpretation of our experiments. It can be ruled out only by a careful comparison of left- and right-injected groups, who receive identically prepared extracts, administered in the same volume of solvent. We believe the explanation of these odd effects may lie in the relative distractability and curiosity of sick and healthy rats, the former being less curious about such minor details as screw-holes, stains on the box walls, or wires visible outside the chamber than the latter. It is noteworthy in this connection that Byrne *et al.* (1966) have reported strong effects from donors trained on the lever closest to the side from which their boxes are opened, while no significant effects are obtained from donors trained on the opposite side, in boxes not radically different from ours except for this lack of symmetry. We suspect that such differences in potency may result from the fact that one group of donors is being trained to enhance and stabilize its natural bias, while the other group must learn to suppress it, making use of different learning mechanisms in what otherwise appears to be an equivalent task.

An effort is made to run all tests at fixed times of the day, with one session beginning about 9:00 A.M., and the second session five to six hours later. Rats are not fed in their home cages until the second session of the day is completed. Correlations are uniformly high between morning and afternoon sessions.

All data are now recorded on standard forms, which include columns for the rat number, treatment given, initial weight and weight at time of injection, number of responses on third preconditioning session, Skinner box number, and numbers of bar presses (left and right) for each pretest and test session. Additional information such as age or shipment of each rat, observations on sick rats, difficulties in injection, etc., may also be recorded. These data are transferred to IBM punch cards for computation.

Other equipment has been used in various experiments, including Y-mazes, a Skinner-box with two foodcups and a single bar (analogous to the two-bar boxes described above, except that the rat must choose the proper cup to approach from a single operant station on the opposite wall), Lashley jumping stands, and boxes with indicator lights (or patterns of lights such as diagonals slanting in opposite directions) to distinguish the correct from the incorrect bar on each trial. All of these types of apparatus have used positive reinforcement. A shuttle box with a white chamber at one end and a dark chamber at the opposite end has been used in attempts at replicating the design of Gay and Raphelson (1967) where the donors are trained to escape a shock administered in one of the two chambers by running to the opposite end. For cats, a large Y-maze with automatic meat feeders has been used. Experiments with these other types of equipment will be described elsewhere.

Data Processing and Computing Techniques

After completion of an experiment, the data for each rat (as specified in the preceding section) are punched on an IBM card, and statistical processing is done by a digital computer. Earlier experiments were processed by hand calculators, but the use of automatic computation has made it possible to compare a number of alternative techniques and transformations of the data, to carry out routine computations of correlations between test results and concomitant variables, and to make comparisons of alternative statistical measures which would be too burdensome to undertake otherwise. A CDC 1604 computer was used for all of the work presented here.

A number of alternative measures of response have been considered in our experiments. Since these measures can not be discussed completely independently of the experimental design, it will be assumed, unless otherwise indicated, that we are talking about the two-bar box with pretesting of the recipient rats. Where pretest data are not available, alternative measures are used which do not require this information. In the following, we are chiefly interested in measures of discrimination between alternative responses, rather than measures of activity, speed of response, or numbers of responses. While it is exceedingly difficult to obtain a measure of discrimination which is completely free of activity effects by using a situation in which there is a minimum "innate preference" for one or the other of the two alternative responses, and computing the differences between the bias shown by each rat on a pretest and the bias shown after injection (as described below), we believe that we have achieved a measure which is almost completely independent of activity effects. Even so, effects which are undoubtedly correlated with basic activity differences between rats (such as the number of responses during preconditioning trials), appear to influence the sensitivity of the rat to our extracts, and may produce shifts in the dose response curve if this is not controlled.

Four alternative measures of effect were studied most intensively:

1. Percentage of left choices;
2. Shift from pretest score in probits;

3. Shift from pretest in scores transformed to angular measure;
4. Shift from pretest, measured as a percentage of the available range (percentage shift).

The percentage of left choices, normally given by

$$\% L = n_L \,/\, (n_L + n_R) \qquad \text{where } n_L = \text{number of left choices}$$
$$\text{and } n_R = \text{number of right choices}$$

is computed in our programs by the modified formula:

$$\% L = (n_L + 1) \,/\, (n_L + n_R + 2)$$

which corrects for bias effects which would otherwise be present with extremely small numbers of responses. Thus, a rat who makes one left choice and no right choices is scored as 67% left; a rat with one left and two right responses is counted as 40% left, etc. Any rat with equal numbers of left and right responses gets a score of 50% left.

Probit shifts are computed by converting the percentage of left responses on the second pretest and on each test to standard normal deviates, or probits, and subtracting the pretest score from the test score. Several test computations were also done with the shift in percentage of left responses, without any sort of transformation; but this fails to give adequate emphasis to the fact that a shift of 5% from 95% left to 100% left is likely to be more significant than a shift from 50% to 55% left. Theoretical discussions of the use of the probit transformation can be found in Finney (1964) or in Li (1964).

The angular transformation (see Finney and Li, for examples) is commonly used to transform data expressed in percentages in order to obtain uniform variance over the range, a necessary assumption for t tests or analysis of variance, as well as for most regression analyses. The transformation is given by

$$\theta = \arcsin \sqrt{P}$$

where P is the percentage of left responses. Shift scores are then computed in the same way as with the probit transformation, by subtracting θ for the pretest from θ for the test score, for each rat.

The fourth measure, the shift in available range, is proportional to the logarithm of the reduction of the error, if "error" is measured in the direction in which the rat shifts from his pretest score. In the form used in our analyses, the shift is given by

$$S = \frac{P\ (\text{test}) - P\ (\text{pretest})}{\text{Range}} \times 100$$

where P (test) = percentage of left responses on test, for a given rat

P (pretest) = percentage of left responses on pretest, for the same rat

$$\text{Range} = \begin{cases} 100 - \text{P (pretest) if P (test)} > \text{P (pretest)} \\ \text{P (pretest) if P (test)} \leq \text{P (pretest)} \end{cases}$$

For example, if a particular rat gives 60% left responses on the pretest and 80% left reponses on a given test, then the S-score for that test is +50, since the rat has shifted upwards by half of the available range. If the pretest score is 80% and the rat moves *down* to 60% left on a given test, then the S-score for that test is −25, since the rat has shifted down by only a quarter of the available range. Note that unlike the other measures described above, this one is asymmetric, in that a shift towards a more extreme score carries more weight than a shift of the same magnitude towards a less extreme score.

Of these four measures, the last is the one that has been adopted for general use. The probit shift, angular shift, and range shift (S-scores) are all very highly correlated, and yield results which are in close agreement in all experiments in which they have been compared. The simple percentage of left responses seems less reliable, frequently disagreeing with the others in small samples, since it is particularly subject to effects of pretest bias or box bias. The S-score and probit scores seem to be the most sensitive measures, occasionally revealing strongly significant effects in cases where the other measures give only weakly significant effects. The S-score seems to be the best by a very slight margin, and has been selected as much for its relative ease of interpretation as for its statistical advantage.

Since an analysis of variance has revealed an extremely significant dependence of results on differential biases among our six boxes, a further correction has been incorporated in all scores for experiments where the number of rats is sufficient to permit an estimate of the box bias to be made. This is done by computing the mean S-score for all rats (both left- and right-injected) for each test box, and subtracting the box mean from the score for each individual rat run in that box. In all recent experiments, care has been taken to balance the assignment of rats to boxes, so that equal numbers of left- and right-injected rats at each dosage are assigned to each test box. The box corrections, however, are still desirable to reduce the variance due to extraneous sources.

A number of methods for controlling the effects of "concomitant variables" have been tried. One method is to perform a multiple regression analysis of the scores for each treatment group in a given experiment, as a function of background variables which are likely to be correlated. These variables include the number of responses on the third preconditioning session (which provides a measure of preinjection activity and speed of learning in the one-bar task) the initial weight of the rat, percentage of weight loss at time of injection, and percentage of left reponses on the second pretest. A linear regression equation is then used to predict the score that would be obtained by a "standard rat" (*e.g.*, a rat weighing 200 gm initially, 160 gm at time of injection, with an activity measure of 30 responses on the third preconditioning session, and 50% left reponses on pretest 2.) This method was used to compare "standard low-activity rats" (activity scores of five reponses) with "standard high-activity rats" (activity scores of 30 responses). Standard errors of estimate were used to permit an evaluation of the significance of differences thus obtained between left- and right-injected groups. The results strongly suggest a shift in sensitivity to the extracts, with the more active rats being about twice as sensitive as the relatively inactive rats; *i.e.*, a feature of the dose-response curve that would appear at a dose of 0.5 brains for the inactive group would appear at 0.25 brains for the inactive group. A pair of curves illustrating this effect, for the pooled data from several experiments, is shown in Fig. 3, where the phase shift is most evident at the lower dosages.

Although this method of "standard rats" is theoretically attractive, and offers a method of correcting for effects of concomitant variables without throwing out any rats from the sample, it becomes quite hazardous when the points for which predictions are to be made do not fall close to the center

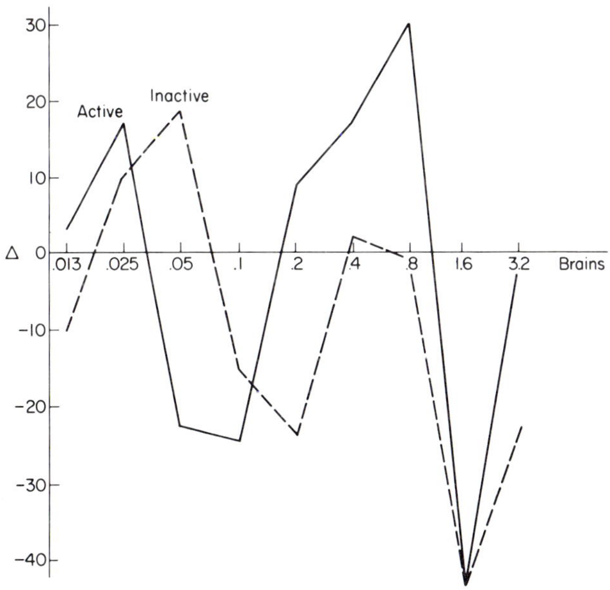

Figure 3 Comparison of active and inactive standard rats (Active = 30 responses, Inactive = five responses in third preconditioning session. Data from means of six tests, on Expt. C24 B and C.)

of the experimental data distribution. The greater the extrapolation, the more hazardous the technique, and the greater the error that might result from departures from linearity. An alternative method has also been used, which eliminates the differences between concomitant variables for the left and right groups at a given dosage by computing scores of "standard rats" situated at the joint means of the two groups. This method may be subject to the same difficulty if the numbers of rats are small in each group, and the differences between group means is large. Consequently, in the data presented for evaluation in the following chapters, linear regression techniques have not been used. Wherever possible, the rats have been sorted according to activity level and results have been tabulated separately for the "inactive" group with 15 or less responses on the third preconditioning session. The active group, although more sensitive, also seems to be more variable, giving lower levels of significance than the inactive rats in many experiments. In the few cases where the numbers of rats tested are too small to permit such a breakdown, the data are presented for pooled rats regardless of activity level.

Similar shifts in sensitivity are obtained as a function of starvation, with more starved rats tending to be more sensitive to the extract than less starved rats. Since starvation has generally been controlled within reasonable limits, and there are generally too few rats to permit a breakdown by both activity and starvation simultaneously, only activity level has been used for tabulation of rats in the data presented in this report, unless otherwise noted. The third main variable which has been studied for correlation effects, percentage of left responses on pretest 2, has high correlations with the percentage of left responses on later test scores, but no systematic correlation with any of the shift measures, where the pretest score has been subtracted out. The S-scores used in this report are particularly free from any correlation with percentage of left responses on the pretest. However, in order to obtain an unbiased starting group of rats, it may be an advantage to delete any rats from the sample with extreme left or right biases on the pretest. This has not been done in any of the analyses given here.

The computer programs used for data analysis have been written to routinely run a number of alternative statistical tests on each experiment. These include t-tests, Mann–Whitney U-tests, Chi-square, and Kruskal–Wallis analysis of variance. The correlation between results obtained by the parametric t-test and the nonparametric U-test (Siegel, 1956) is extremely high, converging to almost perfect agreement when the number of rats is large. The U-test, which is slightly more conservative, is the one used in this report. Little use has been made of the Chi-square analyses, although they tend to support the other tests on the basis of a crude classification of rats whose scores are "above the mean" or "below the mean." The Kruskal–

Wallis test (described in Siegel) is helpful in testing an entire experiment for overall significance, where there may be a large number of different treatment groups at different dosages, and where we are interested in differences between doses as well as differences between left and right injections.

EXPERIMENTS ON LOCATION DISCRIMINATION

Preliminary Methodological Experiments

Previous experiments reported by Rosenblatt and Miller (1966) compared a number of different types of test apparatus and extraction procedures, in an attempt to develop a standard assay technique. It was concluded that a two-bar Skinner box with preconditioning of recipients in a one-bar box was the most satisfactory of the designs studied. Subsequent work has been concerned chiefly with the development of a standard extraction method, and with studies of the dose-response curves obtained with the two-bar box. Before a completely standard test procedure was adopted, twelve more preliminary experiments were completed, which are summarized here without detailed presentation of the data. The subsequent clarification of dose-response relationships shows that many of these preliminary findings (as well as the results of previous studies) could easily be misinterpreted without considerably better calibration of effective dosages than was available at the time. The complexity of the dose-response curve makes it imperative to run a full series of dosage points over a sufficient range to identify key features, if the relative potency of different fractions or extraction procedures is to be compared. Experiments are numbered with a letter to identify the series, a number for the donor group, and a final letter to identify the replication or batch of recipients. Thus, Expt. C-23-C signifies series C (consisting of right/left discrimination experiments), the 23rd batch of donor rats extracted, and replication C. In a given replication, we have tried to keep recipients as uniform as possible (generally from a single shipment), and typically run a number of left- and right-injected groups at different dosages. Unless otherwise indicated, all donors are trained ten days, 15 minutes per day, in two-bar boxes. The preliminary experiments, beyond those previously reported, were as follows:

Expt. F-1. Preliminary study of subcellular fractions, with a two-door box. This experiment was referred to briefly in a note added in proof in Rosenblatt and Miller, and the box used was Box B, shown in Fig. 1 of that paper. Subcellular fractions were prepared after the method of Whittaker (1959), using a sucrose homogenate of the brains. Fractions tested included crude mitochondria, nuclear, microsomal, and soluble fractions. Further subfractions included purified mitochondria, myelin, and nerve-ending particles (NEPs). The final fractions were resuspended in Triton and precipitated with ethanol. The ethanol precipitate was then dissolved as far as possible for injection. Doses of 1.5 brains and 3.0 brains per recipient were used for the nuclear, microsomal, and soluble fractions, with five left-injected and five right-injected rats for each treatment group. A single dose of three brains was used for the remaining fractions, with four left and four right rats in each case. These small numbers and lack of adequate dose-response curves make the results of this experiment marginal, at best, and the design should eventually be repeated with improved extraction techniques. The strongest indication of activity came from the pure mitochondrial subfraction, where the probability was 0.17 for the first test session, and 0.343 for the mean of six tests, with four left and four right recipients. No fraction gave clearly significant results.

Expt. C-13. TCA extraction method, with two-bar box. Male rats were used as donors, females as recipients. A single dosage of 1.5 to 1.6 brains was used for all recipients. Brains were initially homogenized in 0.5% Triton, shaken 24 hours at 0–4°C, and centrifuged at 60,000 *g* for one hour. The precipitate was resuspended and extracted with water. The precipitate was then collected by a second centrifugation and the supernatants were pooled. Two final extracts were obtained: Extract A was precipitated with five percent TCA, neutralized with NaOH, and the soluble supernatant used for injection; Fraction B consisted of a water extract of all of the pellets left over from Fraction A, which was precipitated with ethanol, and the precipitate used for injection. Fraction A (TCA-soluble) was injected into 16 recipients, and Fraction B (TCA-insoluble) into 15 recipients. Testing of recipients was carried out according to the procedures described in Rosenblatt and Miller. Unlike the previous TCA experiment described in that paper, the TCA-soluble fraction gave no significant results, while the TCA-insoluble fraction gave significant inversion effects in the first and sixth test sessions ($P = 0.983$ and 0.968 with one-tailed U-tests).

Expt. C-14. TCA extracts with concentrations of zero, one, and five percent TCA were compared with water extracts of the one percent and five percent TCA precipitates, using the two-bar box. Brains were initially homogenized in Triton, precipitated with ethanol, lyophilized to dryness, and the precipitate then resuspended and treated with TCA in the concentrations indicated. Six left and six right rats were injected with two brains each. The fraction with no TCA treatment gave no significant results. One percent TCA gave marginally significant inversions (P values from 0.8 to 0.97 in the six sessions; 0.934 for the mean of six tests). Five percent TCA gave very similar results (P values from 0.8 to 0.97, with $P = 0.91$ for the mean of six tests). The water extracts of the TCA precipitate gave a slight but nonsignificant positive bias (P = 0.12 for the most significant test session). At the time these results were first obtained, they were taken to indicate that the TCA-soluble fractions were less active than the TCA precipitate, and that the potency of the TCA-soluble fraction reported in our earlier paper (Rosenblatt and Miller, 1966) was a spurious result. In view of the complexity of the dose-response curve revealed by later experiments, this conclusion may be incorrect, and the significant inversions found in the TCA-soluble fractions (and lack of significance from the precipitates) in this and the preceding experiment seem to be consistent with the earlier interpretation.

Expt. C-15. Dose-response curve, with ethanol precipitate in two-bar box. The extract was obtained exactly as in the previous experiment, except that TCA was not used. Doses of one-half, one, two, and four brains were injected into eight left and eight right rats each. We predicted that the lower doses would give inversion effects, and the higher doses positive effects, in accordance with the dose-response curve previously hypothesized (Rosenblatt and Miller, 1966). The results appeared to fit this hypothesis nicely, with Δ-scores (mean of left-injected group minus mean of right-injected group) of −11.7, −18.8, 28.7, and 18.5, for the means of six tests, for the four doses. The P-values (with 1-tailed U-tests) for the pooled low-dosage groups ranged from 0.84 to 0.94, with P = 0.923 for the mean of six tests; for the pooled high-dosage groups P ranged from 0.002 to 0.068, with P = 0.008 for the mean. Successive test sessions were quite consistent, and left and right dose-response curves were roughly symmetric mirror images of one another.

Expt. C-16. This was the last attempt to use a TCA extract. An ethanol fraction, a ten percent TCA-soluble fraction, and extract of the ten percent TCA precipitate were obtained. Each fraction was injected into five left and five right recipients, at each of two doses (one brain and two brains). The ethanol precipitate gave no significant results at the low dose, but gave significant inversions in the early sessions with the high dose (P up to 0.992). The TCA-soluble fraction showed inversion tendencies at both dosages, but with only one significant session (P = 0.96), occurring in the one-brain dosage group. The TCA-insoluble fraction also showed inversion effects, particularly in the first session, although the higher dosage began to show a positive bias in the later tests. The pooled data for all three fractions for Test 1 gave P = 0.875 at one brain, and P = 0.9993 at two brains. At the time this experiment was completed, it was assumed that these strong inversion effects (Δ up to 38%) indicated weak extracts, falling in the left-hand portion of the previous dose-response curve. This interpretation may, of course, still be correct, but since inversions were later found at higher doses we have insufficient data to be sure of this. In general, it is noteworthy that the extra purification which results from the addition of TCA to the ethanol extract in this and earlier experiments does seem to have a possible potentiating effect, since we have frequently obtained levels of significance from the TCA extracts and washes of the TCA precipitate which we have been unable to obtain from the parent fraction.

Expt. C-17. Dose-response curves comparing Triton extract with water extract. Since it is difficult to remove the Triton from the extracts, and it seemed desirable to develop an extraction procedure which would be as free from unnecessary reagents as possible, a water extract was obtained by extracting the brain homogenate successively with 4, 12, and 24 hour water washes, and pooling the three supernatants. The residue was then extracted with Triton, followed by the ethanol precipitation procedure previously employed. Doses of one-half, one, two, and four brains were used, with eight left and eight right recipients at each dose. The Triton extract gave dose-response curves roughly comparable to those obtained before (in Expt. C-15) with a minimum at one brain, and increasing to a maximum at four brains. The water extract, however, started with a slight positive effect at one-half brain, increased to a maximum at one brain, and fell back to zero by four brains. A comparison of starved and unstarved rats (based on weight reduction to 82% or less) showed a strong positive effect over most of the range for the starved rats, and inversion effects for the relatively heavy rats. At present, it seems likely that the differences between Triton and water extracts should be interpreted as an indication that they differed widely in potency, corresponding to different ranges of the full dose-response curve. Without controlling for weight loss, results obtained in this experiment were not statistically significant, but by analyzing the starved rats separately, probabilities down to 0.016 were obtained for the pooled groups.

Expt. C-18. Comparison of preconditioning methods, using ethanol extracts as before. Three groups of recipients were preconditioned: Group A was trained for seven days being reinforced for pushing either bar in the two-bar box; Group B being trained to alternate left and right in the two-bar box for the same period; and Group C being preconditioned in the one-bar box as usual. Doses of 1.5 brains were used in all cases. The only group to show any strong effects was Group C, the others tending to retain their previously acquired bias, and showing a diminished sensitivity to the extract. In Test 1 (the strongest of the six tests) P for Group C was 0.032, and for the pool of all three groups was 0.023. The one-bar preconditioning method has been used in all subsequent experiments with five preconditioning sessions, although beginning with Expt. C-22 a pair of pretests in the two-bar box were introduced following the one-bar training and prior to injection.

Expt. C-19. Preliminary attempt at acetone fractionation procedure, using two-bar box, preconditioning in one-bar box, without pretests. Four fractions were tested, precipitated at 0–10% acetone, 10–40%, 40–60%, and

60–90%. Acetone was added relatively rapidly; slow addition, as prescribed in our standard procedure in the last chapter, was only begun in later experiments. Doses of one-half, one, two, and four were tested, except in the case of the 40% fraction (*i.e.*, 10–40) where the highest dose proved to be lethal and was abandoned. Eight rats in each treatment group were injected with left-trained extract. A group of control rats were given water injections. No right-trained extract was used, all comparisons being based on differences from the control group. This practice is not to be recommended, as differences between rats injected with different volumes of water were discovered in this experiment (as noted previously), and the right vs. left comparison seems to be much more sensitive. The most significant effects were found in the 90% extract, particularly at the highest dose, where probabilities down to 0.003 were obtained.

Expt. C-20. Trial of two-cup box, with single bar at opposite end. Donors were trained to go to right or left cup to obtain pellets after pushing the bar. It was hoped that performance would be improved by associating the location to be discriminated more directly with the food reward. Results with 16 recipients injected with 0.7 brains each gave a P value of 0.07 on Test 1, and did not seem to support the idea that this method would be markedly superior to the two-bar box.

Expt. C-21. Repeat of C-20, with doses of 1.6 brains, and additional water-injected control rats. The most significant result was an inversion effect with a significance level of 0.98 on Test 2, and diminishing in magnitude thereafter. The two-cup box was abandoned after this experiment.

Expt. C-22. Dose-response curves with same fractions as in C-19, but with both left and right-trained extracts. Doses were one-half, one, and four brains. Strong differences (significant at $P = 0.0005$) were found between active and inactive rats (taking 30 responses on third preconditioning session as the cutting point for activity). In this dosage range, inactive rats tended to give positive Δ's, while the active rats showed strong inversion effects. The strongest effects were obtained from the 60 and 90% fractions. This tendency for inactive groups to transfer positively, while active rats show inversions, was found to be present very markedly in earlier experiments as well, when these were re-examined. In later experiments, however, beginning with C-23, in which pretesting in the two-bar box was introduced for the first time, this simple relationship between activity and inversion seemed to disappear. It now appears that a sensitivity difference between active and inactive rats (as described in the last section) more properly accounts for the differences, with active rats consequently being more positive or more inverted than inactive rats depending upon the dosage range. Further experiments to determine whether this correlation of inversions with high activity would reappear more regularly in the absence of pretesting (*i.e.*, with rats who are naive with respect to the test box) might be worth carrying out in the future.

Expt. S-1. Savings method, with one-bar box and Y-maze. Since savings measures are generally used as particularly sensitive measures of retention in conventional learning experiments, and have been widely used by other researchers in "transfer" experiments, this experiment was undertaken to see whether we might be able to use them successfully, in the hopes that this method might reduce or eliminate the "inversion phenomenon." Three groups of 36 donors were used: one group trained in a one-bar box, one group trained to turn left in the Y-maze (described in Rosenblatt and Miller, 1966), and one group of naive rats. Recipients tested in the Y-maze consisted of four groups of 12 rats each: uninjected controls, controls injected with 1.5 brains of naive extract from starved rats kept in their home cages, rats injected with 1.5 brains of Y-maze extract, and rats injected with 1.5 brains of one-bar extract. Recipients tested in the one-bar box consisted of four additional groups of 12 rats with the same four treatments. We hypothesized that if specific transfer effects were present, then the homologous extracts would facilitate learning, while the nonhomologous extracts would do so to a lesser extent or not at all. If the nonspecific activation effects which we believed to be present in the extracts were large compared to task-specific effects, then it was hoped that the two control groups would show this to be the case. In actuality, the results were so complicated as to be difficult to interpret clearly. In the Y-maze the rats injected with the control extract learned significantly faster than any of the other three groups, and the Y-maze extract was the poorest of the four; while in the one-bar box the rats injected with the naive extract were the slowest group, and the Y-maze extract gave the fastest learning. Thus, there seem to be inversion effects superimposed upon a rather complicated activation phenomenon. Activity scores showed that the rats receiving naive extract acted as if they were much hungrier than the others, which helped in the Y-maze (by facilitating exploratory behavior) but may have hindered the learning of the one-bar box since hungry rats discovering a source of pellets refuse to leave the foodcup, and are thus retarded in learning to cross the box and press the bar on the other side. Additional experiments would be necessary to test this conjecture. In any case, this experiment serves as a warning against the assumption that a "naive" extract provides a neutral control.

Experiments with Standard Assay Procedure

The preliminary experiments described above led to the standard procedure which has been largely detailed in the preceding chapter, and which was first tested in its current form in Expt. C-22. In this procedure, left and right extracts are obtained from donors trained for ten days, 15 minutes per day, unless otherwise noted. Recipients are preconditioned for five days, 15 minutes per day, in the one-bar box, and are then given two test sessions (five minutes each) on the sixth day. They are injected on the seventh day, and tested with two five-minute tests per day for the three days following. Data analysis is done by computer, as described in the preceding chapter. Recipients are fed for pushing either bar both on pretests and postinjection tests. Rats are starved and maintained according to the schedule described in Chapter 2.

Table I shows the probabilities and Tables II and III the Δ values obtained in this series of experiments. P-values are obtained by one-tailed U-tests. Δ-scores are always obtained by subtracting the mean S-score for the right-injected group from the mean S-score of the left-injected group. Since the S-scores (defined in the previous chapter) measure shifts from pretest bias ranging from -100 to $+100\%$, the value of Δ can vary, in principle, between $+200$ and -200. In actuality, Δ-values greater than 50 indicate a strong effect, and Δ with magnitude greater than 100 is rare. A Δ-value of 60 would be obtained, for example, if an initially unbiased group of rats had been shifted to 65% left responses by the left extract, and 35% left responses by the right extract, at a given dosage. The probabilities in Table I are shown for all rats in the samples, as are the Δ-scores in Table II. Since more reliable results are obtained by controlling for activity, Table III shows Δ-scores for the "inactive" rats, who make 15 or less responses on the third preconditioning test. In the experiments described in the preceding section, this criterion seemed to give the clearest difference between "active" and "inactive" rats and has been employed ever since, although a better criterion might still be found. Table III reports results only for those cases in which the number of rats was sufficient to permit a meaningful subsample to be obtained.

TABLE I

LEVELS OF SIGNIFICANCE WITH ONE-TAILED U-TEST

Expt.	Acetone Fraction in Percentage	Dose	N (L,R)	Probabilities for Each Test and Mean 1	2	3	4	5	6	M
C-23A	40-60	.5	42,42	.0139	.4064	.0979	.0713	.1011	.0700	.0308
		1.0	42,41	.9092	.6473	.6716	.7107	.5544	.3570	.5544
		2.0	40,42	.2702	.2392	.6378	.7174	.7997	.6343	.6378

TABLE I (*continued*)

Expt.	Acetone Fraction in Percentage	Dose	N (L,R)	1	2	3	4	5	6	M
				Probabilities for Each Test and Mean						
C–23B	60–90	.5	20,20	.0720	.4676	.0583	.0798	.2943	.2244	.0881
		1.0	20,18	.2994	.2415	.0847	.1748	.1673	.3200	.0942
		2.0	19,16	.2332	.0396	.0318	.1103	.2332	.1166	.1041
		4.0	16,11	.7064	.3836	.7231	.5393	.4025	.4803	.5975
C–23C	0–10	.5	17,17	.9415	.9814	.9185	.9076	.9742	.9590	.9491
		2.0	13,19	.5534	.1371	.3577	.2633	.4924	.3435	.3158
	10–40	.5	22,21	.4420	.4903	.4516	.8091	.7518	.7118	.5097
		2.0	13,12	.2073	.6483	.5433	.8224	.6878	.6878	.6878
C–24A	0–60	.05	7,6	.5000	.9773	.7161	.8734	.8413	.9568	.9234
		.1	13,13	.4352	.7392	.6402	.7392	.8649	.8649	.7714
		.2	11,11	.4348	.6410	.5131	.4869	.5131	.4091	.6162
		.4	12,12	.7373	.6888	.7309	.9333	.5460	.9079	.8980
		.8	13,11	.5000	.6888	.9507	.9866	.9215	.8055	.9445
		1.6	12,13	.3928	.5576	.7474	.8789	.5346	.6683	.7252
C–24B	60–90	.05	9,8	.3864	.4789	.6498	.6498	.4237	.2207	.4617
		.1	10,10	.7971	.5601	.9794	.9993	.9937	.9949	.9706
		.2	10,10	.3415	.5649	.9292	.8735	.7162	.6188	.7971
		.4	10,10	.7273	.9904	.6188	.7273	.6473	.8178	.8178
		.8	10,10	.1284	.7273	.4699	.2984	.4103	.2727	.3527
		1.6	10,10	.9224	.8760	.8932	.9863	.9750	.9971	.9750
		3.2	8,10	.2670	.2970	.7614	.3611	.3611	.1871	.5354
C–24C	60–90	.013	15,15	.8315	.7122	.8452	.3622	.7575	.1783	.5741
		.025	14,15	.5000	.7959	.5770	.1971	.5641	.5348	.4479
		.05	15,15	.6200	.6378	.3468	.6378	.5083	.4097	.5248
		.1	15,13	.8779	.9552	.9129	.8816	.9008	.9838	.9266
		.2	15,14	.7436	.5348	.6687	.8203	.6844	.6687	.5693
		.4	15,15	.0274	.1797	.0599	.3308	.3800	.3156	.1548
C–24D	0–10	.4	20,19	.8572	.9357	.9422	.7224	.6093	.1397	.7588
		.8	18,18	.9214	.9396	.8972	.9356	.9066	.9396	.9315
	10–40	.4	18,18	.0690	.1000	.3289	.3758	.1241	.1879	.2238
		.8	17,18	.0716	.3717	.2822	.5395	.5069	.2509	.1954
C–25A	0–50	.013	14,13	.7882	.6373	.6212	.7116	.7371	.9094	.8779
		.025	14,13	.2301	.3317	.6209	.2232	.0867	.4352	.2334
		.05	16,15	.8029	.4917	.7055	.5938	.0946	.2602	.4685
		.1	15,15	.0208	.1791	.7400	.3670	.2047	.6998	.4917
		.2	14,15	.3310	.8217	.9038	.7043	.1085	.5913	.6200
		.4	15,15	.9956	.8894	.7198	.5577	.3110	.4359	.7398

TABLE I (*continued*)

Expt.	Acetone Fraction in Percentage	Dose	N (L,R)	Probabilities for each Test and Mean 1	2	3	4	5	6	M
C–25B	50–75	.013	15,14	.8224	.8358	.9888	.9912	.9836	.9913	.9752
		.025	15,14	.0059	.0202	.0088	.0143	.0073	.0092	.0116
		.05	14,14	.1351	.3566	.4230	.2311	.3739	.2311	.2600
		.1	15,14	.1415	.5387	.4613	.2101	.4479	.2035	.2564
		.2	15,15	.2035	.0248	.0107	.1649	.1754	.4917	.1186
		.4	14,15	.9505	.7709	.7198	.6365	.8718	.9120	.8525
C–25C	75–95	.025	15,14	.5174	.6528	.6351	.5821	.4271	.3717	.6200
		.05	13,12	.3317	.2748	.0514	.1055	.3431	.0785	.1267
		.1	14,12	.8090	.5000	.0825	.0825	.2053	.5648	.1773
		.2	14,13	.4490	.7556	.2777	.1141	.2802	.7481	.3670
		.4	14,13	.4041	.7666	.5959	.5770	.7314	.3854	.6146
C–26A (X)	50–75	.006	4,4	.557	.314	.429	.314	.114	.200	.429
		.013	4,4	.443	.171	.014	.057	.100	.100	.100
		.025	4,4	.657	.757	.557	.557	.557	.829	.657
		.050	4,3	.314	.114	.200	.114	.057	.028	.114
		.100	4,4	.343	.443	.443	.171	.343	.557	.343
		.200	4,4	.829	.557	.571	.829	.829	.657	.829
C–26A (Y)	50–75	.006	5,5	.206	.579	.925	.875	.548	.421	.655
		.013	5,5	.111	.155	.111	.345	.274	.155	.210
		.025	5,5	.655	.952	.972	.972	.972	.726	.972
		.050	5,5	.952	.944	.992	.889	.992	.889	.992
		.100	5,5	.726	.579	.579	.579	.155	.274	.421
		.200	5,5	.210	.500	.210	.421	.155	.274	.274
C–26A (Z)	50–75	.006	5,4	.365	.452	.206	.452	.452	.365	.365
		.013	5,5	.500	.790	.500	.579	.345	.421	.579
		.025	5,4	.365	.452	.095	.143	.095	.196	.206
		.050	5,5	.500	.500	.155	.028	.048	.155	.155
		.100	5,5	.500	.845	.726	.579	.790	.655	.790
		.200	5,5	.421	.111	.210	.500	.421	.274	.210
C–26B	50–75	.0008	10,10	.2568	.2249	.1846	.1680	.5000	.5601	.3251
		.0016	9,10	.4351	.2265	.6585	.9488	.9568	.9292	.9174
		.0031	10,10	.6188	.3121	.0994	.3251	.2838	.6188	.4399
		.0063	9,9	.4824	.7735	.7181	.5000	.7793	.5000	.4126
		.0125	10,10	.0562	.0604	.0028	.0302	.0206	.1443	.0247
		.0250	10,10	.8178	.5151	.6473	.4699	.2984	.2249	.5000
		.0500	8,10	.4646	.0171	.1642	.0549	.0104	.0132	.0165
		.1000	8,9	.4237	.2090	.3763	.3502	.5000	.5000	.4617
		.2000	9,10	.7688	.6879	.7688	.9568	.8557	.7928	.8154

TABLE I (*continued*)

Expt.	Acetone Fraction in Percentage	Dose	N (L,R)	Probabilities for Each Test and Mean 1	2	3	4	5	6	M
C–26C	50–75	.006	27,24	.5400	.7726	.5884	.6805	.8068	.8294	.7207
		.013	24,24	.1325	.0285	.1530	.2326	.0211	.2020	.0168
		.025	24,24	.6480	.6429	.4906	.0868	.0952	.2443	.3031
		.050	24,24	.2315	.8409	.0900	.0989	.3673	.6095	.1821
C–26D (I)	50–75	.006	12,12	.6136	.7181	.8368	.6983	.8873	.7558	.8368
		.013	10,12	.6939	.5308	.3881	.0799	.3951	.0186	.4475
		.025	12,11	.8998	.4267	.6667	.4607	.6888	.5131	.7102
		.050	13,8	.3860	.1233	.0373	.0213	.0962	.0149	.0479
C–27A (6)	50–75	.013	5,6	.535	.835	.535	.604	.669	.331	.396
		.025	6,6	.786	.214	.396	.535	.706	.758	.409
		.050	6,6	.155	.214	.465	.604	.758	.911	.650
C–27A (10)	50–75	.013	6,6	.548	.365	.452	.655	.238	.579	.531
		.025	6,6	.886	.421	.026	.095	.066	.123	.155
		.050	6,6	.331	.650	.531	.531	.409	.409	.531
C–27A (14)	50–75	.013	6,5	.421	.790	.465	.790	.500	.790	.535
		.025	6,6	.469	.350	.409	.469	.531	.469	.294
		.050	6,6	.871	.543	.669	.350	.457	.706	.531
C–27B (6)	50–75	.013	6,6	.214	.669	.619	.604	.824	.824	.531
		.025	6,6	.619	.561	.604	.543	.914	.535	.531
		.050	6,6	.824	.732	.845	.531	.331	.786	.591
C–27B (10)	50–75	.013	6,6	.465	.409	.465	.531	.669	.331	.469
		.025	6,6	.604	.758	.604	.669	.695	.871	.845
		.050	6,6	.274	.066	.086	.604	.409	.197	.350
C–27B (14)	50–75	.013	6,6	.095	.032	.143	.123	.531	.396	.155
		.025	6,6	.268	.591	.604	.669	.695	.238	.469
		.050	6,6	.535	.331	.726	.176	.214	.009	.294
C–27C (10)	50–75	.013	5,6	.810	.700	.535	.535	.759	.543	.786
		.018	5,6	.548	.365	.786	.925	.786	.500	.769
		.025	6,6	.268	.220	.531	.138	.090	.066	.155
		.035	6,6	.439	.176	.027	.120	.066	.120	.066
		.050	6,5	.956	.732	.622	.500	.535	.457	.619

TABLE I (*continued*)

Expt.	Acetone Fraction in Percentage	Dose	N (L,R)	Probabilities for Each Test and Mean 1	2	3	4	5	6	M
C–27C (14)	50–75	.013	6,6	.220	.650	.439	.409	.197	.120	.242
		.018	5,6	.758	.048	.155	.190	.758	.396	.331
		.025	5,5	.635	.365	.655	.655	.794	.889	.889
		.035	6,6	.535	.732	.500	.786	.591	.350	.531
		.050	6,6	.937	.669	.793	.862	.967	.695	.845

TABLE II

DIFFERENCES BETWEEN LEFT AND RIGHT GROUPS (Δ) FOR ALL RATS TESTED

Expt.	Acetone Fraction in Percentage	Dose	Δ For Each Test and Mean 1	2	3	4	5	6	M
C–23A	40–60	.5	17.9	3.6	10.9	14.8	13.9	17.0	13.8
		1.0	−5.5	1.6	−.4	1.4	.4	8.2	1.6
		2.0	7.3	9.4	2.7	−1.9	−5.8	.7	2.1
C–23B	60–90	.5	18.2	2.1	16.1	13.0	5.5	7.0	10.3
		1.0	2.6	3.1	12.0	12.7	12.9	12.8	9.3
		2.0	8.1	17.0	20.3	16.5	14.3	17.8	15.7
		4.0	−11.3	−.4	−19.6	−8.2	−1.9	−5.7	−7.9
C–23C	0–10	.5	−21.1	−29.9	−20.7	−23.4	−40.1	−35.2	−28.4
		2.0	−.7	21.8	7.6	13.1	.3	9.2	8.5
	10–40	.5	−2.4	−7.3	−7.0	−16.7	−14.4	−9.9	−9.6
		2.0	10.7	−7.2	−1.1	−12.6	−4.7	−5.5	−3.4
C–24A	0–60	.05	−6.6	−40.6	−11.7	−16.9	−15.4	−26.5	−19.6
		.1	9.5	−9.0	−2.2	−6.9	−6.5	−14.4	−5.0
		.2	3.5	−5.7	−1.5	−3.5	−8.3	−1.2	−2.8
		.4	−13.1	−1.1	−7.2	−21.7	−.2	−21.9	−11.0
		.8	−1.8	−7.6	−25.0	−36.5	−16.8	−15.4	−17.6
		1.6	3.1	−2.0	−10.9	−13.5	−7.0	−8.0	−5.7
C–24B	60–90	.05	12.3	12.5	2.9	3.3	5.2	14.5	8.9
		.1	−7.2	−2.5	−36.4	−49.7	−46.8	−50.4	−32.3
		.2	6.3	−3.3	−15.9	−20.4	−14.5	−8.2	−12.7
		.4	−16.4	−29.5	−5.2	−5.4	−3.7	−8.7	−11.5
		.8	9.3	−7.9	3.7	12.2	9.5	11.5	6.4
		1.6	−14.7	−24.0	−22.6	−41.3	−21.7	−44.4	−24.6
		3.2	6.2	9.0	−12.5	11.5	4.6	11.9	5.1

TABLE II (*continued*)

Expt.	Acetone Fraction in Percentage	Dose	Δ for Each Test and Mean 1	2	3	4	5	6	M
C–24C	60–90	.013	−4.6	−7.4	−16.3	2.7	−17.3	10.6	−5.8
		.025	−2.7	−7.9	−6.1	16.5	−2.2	1.7	1.4
		.05	−2.6	−2.6	12.1	−11.3	.5	5.0	−.4
		.1	−8.0	−23.9	−17.4	−25.9	−24.6	−33.9	−19.7
		.2	−7.6	−.3	−14.7	−19.0	−12.0	−10.2	−10.7
		.4	23.0	18.5	29.6	7.2	11.8	13.3	18.6
C–24D	0–10	.4	−10.6	−15.1	−14.6	−6.7	.9	13.2	−5.6
		.8	−16.5	−14.5	−16.4	−25.5	−22.8	−31.6	−20.1
	10–40	.4	15.4	18.7	14.1	9.9	19.7	16.9	15.8
		.8	20.9	10.7	11.1	2.5	.6	11.6	8.4
C–24 B & C Pooled	60–90	.05	2.9	2.3	8.5	−6.3	1.9	8.1	2.7
		.1	−7.9	−15.0	−25.9	−36.7	−34.7	−41.2	−25.4
		.2	−2.0	−1.4	−15.2	−19.5	−13.0	−9.5	−11.7
		.4	6.9	−1.2	15.7	1.5	5.3	4.2	6.6
C–25A	0–50	.013	−6.3	2.4	3.7	−7.8	−5.3	−18.8	−11.0
		.025	6.5	.2	−1.0	10.6	29.0	4.7	5.4
		.05	−10.4	.0	−8.8	−4.6	17.7	16.3	.5
		.1	23.9	16.8	−5.1	3.4	10.5	−.1	4.2
		.2	15.0	−6.5	−10.8	6.2	20.0	1.9	5.5
		.4	−26.9	−17.0	−12.1	−2.2	9.7	6.9	−8.7
C–25B	50–75	.013	−11.5	−10.6	−45.3	−45.7	−45.1	−48.0	−29.7
		.025	27.3	34.2	36.9	44.5	49.0	47.9	37.1
		.05	13.5	3.5	4.2	4.3	6.0	11.7	7.0
		.1	10.2	−.7	3.9	7.8	−2.1	19.6	6.4
		.2	12.0	27.4	30.4	13.7	16.1	8.3	17.9
		.4	−12.8	−9.7	−8.3	−4.8	−14.2	−22.1	−11.8
C–25C	75–95	.025	−1.4	.4	4.5	6.6	11.7	10.8	4.0
		.05	7.4	16.3	19.1	20.8	16.1	33.4	18.3
		.1	−12.9	−.8	11.3	17.3	11.1	−8.3	4.2
		.2	3.2	−8.4	9.5	25.1	2.0	−6.9	2.0
		.4	6.0	−2.6	−4.5	−2.5	−5.8	8.3	.0
C–26A–X	50–75	.006	2.8	3.0	4.9	26.5	39.1	25.3	20.3
		.013	2.9	33.2	62.6	62.0	43.1	37.9	40.3
		.025	−17.4	−20.5	−19.7	−20.2	−13.2	−39.7	−21.8
		.050	33.8	45.6	40.2	60.7	88.0	43.2	51.9
		.100	27.2	11.9	4.6	43.9	13.0	1.9	17.1
		.200	4.0	7.0	3.4	−30.3	−30.1	−20.2	−11.9

TABLE II (*continued*)

Expt.	Acetone Fraction in Percentage	Dose	Δ For Each Test and Mean 1	2	3	4	5	6	M
C–26A–Y	50–75	.006	17.7	−1.1	−31.3	−34.8	13.4	15.4	−3.1
		.013	30.7	32.8	38.2	26.2	30.5	47.9	34.4
		.025	−3.2	−45.5	−43.6	−57.2	−53.1	−26.3	−38.2
		.050	−26.9	−48.3	−55.0	−32.5	−46.4	−29.4	−39.1
		.100	−22.0	−12.1	−3.2	4.8	33.4	31.0	5.3
		.200	11.3	4.4	28.6	18.9	40.3	41.5	24.2
C–26A–Z	50–75	.006	10.9	6.5	18.9	6.2	.1	13.1	9.3
		.013	12.1	−30.6	17.6	.8	12.9	15.2	4.7
		.025	8.4	−3.7	37.3	37.9	42.1	31.6	26.5
		.050	6.5	6.8	32.0	55.1	57.7	37.3	32.6
		.100	−11.5	−41.2	−17.1	−31.8	−12.6	−14.2	−21.5
		.200	1.2	25.1	27.7	13.8	28.7	30.5	21.1
C–26B	50–75	.0008	6.2	12.3	13.0	27.4	4.1	2.6	11.6
		.0016	.7	12.5	−7.2	−26.3	−32.4	−27.0	−13.6
		.0031	−4.3	11.0	18.9	9.9	16.6	−5.2	6.5
		.0063	4.5	−9.2	−12.2	−.7	−9.7	−3.0	−8.0
		.0125	19.8	22.1	49.8	33.6	47.2	24.4	33.5
		.0250	−11.1	2.1	−1.8	−.6	8.6	15.0	2.0
		.0500	.9	42.3	14.8	27.7	49.4	60.8	32.7
		.1000	.4	12.7	1.8	6.0	−1.8	.3	2.0
		.2000	−12.9	−5.8	−13.8	−27.7	−21.6	−16.9	−16.5
C–26C	50–75	.006	.1	−8.0	.1	−5.7	−9.3	−12.4	−5.4
		.013	10.5	25.2	12.1	12.2	23.6	13.1	16.7
		.025	−3.9	−4.2	3.0	19.3	15.3	9.6	4.1
		.050	10.0	−17.0	12.7	13.5	5.9	6.6	5.6
C–26D (I)	50–75	.006	−4.0	−11.5	−23.7	−10.8	−33.5	−22.8	−17.7
		.013	6.5	−8.6	3.2	17.3	6.6	33.8	4.8
		.025	−3.4	1.8	−3.1	3.6	13.4	−2.4	−3.5
		.050	−.3	19.5	35.6	28.8	26.7	48.2	25.3
C–27A (6)	50–75	.013	3.8	−33.3	−8.4	−5.4	−9.5	6.1	−7.8
		.025	−4.3	20.4	−5.4	−.9	−8.0	−29.5	−5.6
		.050	22.4	12.6	−5.1	−18.2	−22.8	−40.9	−9.6
C–27A (10)	50–75	.013	−8.0	9.6	−12.8	−9.6	34.4	4.1	2.8
		.025	−11.4	12.8	21.1	25.8	32.2	27.0	13.3
		.050	−1.0	−6.9	−6.8	.2	3.7	11.0	−1.2

TABLE II (*continued*)

Expt.	Acetone Fraction in Percentage	Dose	Δ For each Test and Mean 1	2	3	4	5	6	M
C–27A (14)	50–75	.013	.1	−24.6	1.7	−19.0	8.7	−6.4	−4.0
		.025	2.1	10.9	13.6	4.7	1.7	13.7	7.8
		.050	−15.6	12.5	−9.6	22.9	3.9	−29.3	−.8
C–27B (6)	50–75	.013	19.9	−10.8	−13.8	−16.2	34.8	−34.1	−6.8
		.025	−3.1	1.0	−4.0	−4.2	−36.9	−12.0	.8
		.050	−10.0	−12.5	−14.2	.4	13.8	−18.9	−7.6
C–27B (10)	50–75	.013	10.3	13.2	11.5	−3.4	−12.6	10.3	3.1
		.025	−9.0	−12.4	−12.3	−15.6	−7.5	−48.2	−16.1
		.050	14.6	25.2	33.5	1.7	8.6	22.7	14.7
C–27B (14)	50–75	.013	31.9	19.6	26.8	40.8	2.8	3.7	19.1
		.025	14.0	−5.6	−9.8	−9.4	−9.9	12.3	6.4
		.050	−7.4	14.9	−7.9	19.9	30.1	52.9	9.5
C–27C (10)	50–75	.013	−19.5	−22.9	−5.2	−8.2	−23.3	−11.7	−22.1
		.018	−4.4	20.0	−11.5	−45.8	−14.3	4.0	−2.7
		.025	5.9	12.0	−7.5	12.5	26.8	46.1	10.5
		.035	7.6	29.5	32.3	33.8	42.2	35.0	28.7
		.050	−33.1	7.9	−10.6	−4.5	−14.8	−22.7	−13.4
C–27C (14)	50–75	.013	22.6	−.7	−4.6	−1.3	13.1	21.5	12.7
		.018	−3.8	38.8	23.6	25.0	−2.2	12.8	12.4
		.025	−7.5	−4.0	−6.2	−7.2	−25.2	−48.4	−21.5
		.035	6.9	−13.3	9.0	−12.3	−11.3	17.0	−6.0
		.050	−51.4	−21.8	−15.7	−23.2	−39.8	−15.8	−16.9

TABLE III

DIFFERENCES BETWEEN LEFT AND RIGHT GROUPS (Δ), FOR INACTIVE RATS (ACTIVITY <16)

Expt.	Acetone Fraction in Percentage	Dose	Δ For Each Test and Mean 1	2	3	4	5	6	M
C–23A	40–60	.5	18.4	−2.7	21.0	18.4	14.6	20.5	16.1
		1.0	3.4	11.6	14.2	11.5	10.8	20.7	12.7
		2.0	10.2	11.8	1.9	−8.7	−13.6	−4.9	−.4

TABLE III (*continued*)

Expt.	Acetone Fraction in Percentage	Dose	Δ For Each Test and Mean						
			1	2	3	4	5	6	M
C–23B	60–90	.5	17.6	11.2	22.0	9.4	10.0	7.8	13.0
		1.0	12.9	18.3	21.9	22.0	8.6	19.3	17.2
		2.0	9.4	21.4	29.8	22.8	22.3	25.4	21.9
		4.0	−20.0	−7.7	−20.0	−2.7	−11.5	−1.7	−10.6
C–23C	0–10	.5	−8.6	−26.1	−.1	−6.9	−30.7	−26.8	−16.5
		2.0	−11.9	5.6	−4.7	−10.2	−16.1	−1.6	−6.5
	10–40	.5	−1.3	−9.0	−19.8	−22.1	−24.2	−18.6	−15.8
		2.0	14.4	.3	−6.9	−15.3	−8.1	−2.2	−3.0
C–24A	0–60	.05	−53.1	−54.6	−25.8	−15.4	−14.4	5.5	−26.3
		.1	4.2	−9.1	−7.3	−14.9	−11.3	−25.4	−10.6
		.2	−8.1	−11.3	.9	1.4	−12.2	−22.1	−8.6
		.4	−7.5	6.3	−6.8	−24.0	1.1	−14.8	−7.9
		.8	3.4	−10.0	−21.9	−36.9	−13.6	−15.5	−17.2
		1.6	11.7	4.8	−11.5	−21.1	−19.1	−16.3	−8.0
C–24B	60–90	.05	34.1	42.6	30.2	30.2	36.4	53.5	38.0
		.1	−11.9	−7.6	−44.0	−64.4	−59.9	−53.8	−40.3
		.2	7.6	−5.6	−17.0	−22.1	−21.8	−12.8	−16.3
		.4	−12.3	−30.0	−6.6	−9.4	−7.2	−12.2	−13.0
		.8	7.6	−11.5	2.9	10.1	6.7	8.1	4.0
		1.6	−15.2	−26.8	−22.9	−43.3	−25.3	−55.4	−26.7
		3.2	22.3	−11.9	−48.4	−10.9	11.5	31.5	−1.0
C–24C	60–90	.01	−20.0	−13.0	−35.5	−9.5	−37.3	13.5	−16.8
		.025	3.5	−2.3	−3.2	1.5	−24.9	−8.7	−2.5
		.05	31.7	21.8	33.5	−6.0	12.7	11.5	15.8
		.1	17.3	−18.5	2.6	−8.6	−6.8	−23.3	−3.0
		.2	−44.0	−25.1	−56.2	−57.6	−31.0	−44.4	−43.1
		.4	18.4	14.1	41.0	28.1	15.2	11.9	23.5
C–24D	0–10	.4	−10.7	−12.4	−4.7	1.5	26.1	28.1	4.0
		.8	−12.5	−14.8	−18.0	−27.3	−27.6	−28.8	−20.6
	10–40	.4	26.7	24.2	10.2	7.1	17.8	11.3	16.2
		.8	26.9	26.6	14.7	6.8	−1.7	.7	10.9
C–24 B & C Pooled	60–90	.05	31.7	27.5	32.6	5.3	20.3	24.8	22.8
		.1	4.2	−11.9	−16.4	−33.4	−30.9	−34.4	−18.0
		.2	−20.3	−16.1	−38.6	−41.4	−26.9	−29.6	−30.6
		.4	5.8	−4.7	21.3	10.1	4.5	.4	8.3

TABLE III (*continued*)

Expt.	Acetone Fraction in Percentage	Dose	Δ For Each Test and Mean 1	2	3	4	5	6	M
C–25A	0–50	.013	−19.8	−5.0	−12.7	−20.6	−40.8	−25.1	−24.5
		.025	30.4	12.3	9.7	6.2	24.5	5.2	7.7
		.05	−21.5	5.4	−20.7	−7.2	25.4	28.7	.3
		.1	30.9	15.1	−4.4	8.5	16.3	1.1	5.8
		.2	4.1	−11.2	−30.5	8.5	3.0	−8.7	−1.5
		.4	−30.9	−12.0	−12.8	−7.5	5.7	6.2	−11.7
C–25B	50–75	.013	−10.6	−10.0	−44.8	−43.8	−42.7	−47.4	−28.3
		.025	20.1	32.7	33.9	56.8	51.4	52.0	37.5
		.05	23.5	7.1	8.8	14.0	21.3	24.6	16.2
		.1	−10.5	−17.3	−13.2	−4.4	−16.5	18.3	−6.1
		.2	6.1	24.3	26.6	8.3	11.1	6.9	13.9
		.4	−9.8	−4.0	.8	−.4	−9.5	−19.6	−7.1
C–25C	75–95	.025	6.2	−4.0	−2.9	−3.1	4.8	.9	−.9
		.05	15.8	30.5	32.1	39.8	34.0	52.5	33.3
		.1	−6.9	11.4	19.8	22.1	22.7	1.4	13.2
		.2	23.9	8.0	18.5	24.3	15.3	4.5	13.1
		.4	3.2	10.0	4.8	28.1	29.8	44.3	20.1
C–26A (X)	50–75	.006	−8.5	−6.2	−13.7	−6.5	21.4	3.7	6.0
		.013	9.7	2.6	62.8	46.4	31.7	11.9	27.5
		.025	−39.2	−19.1	−15.1	−38.4	−43.0	−66.4	−36.9
		.050	26.1	17.4	33.5	36.7	66.8	25.1	34.3
		.100	−40.0	−66.7	−60.7	−13.4	−33.3	−.7	−35.8
		.200	18.9	10.0	42.6	−18.1	−36.5	−17.4	−3.9
C–26A (Y)	50–75	.006	27.5	28.9	−16.6	−16.8	16.0	22.4	10.3
		.013	42.8	57.5	79.8	71.0	77.4	80.3	68.2
		.025	−8.5	−41.3	−30.8	−27.6	−21.0	14.2	−19.1
		.050	−10.9	28.3	−42.9	−23.5	−18.4	−6.8	−25.2
		.100	−13.1	−1.5	.3	−7.6	21.4	12.4	2.0
		.200	12.9	−20.4	−2.6	−10.1	12.9	22.9	2.6
C–26A (Z)	50–75	.006	7.6	−5.2	3.7	−22.0	−13.1	−15.3	−7.4
		.013	15.1	−13.3	22.1	2.7	14.9	27.3	11.5
		.025	−25.4	−31.7	6.8	12.6	24.9	22.4	4.1
		.050	5.2	26.6	42.2	42.9	51.1	36.7	34.1
		.100	29.2	−14.9	7.6	−26.4	15.7	8.2	2.4
		.200	21.9	65.8	66.8	46.2	39.4	53.2	49.0

TABLE III (*continued*)

Expt.	Acetone Fraction in Percentage	Dose	Δ For Each Test and Mean 1	2	3	4	5	6	M
C–26B	50–75	.0008	4.2	.7	−4.5	10.5	−15.5	−7.4	.2
		.0016	−8.1	5.0	−7.8	−30.5	−40.9	−34.5	−19.8
		.0031	−7.5	3.3	10.1	6.1	8.9	−12.6	.1
		.0063	7.7	−16.4	−4.1	8.3	−3.5	—.3	−5.7
		.0125	14.0	12.5	38.9	18.9	38.7	19.4	24.7
		.0250	−11.6	−1.6	−5.1	−9.3	−1.3	8.7	−3.4
		.0500	−8.3	39.6	−7.1	27.1	44.8	65.6	27.1
		.1000	2.2	15.3	−3.4	.6	−8.3	−3.4	−.8
		.2000	−20.6	−4.2	−10.0	−21.6	−12.2	−3.9	−12.1
C–26C	50–75	.006	.0	−5.4	.1	−3.0	−8.9	−11.7	−4.4
		.013	13.2	22.1	13.2	11.8	12.9	9.4	16.1
		.025	−1.0	−1.0	6.2	22.2	18.8	21.1	6.8
		.050	12.1	−9.6	18.6	19.7	15.6	1.5	11.3
C 26D (I)	50–75	.006	−.5	−13.9	−31.5	−17.2	−40.9	−34.5	−23.1
		.013	4.9	4.5	10.3	30.6	30.1	42.1	11.9
		.025	−13.5	.4	8.7	.9	−15.4	−6.7	−4.4
		.050	−10.1	29.9	53.2	27.4	36.6	50.0	28.4

In experiment C-23, acetone was still being added quickly, so that the fractions are not as clean as they are in later experiments. From C-24 on, the acetone fractionation procedure is exactly as described in the last chapter. Since the early experiments seemed to indicate that most activity occurred in the 40–60% and 60–90% fractions, the initial four fractions were modified in C-25 to give a fraction from 50–75% acetone, which was predicted to be the active fraction. Since this did indeed seem to be the case, as can be seen from the tables, we have concentrated entirely on this fraction in subsequent experiments. In C-26, a further subfractionation was attempted on the basis of molecular weight, using membrane filters, as described in the section on Preparation and Storage of Extracts. Fraction "X" was supposed to contain molecular weights over 50,000, fraction "Y" 10,000–50,000, and fraction "Z" less than 10,000. Later attempts at calibrating the filters suggest that the actual molecular weights may be substantially less than expected on the basis of the advertised properties of the filters. This might account for the discrepancy between the appearance of the bulk of activity in the middle molecular weight range (fraction Y, as shown in the tables), and our earlier results with Sephadex which suggested a molecular weight of about 5,000 for the active component (Rosenblatt *et al.*, 1966). Experiments C-26B, C, and D were done with the entire filtrate under 50,000 (nominally) without removing the low molecular weight fraction. Based on the shift of key features in the dose-response curves, it appears that about half of the total activity of the 75% acetone fraction was present in the middle-range filtrate. In C-27, we returned to the crude 75% fraction, without filtration, until a better calibration could be obtained for the filters. Expt. C-27 also contains three subgroups: the first uses extract from donors trained for six days, the second from donors trained ten days, and the third from donors trained for 14 days. The donor group is indicated by the numbers in parentheses in the designations of these experiments in the tables.

The group reported for C-26D (designated with the appended Roman numeral, I) was injected and tested by our usual procedure. The same experiment included a second group, C-26D (II), which was run with a double crossover design, half of the rats being injected with the sequence LRL, with one day of testing (two sessions) after each injection, and the other half receiving the sequence RLR. While a crossover design of this type can greatly increase the sensitivity of an assay where, in an additive fashion, there is a large component of individual variability superimposed upon a consistent drug effect, it does not help appreciably in cases where the individual differences have a strong interaction with the drug effect, or where the effect itself may be opposite in different individual cases. Since the latter case seems to be closer to what is happening in our situation, it is unlikely that the crossover design (which presents formidable problems in interpretation of persisting bias effects from one injection to the next, in any case) will be very helpful here. The data from C-26D (II) have not been fully analyzed yet, but preliminary examination does not show any strong effects.

It will be noted from the tabular material that as we continued this series of experiments, the dosage range was lowered progressively, the minimum dose used being 0.0008 brains per recipient. We are still not sure that we have reached the bottom of the dynamic range over which the extract is effective, since each examination of a lower dosage region so far has revealed either a maximum or a minimum, which disappears as the dosage is increased. There is no indication, however, of strong effects at our lowest dosages, so that we may have finally found the "bottom" of the dose-response curve. The conviction that these multiple features of the curve are genuine effects, and not artifacts or statistical flukes, comes from the following main observations:

1. The individual maxima and minima are often highly significant: see, for examples, the 0.1 brain minimum and 1.6 brain minimum in C-24B, the

0.1 brain minimum in C-24C, and the 0.013 brain minimum and 0.025 brain maximum in C-25B, and analogous features in C-26.

2. Maxima and minima have repeatedly occurred at predicted dosage points when we have replicated the same experiment with identical extracts, or extracts obtained by the same procedure. The 50–75% acetone fraction, which was expected to straddle the activity found in the 40–60 fraction and the 60–75% fraction of the earlier experiments, did indeed turn out to be the strongest fraction obtained, with a left-shift of about a factor of two for corresponding features of the dose-response curve, compared to a composite curve obtained from earlier experiments. This shift had been predicted on the assumption that about half of the total activity was turning up in the 40–60 fraction, and about half in the 60–90 fraction. A similar right-shift of the curve for molecular weight fractionation with the filters, was also anticipated and observed. It might be argued that with curves as complex as this, maxima and minima could be matched up regardless of the existence of an actual effect. To show that this is not so, we have tried to form composite curves by turning randomly selected curves backwards or upside down, in order to interchange maxima and minima, or low and high doses. In no such case has it been possible to form a satisfactory composite curve for the full set of experiments. The composite actually obtained by superimposing all curves from "potent" fractions (60, 90, and 75% acetone fractions) is shown in Fig. 4. This composite was made by first identifying a maximum which we have called "omega" on each of the dose-response curves. The

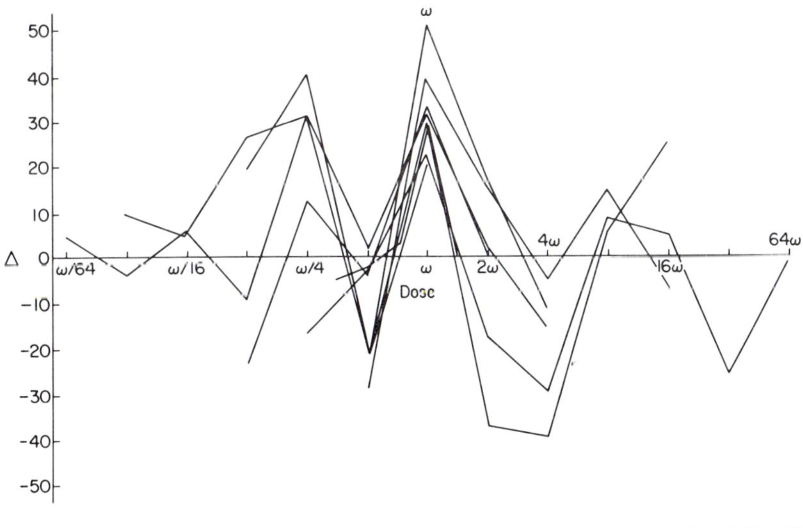

Figure 4 Composite dose response curve. Curves for Expts. C24 B + C, C25B, C26 A, B, D, C27(10) A + B + C, for means of six tests, are placed to coincide at omega.

low-activity rats were used wherever activity controls had been made. The omega-maximum normally occurs at or close to a dose of 0.025 brains in the 50–75% (unfiltered) fraction, and is shifted according to the relative concentration of extracts in other fractions. C-23, which was obtained with imperfect methods and probably differs so greatly in potency that its features could not be identified reliably, was omitted from this series. C-24B and C were pooled since they employed the identical extract and jointly give a greater range than either experiment alone. C-27A, B and C (for the ten-day donor extract) were also pooled, since they again employed the identical extract, and the omega-maximum seems to fall at a point which was not included in C-27A or B (0.035 brains, in C-27C). Having identified this feature, the curves are then traced on a common set of coordinates with omega aligned as an arbitrary origin. The agreement at the location of omega itself is, of course, generated by this procedure, and should not be considered as evidence for the "reality" of the effect at this point. Note, however, the highly consistent minima at a factor of two below and a factor of four above, and the consistent maxima at a factor of four below and a factor of eight above omega. In addition, it should be noted that both C-26 and C-27 were run after the composite was first plotted, with the omega feature falling at or extremely close to the predicted point in each case.

3. Similar features are not found in any of the other acetone fractions, or in control data. While there are many statistical irregularities, as one might expect in a sample of this many experiments, there is no evidence of any regularly recurring features, such as are found in the 50–75% acetone fraction.

4. An analysis of variance was computed for the 0.013 brain and 0.025 brain doses in Expt. C-25B, which were suited to a factorial design. The variables were treatments (left or right), doses, and test boxes. Since the range selected for this study includes a minimum and maximum where doses have opposite effects, the differences between treatments and between doses were not significant at five percent. The treatment-dose interaction effect was significant at $P<0.0001$, the treatment-box interaction at about $P=0.01$, and the dose-box interaction at about $P=0.005$. Thus, there is a clearly significant effect of both dosage and treatment, provided the other variable is controlled properly. The complicating effect of box differences is shown by the significant dose-box interaction, which suggests a differing sensitivity to dosage for rats tested in different boxes.

The most recent experiment, C-27, was performed partly to test the effect of different training times of donors, and partly to reconfirm the presence and shape of the omega feature at the predicted location. In no case were significant effects obtained from the six-day group, but both 10 and 14 days seemed to give an effect. In C-27A the maximum, flanked by an inversion on either side, came at the predicted point for the 10 and 14 day curves, although the maximum was slightly lower in magnitude than had been expected. This suggested that the maximum might be slightly to the right or left of the observed points, since it is clear that these features are sharp enough so that a slight displacement of the curve could miss them. In C-26B, where the identical points were run a second time, this does in fact seem to have occurred, but we felt it was essential to run a third test with closer spacing of the dosage points to confirm this. C-27C does indeed show the expected maximum at a dose of 0.035 brains. The shape of the omega feature, as shown in the pooled data from these three replications, is shown in Fig. 5. There is a suggestion that the same feature in the 14-day curve may have been shifted to the left (indicating a more potent extract), although the shape of the curve is not identical. The negative effects found with the six-day extract are not significant.

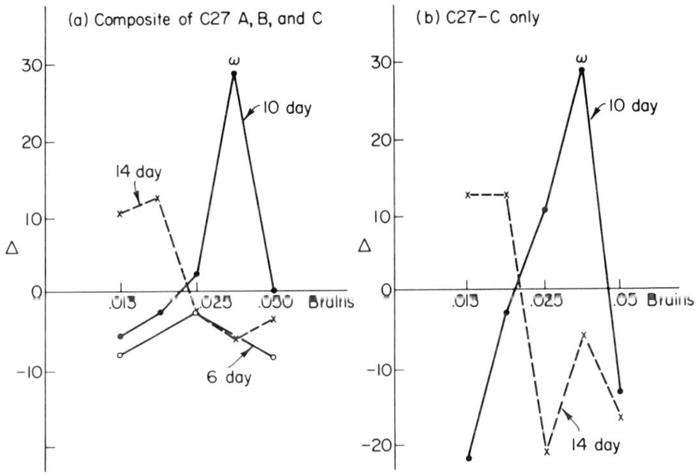

Figure 5 Dose response curves for expt. C27.

CONCLUSIONS, THEORIES, AND FUTURE PLANS

We believe that we have now completed the first exploratory phase of these studies and are ready to begin a more intensive analysis of the chemistry and localization of the discovered effects. The main conclusions which we draw from the data in the preceding chapters are the following:

1. Both left-right discriminations (in various types of apparatus) and

visual discriminations can be affected by extracts obtained from brains of trained animals. These induction effects occur in the rat, and probably to a similar degree in the cat, although the cat data are still insufficient to permit a firm conclusion.

2. The effects obtained depend strongly upon both the dosage and the training of the donor animals. All aspects of the donor experience are likely to be important, with extracts from "naive" animals showing effects which probably depend upon handling and degree of starvation, etc., of the "naive" donors. In left-right discrimination experiments differences between left and right groups are often seen superimposed upon common effects which may vary with the dosage, and which may depend upon the common experience of both left- and right-trained donors (exposure to the same training routine, the same equipment, etc.). Thus, we believe that the induction phenomena are specific, but complex; a single extract frequently conveying a number of different effects which reach their threshold at different dosages.

3. The dynamic range of the dose-response curve is great, and has probably not yet been fully explored. Significant features covered in this report have been found ranging from 0.0008 to 4.0 brains. One would expect that the strongest components are those which are found at the low-dosage end of the curve. We are still hopeful of finding a single, clearly-marked initial feature at the beginning of the curve (*e.g.*, a "first maximum") which could be used as a convenient reference point for assay work.

4. Each feature of the dose response curve typically consists of a sharp increase or a sharp decrease in the percentage of "correct" responses as a critical threshold is reached. As this threshold dosage is passed, the curve goes quickly to a maximum or minimum, and then tends to return to the zero level. A factor of two to four in dosage is generally sufficient to encompass any of the features discovered, giving the entire dose-response curve the aspect of a spectral profile.

5. Particular maxima and minima are reproducible, provided all aspects of extraction procedure, properties of the rats, training of donors, and preconditioning and testing of recipients is held constant. Large samples of rats are necessary to obtain reliable results because of the great individual variability in sensitivity to the extract. Doses should be interpolated as closely as possible to avoid missing an important feature between dosage points. All assays should be carried out with a dose-response curve which can be matched against a standard.

A tentative explanatory hypothesis has been developed to account for this strange set of phenomena. We assume that in a symmetric choice situation, such as the left/right discrimination task, there exist different brain tracts, some of which will favor a right response when activated (or suppressed), and others that would favor a left response under the same conditions. We assume that our extracts contain components which can selectively block or enhance *either* type of tract, but with different efficiency. The binding coefficient of component i in tract x could be designated B_{ix}, the binding of the same component on tract y could be designated B_{iy}, etc. Now suppose the induction effect of a given component increases as the amount bound from a threshold to a saturation level, at which its effect (or rate of induction) is maximum. Suppose tracts x and y have opposite effects (*e.g.*, making the rat prefer left and right bars, respectively), and suppose $B_{ix} > B_{iy}$. Then, as the dosage is increased, we first expect to see an induction of the x response, when component i crosses the concentration threshold in the neighborhood of tract x, but is still below the threshold for tract y. Further increases in dosage will continue to increase this effect, until the saturation level is reached, or the binding to tract y begins to occur. As soon as the y threshold is crossed, the y binding begins to induce the opposite reaction to x, and when both tracts have reached their respective saturation levels, they may be expected to cancel with no behavioral effect being observed. Since the learned task consists of a wide variety of excitatory and inhibitory effects (for example, inhibiting bar-pressing for the wrong bar as well as promoting it for the proper bar) it seems plausible that some of the effects which are thus "teased apart" by the different ranges of the dose-response curve are suggestive of "positive transfer" and others of "negative transfer" or "inversion." It should again be emphasized, however, that according to this theory, it is misleading to speak of "transfer" as if we expected the recipient to duplicate the behavior of the donor under similar circumstances: what we seem to have done is to redistribute various active factors, presumably highly localized in the donor brain, throughout the brain of the recipient, in such a way that the same factor could produce opposite effects depending upon where it is being taken up most readily.

This means that, although there are differences in extracts which depend upon donor training (*e.g.*, component j, with $B_{jx} < B_{jy}$ may be less concentrated than component i in left-trained extract, and more concentrated than i in right-trained extract), the whole question of task-specificity must be considered by studying all of the consequences of injecting different extracts at different doses. An effect that is characteristic of extract A may be absent in extract B, or may simply require a different dosage to bring it out. The bulk of our evidence at this time does suggest that extracts have specific effects on the performance of homologous tasks, but there is little evidence to reject the possibility that they may have equally strong effects on a number of nonhomologous tasks as well.

We still believe that the most plausible site of action of the injected chemical inducers is at the synapse, but whether the synaptic effect (facilitation or blocking) is produced directly by an action of the injected molecules upon the synapse itself or by means of a sequence of intermediary reactions is not yet clear. The timing of the observed effects, with a maximum effect typically occurring at about the third test session, 36 hours after injection, suggests a slow process, possibly involving local morphological changes or protein production. A possible model for this process has been proposed by Rosenblatt (1967), but this remains highly conjectural. The differences in binding suggested by the above description could be accounted for by any differences in steric specificity of inducer molecules coming from different tracts. The degree of differentiation implied for the nervous system seems increasingly plausible as new evidence on specific immunological reactions, regeneration phenomena, and specificity of embryological connection patterns comes to light. It may be that studies in tissue culture will ultimately be capable of revealing the mechanism at work in these phenomena.

At the present time, we are still continuing to explore new assay procedures, particularly for visual discrimination. We are satisfied, however, that the left-right assay is sensitive and repeatable enough to permit us to go on to the next stage of chemical purification, and to further studies of subcellular and regional localization. We plan to subfractionate the 50–75% acetone fraction either by molecular filtration or by Sephadex, according to molecular weight, and then to attempt electrophoretic separation of the resulting fraction. At that time, we hope to have an extract which is pure enough to permit meaningful studies of its chemical composition, and to attempt incorporation of a radioactive label for tracer experiments. Chemical analysis which is currently carried out on a routine basis on all of our extracts shows both protein and RNA to be present. We are probably now in a position to undertake enzymological experiments, and examine the effects of both ribonuclease and various proteolytic enzymes upon the dose-response curve. The subcellular studies will follow the same design as before (in Expt. F-1), except that the standard acetone fraction will be obtained from each subcellular component.

There is no doubt that an experimental design yielding a monotone dose-response curve would enormously facilitate all work in this field. Ungar has demonstrated such curves in a number of cases, and it may be that this is a matter of appropriate choice of task. The experiments in which monotonicity is most readily demonstrated, however, seem in all cases to be suspect as possible studies of activity induction, rather than of discrimination or choice of a specific alternative. The Gay–Raphelson shuttle-box perhaps comes closest to meeting this objection, provided an oppositely trained extract (in which donors are shocked in the light box instead of the dark one) can be clearly demonstrated to produce an opposite effect from the dark-shocked extract. In its initial form, however, this method is highly asymmetric, and fails to satisfy this criterion. Even the visual tasks in which different donor groups are required to press bars marked by light-on or light-off are subject to strong activity effects. This can be seen by the difference of several days in training time required to bring the two groups to a criterion. It is likely that one group is being taught to enhance a natural response tendency (exploration of the light) while the other group must specifically learn to suppress it. Thus, the two groups are actually learning entirely different kinds of tasks in a situation which superficially appears to be balanced, and there is little reason to think that these tasks would "transfer" in a symmetrical fashion.

Thus, despite the complexities of the right-left procedure, we are inclined to follow it further, probably by concentrating on a single feature of the dose-response curve and attempting to discover what factors affect it in the training of the donor rats, the selection and treatment of the recipients, and the composition of the extracts. If this can then be done for several additional features as well, we may be in a better position to understand what we are dealing with.

REFERENCES

Agranoff, B. W. and Klinger, P. D. Puromycin Effect on Memory Fixation in the Goldfish. *Science*, **146** (1964) 952–953.

Babich, F. R., Jacobson, A. L., Bubash, S. and Jacobson, A. Transfer of a Response to Naive Rats by Injection of Ribonucleic Acid Extracted from Trained Rats. *Science*, **149** (1965) 656–657.

Byrne, W. L. *et al.* Memory Transfer. *Science*, **153** (1966) 658–659.

Byrne, W. L. and Samuel, D. Behavioral Modification by Injection of Brain Extract Prepared from a Trained Donor (abstract). *Science*, **154** (1966) 418.

Corning, W. C. and John, E. R. Effect of Ribonuclease on Retention of Conditioned Response in Regenerated Planarians. *Science*, **134** (1961) 1363–1365.

Finney, D. J. *Statistical Method in Biological Assay*, Griffin, London, 1964.

Fjerdingstad, E. J., Nissen, T. and Røigaard-Petersen, H. H. Effect of Ribonucleic Acid (RNA) Extracted from the Brain of Trained Animals on Learning in Rats. *Scand. J. Psychol.*, **6** (1965) 1–6.

Flexner, J. B., Flexner, L. B. and Stellar, E. Memory in Mice as Affected by Intracerebral Puromycin. *Science*, **141** (1963) 57–59.

Gaito, J. *Molecular Psychobiology*, C. Thomas, Springfield, Ill., 1966.

Gay, R. and Raphelson, A. "Transfer of Learning" by Injection of Brain RNA: A Replication. *Psychon. Sci.*, **8** (1967) 369–370.

Grossman, S. *Textbook of Physiological Psychology*, Wiley, New York, 1967.

Hartree, A. Keith-Lee, P. and Morton, W. D. Planaria: Memory Transfer through Caniballism Reexamined. *Science*, **146** (1964) 274–275.

Hydén, H. and Egyházi, E. Nuclear RNA Changes of Nerve Cells During a Learning Experiment in Rats, *Proc. NAS*, **48** (1962) 1366–1373.

John, E. R. *Mechanisms of Memory*, Academic Press, New York, 1967.

Li, C. C. *Introduction to Experimental Statistics*, McGraw Hill, New York, 1964.

McConnell, J. V. *J. Neuropsychiat.*, **3** (1962) 542.

Nissen, T., Røigaard-Petersen, H. H. and Fjerdingstad, E. J. Effect of Ribonucleic Acid (RNA) Extracted from the Brain of Trained Animals on Learning in Rats, II. *Scand. J. Psychol.*, **6** (1965) 265–272.

Reiniš, S., The Formation of Conditioned Reflexes in Rats after the Parenteral Administration of Brain Homogenate, *Activitas Nervosa Superior*, **7** (1965) 167–168.

Rosenblatt, F. Recent Work on Theoretical Models of Biological Memory, in Tou, J. (Ed.) *Computer and Information Sciences, II*, Academic Press, New York, 1967.

Rosenblatt, F., Farrow, J. T. and Herblin, W. Transfer of Conditioned Responses from Trained Rats to Untrained Rats by Means of a Brain Extract, *Nature*, **209** (1966) 46–48.

Rosenblatt, F., Farrow, J. T., and Rhine, S. The Transfer of Learned Behavior from Trained to Untrained Rats by Means of Brain Extracts, I, II. *Proc. Nat. Acad. Sci.*, **55** (1966) 548–555 and 787–792.

Rosenblatt, F. and Miller, R. G. Behavioral Assay Procedures for Transfer of Learned Behavior by Brain Extracts, I, II. *Proc. Nat. Acad. Sci.*, **56** (1966) 1423–1430 and 1683–1688.

Siegel, S. *Nonparametric Statistics*, McGraw–Hill, New York, 1956.

Ungar, G. and Oceguera-Navarro, C. Transfer of Habituation by Material Extracted from Brain. *Nature*, **207** (1965) 301–302.

Whittaker, V. P. The Isolation and Characterization of Acetylcholine Containing Particles from Brain. *Biochem. J.*, **72** (1959) 694–706.

COMMENTS

W. F. Herblin

Central Research Department
E. I. Du Pont de Nemours & Co.

There has been a great deal of discussion as to the validity of certain statistical methods used in studies of the transfer phenomenon. If the total number of experiments which have been performed is considered, including those which yielded negative results, a probability of 0.05 can be questioned, regardless of the procedure by which it was obtained. We therefore decided that statistical significance was a necessary but not a sufficient condition for replication, and looked for other features of the phenomenon which might be used.

Suitable features were readily found in the work of the Cornell group. The active material had been found in a particular fraction derived from an acetone fractionation procedure, and a complex dose response curve had been determined with maxima and minima at definite dose levels. We set out to replicate these features.

With a few specific exceptions, the conditions of the experiments conducted in our laboratory were as closely identical with those of Rosenblatt and his group as we could make them. We have used only one box for all our studies, and therefore have not needed to correct for any box bias. We have kept the injection volume below 0.50 ml, but have varied dosage by varying the injection volume. Finally, our data analysis, while quite similar to the methods reported by Rosenblatt, are not identical.

Our first experiment used 200 animals to investigate the activities of the various acetone fractions. Each fraction was tried at two doses, 0.25 and 0.50 brains/rat, using ten rats per dose. We found activity only in the 60–90% fraction. We also found that this fraction was highly toxic at the 0.50 brain/rat dose. The results obtained from the animals at the lower dose indicated positive transfer that was statistically significant ($p < 0.01$).

Our second experiment, involving 120 rats, used a fraction isolated between 50% and 75% acetone and examined the effect of six different dose levels of the same extract. Statistically significant positive results were obtained at 0.013, 0.025, and 0.25 brains/rat. These maxima are very nearly identical with those reported by Rosenblatt. The results obtained at 0.04, 0.07, and 0.13 were not statistically significant but showed a slight tendency toward inversion. This correlates well with the variable inversion which Rosenblatt has found in this range.

In short, we have been able to replicate not only the transfer phenomenon but also certain additional characteristics of the response. It is the reproducibility of these extra features which convinces us of the validity of the phenomenon under study.

ATTEMPTS TO TRANSFER APPROACH AND AVOIDANCE RESPONSES BY RNA INJECTIONS IN RATS*

James V. McConnell, Tsuyoshi Shigehisa and Harold Salive

Mental Health Research Institute
The University of Michigan
Ann Arbor, Michigan

INTRODUCTION

The first studies suggesting that an interanimal "transfer of acquired behavioral tendencies" was possible were conducted in this laboratory in 1960 (Humphries and Jacobson, 1961; McConnell *et al.*, 1961; McConnell, 1962). Planarians given classical conditioning training using light as the conditioning stimulus (CS) and electric shock as the unconditional stimulus (UCS) were, following training, chopped in pieces and fed to untrained, cannibalistic flatworms. The latter were then given conditioning trials. The response rates to light (CS) of these experimental animals were, from the first set of training trials, significantly greater than the response rates of control cannibals fed the bodies of untrained victims. In 1962, we were the first to show that a crude RNA mix extracted from the bodies of conditioned planarians and injected into the bodies of untrained flatworms apparently

*Various parts of the research described in this paper were supported by contract AT(11–1)–825 from the U.S. Atomic Energy Commission, by grants NIH 5 K3 MH 16697 and MH–13151 from the National Institute of Mental Health, and by grants from the Michigan–Memorial Phoenix Project and the Mental Health Research Institute at The University of Michigan, Ann Arbor.

caused an elevation in response rates in the experimental animals, as compared with the behavior of several groups of control animals injected with a crude RNA mixture extracted from the bodies of various groups of untrained planarians (Zelman *et al.*, 1963). These early experiments have been replicated in recent years in several other laboratories and under more controlled conditions than obtained in the original research (Fried and Horowitz, 1964; Fuchs *et al.*, 1966; Jacobson *et al.*, 1966b, 1966c; John, 1964; Kabat, 1964; McConnell, 1966, 1967a, 1967b; Pickett *et al.*, 1965; Ragland and Ragland, 1965; Rieke and Shannon, 1964; Westerman, 1963a, 1963b). While there remains some argument as to what it is (behaviorally) that transfers in the case of planarians, it seems relatively clear that the "transfer effect" itself is a valid one, at least in flatworms.

The first experiments suggesting that a similar "transfer effect" could be obtained in higher organisms, such as rats and hamsters, were performed in Czechoslovakia by Reiniš (1965, 1966), in Copenhagen by Fjerdingstad, Røigaard–Petersen, and Nissen (1965), in Houston by Ungar and his associates (Ungar and Oceguera–Navarro, 1965), and in Los Angeles by Jacobson and his colleagues (Babich *et al.*, 1965a, 1965b, 1965c; Jacobson *et al.*, 1965, 1966a). Our own work with rats began in 1965, when we attempted to repeat the experiments reported by Jacobson and his colleagues, showing that if one extracted an RNA mixture from the brains of rats trained to approach a food cup when a click sounded, and then injected this mix intraperitoneally into untrained animals, the recipients tended to approach the food cup more than did control animals when presented with the click. Our work yielded statistically significant results, and the use of additional control groups indicated that the "transfer molecule" was probably RNA.

In our first experiments, we used a training apparatus built as closely as possible to the one employed by Jacobson and his associates (Babich *et al.*, 1965a), consisting of a masonite box ($26 \times 26 \times 26$ cm^3) with a grid floor. In one corner of the box was a metal food cup into which food pellets could be dropped from a tube outside the box. A buzzer was mounted in the exact center of the bottom of the box, beneath the grid floor. The walls of the box were painted gray, but an area of 63 cm^2 around the foodcup was white and constituted the "food cup area." The subjects were 64 male hooded rats (Long–Evans) from Maxfield Farms in Cincinnati. The animals were approximately 60 days old, and weighed approximately 250 gm, when used. Both donor and recipient animals were put on a food deprivation schedule and maintained at approximately 90% body weight. The recipients were 48 hours deprived when tested. Twenty-four of the donors were trained five days by hand-shaping to approach the food cup to get a pellet of food whenever the buzzer sounded. By the end of training, all the animals were responding appropriately at least 95% of the time. The eight untrained donors were handled and deprived in the same fashion as were the trained donors, but were not put in the training box. Immediately after their last training session, the donors were dropped into liquid nitrogen and frozen. The whole cerebrum (*i.e.*, the portion of the brain posterior to the olfactory lobes and anterior to the cerebellum) was used for extraction purposes. The brains were individually placed in a cold homogenizer and ground with ten volumes of phosphate buffer for approximately two minutes. An equal volume of 88% phenol was added to the homogenate, and the mix stirred for one hour at 2°C. The mixture was then centrifuged at 20,000 *g* for five minutes at 2°C. The phenol layer was carefully removed and discarded, an additional half-volume of phenol added, the mix stirred for 30 minutes, and then centrifuged at 20,000 *g* for ten minutes. The aqueous phase was carefully drawn off and centrifuged at 3500 rpm for 15 minutes. The RNA was precipitated by addition of 0.1 volume of 20% potassium acetate and 2.5 volumes of 95% ethanol. The RNA was separated by centrifuging at 3000 rpm for 15 minutes, washed with ethanol-water (3:1) twice, and was then dissolved in one ml of isotonic saline after the remaining ethanol was evaporated off. This preparation constituted the "unpurified" or "crude" RNA mixture.

Some of the "crude" RNA was then hydrolyzed with RNase. This RNA was dissolved in a solution of (1 mg/ml) ribonuclease (bovine pancreatic, phosphate free, and lyophylized) instead of the isotonic saline, and was incubated at 37°C for five hours. Ten μg/ml bentonite were added to the incubated solution, and the clear supernatant was drawn off after the mixture had been centrifuged at 10,000 *g* for one hour. This preparation constituted the hydrolyzed "crude" RNA mixture.

For the preparation of "purified" RNA from both trained and untrained donors, the "crude" precipitated RNA was dissolved in three ml of distilled water. To this solution, one volume of 2.5 M K_2HPO_4, 0.05 volume of 33.3% H_3PO_4, and one volume of 2-methoxyethanol were added, in that order. The mixture was shaken vigorously and allowed to stand at room temperature for 15 minutes; it was then centrifuged at 3000 rpm for 15 minutes. The top layer (containing most of the RNA) was then carefully drawn off. A small amount of the top layer was poured back into the lower layer and this mixture was shaken vigorously and then allowed to stand for five to ten minutes so that separation into two layers again occurred. This procedure was repeated three times. Then the combined top layers were centrifuged at 10,000 *g* for one hour to remove any remaining 2-methoxyethanol. The RNA was precipitated from the clear supernatant obtained by this centrifugation. RNA yields were determined from the optical density at 260 mμ with ten mm cell length. For the "crude" RNA preparation, the average yield was 0.96 mg/gm of brain tissue; the yield was 0.62

mg/gm for the "purified" RNA from both trained and untrained donors.

Four groups of recipients were used. The nine animals in Group I were injected with the "purified" trained RNA, the eight animals in Group II received "crude" trained RNA, the seven animals in Group III received hydrolized "crude" trained RNA, while the eight animals in Group IV received "purified" untrained RNA. Each recipient was injected intraperitoneally (without anesthetic) with an amount of the appropriate preparation that represented the RNA from one donor brain. The animals were assigned code numbers, and injection and testing were performed "blind."

Six test sessions of five trials each were given. The first such session occurred six hours after injection; additional test sessions were given once each 24 hours thereafter for five days. Each animal thus received a total of 30 test trials. A test trial was not begun unless (1) the animal was at least one body length distant from the food cup, (2) unless the animal was turned at least 90° away from the cup, and (3) unless the animal was not making any gross locomotor movements. At the start of a test trial, the buzzer was sounded and if any part of the animal's body crossed into the white "food cup area" within five seconds, the response was scored as an "approach." No food or other reward was given to any of the recipients while they were in the test chamber. Two observers recorded all responses independently.

One week after the recipients had been given the 30 unreinforced test trials, each animal was given 100 reinforced trials during which a pellet of food was dropped into the food cup immediately after the buzzer sounded. If the animal reached the food cup in five seconds after the buzzer sounded, the response was considered "correct."

The results of this first set of experiments are shown in Tables I, II, and III. As the data in Table I indicate, the animals that received either "crude" or "purified" RNA from the brains of trained donors were significantly superior to the animals in both control groups in terms of the number of approaches they made to the food cup during the 30 unreinforced test trials. The data in Table II show that the animals receiving the "purified" untrained RNA made significantly fewer errors during the subsequent 100 reinforced trials than did the animals in any other group, while the animals receiving the hydrolyzed "crude" trained RNA were significantly the poorest group. If we look at the last trial on which each animal failed to respond correctly during these 100 reinforced trials (Table III), we find much the same picture.

These data suggest to us that one must measure several different "transfer" effects: There seems to be a "nonspecific," excitatory, perhaps pharmacological effect on performance from injecting any kind of an RNA mixture into untrained animals. This nonspecific effect shows up during reinforced training trials, but is not apparent during unreinforced test trials. In addition, it would seem that one can obtain both facilitatory and inhibitory "transfer"

TABLE I

NUMBER OF APPROACHES TO THE FOODCUP FOR FOUR GROUPS OF RATS INJECTED WITH VARIOUS SUBSTANCES[a]

	I Purified Trained RNA	II Crude Trained RNA	III Hydrolyzed Crude Trained RNA	IV Purified Untrained RNA
Ss				
	1	1	0	1
	3	2	0	1
	3	3	2	2
	5	3	2	3
	5	4	3	3
	5	5	3	4
	6	5	3	4
	6	7	—	4
	7	—	—	—
Mean	4.6	3.8	1.9	2.8

[a] By the Mann–Whitney U-Test,
Group I is significantly superior to Groups III and IV.
Group II is significantly superior to Group III.

TABLE II

NUMBER OF INCORRECT RESPONSES OUT OF 100 REINFORCED TRIALS FOR FOUR GROUPS OF RATS INJECTED WITH VARIOUS SUBSTANCES[a]

	I Purified Trained RNA	II Crude Trained RNA	III Hydrolyzed Crude Trained RNA	IV Purified Untrained RNA
MEAN	4.0	3.6	5.1	2.0

[a] By the Mann–Whitney U-Test,
Group IV is significantly different from all three other groups.
Group II approaches significant difference from Group III ($p = 0.076$).

effects from injecting material from trained brains into recipients. The facilitatory effect shows up most clearly during unreinforced trials, but to some extent also during subsequent reinforced trials, while the inhibitory effect shows up most clearly during the reinforced trials.

These findings should not be too surprising. For decades now, learning

TABLE III

THE LAST TRIAL ON WHICH EACH ANIMAL FAILED TO RESPOND CORRECTLY DURING 100 REINFORCED TRIALS FOR FOUR GROUPS OF RATS INJECTED WITH VARIOUS SUBSTANCES[a]

	I Purified Trained RNA	II Crude Trained RNA	III Hydrolyzed Crude Trained RNA	IV Purified Untrained RNA
MEAN	36	27.9	46	7.7
MEDIAN	31	27.5	44	6

[a] By the Mann–Whitney U-Test,
Group IV differs significantly from all other Groups.
Group II differs from Group III at the $p = 0.168$ level of confidence.

theorists have emphasized that in any learning situation, the animal surely acquires both approach and avoidance tendencies, whether the reinforcement used to control the animal's behavior is positive or negative. Thus the animal's behavior is seen as reflecting some kind of "algebraic sum" of several different excitatory and inhibitory processes. In our opinion, the outcome of any "transfer" experiment hinges on the balance obtained among several different "transfer" factors, at least one of which is inhibitory. We believe it likely that one reason that several laboratories have been unable to replicate the many successful "transfer" experiments now in print, stems from their use of procedures that maximize the "transfer" of inhibitory factor(s) rather than excitatory or facilitatory factors.

OPERANT CONDITIONING STUDIES

Although we were successful in replicating the work of Jacobson and his colleagues on our first experiments, several considerations prompted us to switch to a different experimental paradigm. To begin with, the Jacobson procedure involved the use of observers who watched the animal and merely recorded whether or not it moved to the food cup within five seconds after the buzzer had sounded. While we obtained high interjudge reliability in these studies (96%), we believed that use of a procedure in which the recipient's responses were automatically recorded might free us from any lingering suspicion that "experimenter bias" was somehow affecting the results of our experiment. Then, too, we wished to find a paradigm that would let us determine what experimental conditions would maximize the facilitatory factors and minimize the inhibitory factors described above. The use of an operant conditioning chamber (Skinner box) and a specific experimental technique called "magazine training" seemed best to fulfill our requirements. The next eight studies were then begun using this technique.

In the standard operant conditioning apparatus, the lever or bar that the rat must depress is always located near the food cup that delivers the positive reinforcement. In general, the further away the food cup is from the lever, the more difficult it is for the animal to associate pressing the bar with the appearance of the food. If the cup is directly opposite the lever in a fairly large chamber, most rats learn the problem fairly slowly. One may make the task even more difficult by substituting a dipper arrangement of some kind for the food cup. Usually, when the animal presses the bar, a pellet of food drops into the cup and remains there until the animal finds it; the experimenter has no way of removing the pellet once it has been dropped into the cup. The dipper, which typically contains a liquid reinforcer such as sweetened milk, can be elevated into the chamber automatically, but if the animal does not find and drink the milk within a period of a few seconds (the exact time being specified by the experimenter), the dipper is automatically withdrawn until the animal presses the bar again. The solenoid operating the dipper typically makes a loud and rather distinct "click" noise as it elevates the dipper, a fact that makes most rats avoid the vicinity of the dipper at least at the beginning of training. Under these conditions, a great many rats fail to learn to bar-press since they must somehow learn to rush from one side of the chamber to the other in order to claim the reward.

If one wishes to teach a rat to bar-press under these conditions, one must typically break the training into two parts. At the beginning of the experiment, the rat is put in the chamber and the machinery set so that the lever is nonfunctional and the dipper is automatically elevated into the chamber at irregular intervals. The rat soon learns to sit by the dipper and wait for its daily ration of "manna from heaven" to appear. This first part of the procedure is called "magazine training," because what the animal learns to do is to approach the food magazine on signal. Once the animal knows the significance of the dipper, and associates the loud "click" with the presence of a reward, the second stage of training begins. In the second part of training, the dipper is set to operate only when the animal presses the bar. A rat beginning this part of training typically sits quietly by the dipper for a while, waiting for it to operate automatically. When, after a few minutes, no food appears, the rat often begins wandering about the chamber and, presumably by accident, presses on the lever. The dipper elevates into the chamber, and the loud "click" made by the apparatus reminds the rat that food is now present. The second stage of training usually proceeds very rapidly, and the animal learns to press the bar to get the milk without further difficulty. However, if the "magazine training" sessions are omitted, the rat learns the task slowly, if at all.

In the present series of experiments, the donor animals were always trained to go to a dipper to obtain milk when a click sounded; the recipients were forced to learn to press a bar in order to activate the dipper, but the recipients were never given the all-important first-stage "magazine training." It was our belief that if the tendency to approach the milk dipper when the click sounded could be transferred chemically, then the recipients that were injected with "trained" brain materials would learn the bar-press task rapidly, while the control recipients injected with "untrained" brain materials would learn slowly, if at all.

It was also our opinion that injecting the RNA directly into the brains of the recipient animals was more likely to yield success than the intraperitoneal injections used in earlier experiments. We chose the technique of injecting into the subarachnoid space of the forebrain described by Lindberg and Ernster (1950), since these authors present evidence that materials injected by their method diffuse rapidly to different parts of the brain, and because the RNA thus injected is presumably "inside" the "blood-brain barrier."

An extraction procedure similar to the one that yielded the "crude" RNA in the first experiments was used throughout this second set of studies. Some modifications were introduced, however: During homogenization of the brains, isotonic saline was used rather than phosphate buffer, and the homogenate was stirred with phenol only once, for 30 minutes, then centrifuged at 20,000 *g* for 60 minutes; 0.1 vol. $MgCl_2$ (1 M) and ethanol were added and the RNA was precipitated overnight in a freezer. The RNA was collected after centrifugation at 2500 rpm for 30 minutes and was washed with 75% ethanol three times or more. Yields of RNA were approximately 1.5 mg/gm of brain tissue, but measurable amounts of DNA, protein, and polysaccharides were present in the mix. Rather than freezing the whole animal in liquid nitrogen and removing the brain afterwards, as we did in the first studies, we decapitated the animals and removed the brain and froze it by placing it in powdered dry ice. Decapitation took place eight hours after final training of the donors. All experiments, save the first one reported below, were performed "blind"; in all experiments, however, the data were recorded automatically.

The subjects were male hooded rats of the Long–Evans strain (obtained from Maxfield Animal Supply Co., Cincinnati, Ohio; and from Blue Spruce Farm, Inc., Altamont, New York), having an average body weight of 250 gm, and were 50–60 days old at the beginning of the handling and deprivation schedules. The animals were maintained in group cages and received food (Rockland Diet) and water *ad lib.* for five days. On the sixth day after arrival, each animal was moved to an individual metal cage and handling and deprivation begun. Except as noted below, all animals were brought down to 80% of their body weight and maintained there for the balance of the experiment. The handling schedule was maintained for more than ten days in the donors before their training began, and for more than 20 days in the recipients before pre-injection testing (except for Experiment MT-2a, described below).

Apparatus for Training and Testing

All training (donors) and testing (recipients) was conducted in two identical wooden, two-bar, operant conditioning chambers (Fig. 1) with a 16 × 14 in. mesh floor and with seven in. high walls, one wall consisting of a sliding door of one-way glass. The chambers were housed in separate sound-insulated enclosures covered with fiber glass. The interior walls and floors of the chambers were painted flat black. The ceilings were glass through which the inside of the chambers were illuminated by 14W cool white bulbs (fluorescent).

The manipulanda, two square bars (1 × 1 × 0.25 in.) protruded, eight in. apart, into the chambers through vertical slits in one of the walls, two in. above the floor. A vertical force on each bar of approximately 24 gm triggered a solenoid that raised a metal dipper which delivered 0.08 ml of milk through the hole in the bottom of the dipper area. A cavity was made two in. above the floor, so as to form the dipper area (2.5 in. in diameter, 1.25 in. deep) in the wall opposite to the one in which the bars were set. A solenoid, placed behind the dipper, generated a distinct click (which was much louder than the usual clicks in conventional Skinner boxes) and lifted up the dipper through the hole. A milk dish which was placed directly under the dipper area was sufficiently large to contain enough milk for 180 continuous reinforcements and deep enough to prevent the recipients from stealing the milk by stretching their forelegs through the hole.

In our later experiments, a photocell device was attached to the dipper area to record the number of times the animal stuck any part of its body inside the dipper area.

A jet of air was introduced into the sound-insulated enclosure through a small hole (0.5 in. in diameter) made in an upper part of a wall. Another small hole was connected with a vacuum device. These two attachments provided cool fresh air, constant white noise, and removed warm stale air from the inside of the chamber throughout the experiment. The temperature inside the chambers was kept at approximately 70–75°F.

The milk dish, dipper, the dipper area (inside and outside of the chamber), and bars and bar area were cleaned, and fresh sawdust was placed in the

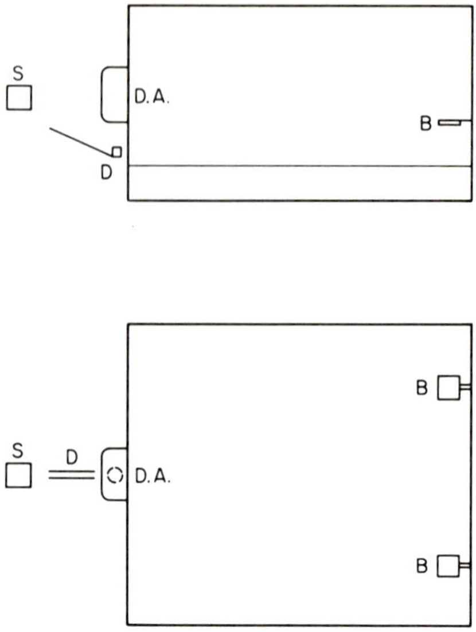

Figure 1 Apparatus used in the magazine training and the pre- and post-injection tests. (S = solenoid; D = dipper; D. A. = dipper area; B = bar.)

tray under the mesh floor of the chamber whenever necessary (usually every four hours). Condensed milk to which was added ten percent sucrose was used as a reinforcer.

During the donor training, both bars were disconnected from the dipper mechanism so that a bar-press did not activate the click noise or dipper. During the recipient testing, the bars were connected to the dipper mechanism so that each bar-press activated the dipper (which made a "click" as it elevated into place).

Training and testing of both donors and recipients proceeded automatically throughout all experiments. The animals were put into the chamber when it was dark. Immediately after the door was closed, the house light was turned on and the first dipper presentation occurred (in donor training).

Training of Donors

Each donor animal was trained to approach the dipper under a variable interval (VI–45 seconds) schedule. The animal was given 60 trials per session for five days (plus one extra session at the start for hand-shaping). During the first two days, the dipper was elevated for 20 seconds during each training trial. On the third day, the time was cut to ten seconds for each presentation, to 7.5 seconds on the fourth day, and to five seconds on the final day of training. The donors were maintained at 80% body weight and were trained when 23-hours deprived. After each training session, the animal was weighed and given enough Rockland Diet to bring it back up to 80% of body weight.

During the preliminary shaping session, the dipper (with milk in it) was held up by E's hand from the beginning of the session until the animal found it and consumed the milk. The animal was given the milk ten times in this manner whenever it approached the dipper. No click was sounded at this stage of shaping. Next, the click was sounded while the animal was sticking its nose inside the dipper area deep enough so that it could very easily find the dipper with milk, immediately after the click, in spite of any emotional response the animal might evince to the onset of the click. This click-dipper presentation was repeated successively more than ten times until the rat consumed the milk within 20 seconds, ten times in succession. Then, the click-dipper presentation was withheld until the animal moved an inch or two away from the dipper area. After the click-dipper presentation was repeated for five to ten times under this condition, the intertrial interval was gradually increased up to 60 seconds, and then it was gradually shifted to the VI–45 seconds dipper-presentation schedule. Total number of dipper presentations per day during this and subsequent days was approximately 60. Almost all animals established a dipper-approach response during this shaping session.

Untrained donor animals were treated similarly to the trained donors outside of the chamber situation. These animals had been handled, deprived, and maintained at 80% body weight throughout. During the last six days prior to sacrifice, untrained donors were given extra handling equivalent in time to the trained donors and weighed, fed, and maintained at 80% body weight. The untrained donors did not experience the chamber situation, however, and were not fed milk.

Trained and untrained donors were sacrificed in pairs.

Injection Procedure

Fourteen hours prior to its first postinjection test session, the recipient animal was injected with RNA from a trained or untrained donor. The injection was made into the space between the olfactory bulbs to facilitate rapid diffusion of the injected materials into the subarachnoid space of the brain.

Prior to injection of the RNA, each animal was anæsthetized with an intraperitoneal injection of sodium-pentobarbital in a manner that minimized emotional disturbance. A cut was made along the line of suture sagittalis between the animal's eyeballs with a sharp knife, the remaining muscles were pushed aside, and the skull was exposed. The incision area was thoroughly sterilized (with zephiran-chloride: ca. 0.2%). A dental drill with a one-mm-wide tip was used to make a hole in the suture sagittalis directly between the eyeballs (*i.e.*, above the space between the olfactory bulbs). The drill tip was placed on the suture sagittalis to form an angle of 45° with the frontal plane (Lindberg and Ernster, 1950).

The hole was drilled gently only about two-thirds of the way through the skull, to prevent mechanical leakage of the RNA solution after injection. A 27-gauge needle was placed in the hole and was rotated several times under moderate pressure for 10–15 seconds to slightly pierce the remaining bone. This needle was then removed, and a 250 μliter tuberculin syringe with a 30-gauge 0.25-inch-long needle was used to inject the RNA. This syringe had been kept in a refrigerator and was loaded with 200 μliter of cool RNA solution immediately before the injection. The tip of the needle was introduced by rotation under minimum pressure. The needle was kept at a 45° angle with the frontal plane, two mm deep from the outside surface of the skull. Two hundred μliter of the cool RNA solution were injected into each recipient over a five-minute period.

The needle was withdrawn slowly after the injection was made so that leakage of the injected solution was minimized. After complete withdrawal of the needle, the incision was sterilized and the epidermal tissues were dried slightly with cotton for 30 seconds. The incision was closed without suturing, by pressing the skin edges together with moderate pressure from a pair of forceps.

Testing Procedure

(a) Pre-injection test: Each recipient animal was tested ten hours before injection (*i.e.*, 14 hours before the first session of the postinjection test) under the same experimental conditions as obtained during the postinjection test. Animals were then assigned to experimental or control groups according to baseline response tendencies during the pre-injection test.

The pre-injection test lasted for 60 minutes, during which time a bar-press elevated the dipper for 20 seconds. However, it was typically the case that few, if any, animals actually obtained the milk during the pre-injection test.

In several of our early experiments in this series, we found a significant correlation between the number of bar-presses the rat made during pre-injection testing (regardless of whether or not it obtained reinforcement) and the probability that it would eventually learn to press the bar to obtain the milk. In general, it appears that animals with a high pre-injection response rate (more than ten bar-presses during the hour's test) will learn the task no matter what one injects them with, while animals with a very low response rate (zero to two bar-presses) fail to learn no matter what one injects them with. The reasons for this fact are obvious and have nothing to do with "memory transfer" *per se.* Rats with a very high initial response rate are highly active animals; hence, much more likely to explore the box immediately after a bar press, and more likely to discover the connection between the milk and pressing the bar. Rats that are (for some reason) low responders tend to be frightened animals that cower in the corner of the box and do not explore; this high emotionality appears to overwhelm the effects of injection of trained material.

It is important to note, then, that the "transfer effect" can best be demonstrated in this paradigm when one uses animals with a "medium" pre-injection response rate (two to ten bar-presses) and when one makes sure that experimental and control groups are balanced in terms of this factor.

It has been our experience that the average pre-injection response rate varies widely in animals obtained from the same supplier at different times. We found it expedient to order at least twice as many recipient animals as we planned to use so that we could discard the very high and low responders prior to injection.

(b) Postinjection test: Following the pre-injection test, each animal was weighed but was not fed; thus, when it was given its first postinjection test 14 hours after the injection, the animal had been food-deprived for 48 hours. The postinjection testing procedure was the same as was used for the pre-injection test. The dipper was elevated for 20 seconds after each bar-press, the intersession interval was 24 hours, the animals were food-deprived for 23 hours (after the first session), and were maintained at 80% of body weight across sessions.

Criterion for learning was specified as being 60 "correct" bar-press responses within 30 minutes. "Correct" bar-press responses were defined as those responses that actually actuated the dipper. Any responses the animal made during the 20 second dipper-holding period (after a "correct" response) were thus not counted. The animals were tested for ten sessions, or until they met the criterion.

Results

Experiment MT-1. The results of this experiment are shown in Table IV. Although the data were recorded automatically, the experiment was not run "blind." All of the rats injected with "trained" RNA learned within nine days; none of the rats injected with RNA from untrained donors learned within ten days. Rat 7 in the experimental group was in poor health during the first five days following injection. The two groups differ significantly ($p < 0.01$).

Experiment MT-2. Because Experiment MT-1 was not performed "blind," we attempted an immediate replication using "blind" techniques.

TABLE IV

NUMBER OF BAR-PRESS RESPONSES DURING EACH ONE-HOUR DAILY TEST SESSION FOR NAIVE RATS INJECTED WITH RNA FROM TRAINED AND UNTRAINED DONORS (EXPERIMENT MT–1)

	Injected With RNA From Brains of Trained Donors					Injected With RNA From Brains of Untrained Donors				
Day	Rat 1	2	5	7	10	3	4	6	8	9
1	11	6	26	9	2	2	2	4	2	0
2	55	13	55	2	46	0	3	0	0	1
3	60	60	60	5	60	3	3	2	1	4
4	60[a]	34	60	3	60	1	1	0	1	6
5	—	60[a]	60[a]	5	60[a]	1	4	2	2	4
6	—	—	—	28	—	1	4	1	0	0
7	—	—	—	60	—	1	4	2	0	0
8	—	—	—	60	—	0	4	2	0	0
9	—	—	—	60[a]	—	0	1	8	0	2
10	—	—	—	—	—	0	3	6	1	4

[a] Criterion is 60 responses in less than 30 minutes.

In our rush to get the experiment underway, however, we gave the recipients only three days' handling prior to injection, and did not bring them down to 80% body weight. They were not 48-hours deprived when given their first test, but were maintained on a 24-hour deprivation schedule thereafter. The experiment (MT-2a) was halted after four days' training for obvious reasons, and all the animals were given a week of handling and were brought down to 80% body weight. They were then re-injected with additional "trained" and "untrained" RNA. Since one of the control animals had reached criterion and two others died, two of the animals that were injected with "trained" RNA the first time were injected with "untrained" RNA the second time. After the second injection, the animals were again assigned code numbers and the data gathered "blind" (MT-2b).

The results of Experiment MT-2, shown in Table V, strongly suggest that the recipients must be well-gentled and deprived if the "transfer" phenomenon is to be demonstrated in a food-approach situation, a fact first pointed out by Reiniš (1966). The fact that animals 14 and 18, which received "trained" RNA on the first injection and "untrained" RNA the second, did not reach criterion rapidly, suggests that there were few if any "lingering" effects of the first injection still present by the time the second injection was made. During the second part of Experiment MT-2, the experimental animals were significantly superior to the controls ($p < 0.01$).

TABLE V

NUMBER OF BAR-PRESS RESPONSES DURING EACH ONE-HOUR DAILY TEST SESSION FOR NAIVE RATS INJECTED AND THEN RE-INJECTED WITH RNA FROM TRAINED AND UNTRAINED DONORS (EXPERIMENT MT–2a AND MT–2b)

	Injected with RNA From Brains of Trained Donors									Injected with RNA From Brains of Untrained Donors					
Day	Rat 1	3	6	11	13	14	18	19	20	2	4	10	15	16	17
1	12	7	4	7	10	3	3	5	3	3	3	6	5	6	41
2	3	2	3	1	5	0	1	6	0	7	2	3	4	3	60
3	3	2	0	0	2	2	1	2	1	1	0	1	9	3	60[a]
4	7	3	2	12	1	1	3	6	3	2	2	0	—	5	—

	Re-Injected with RNA From Brains of Trained Donors					Re-Injected with RNA From Brains of Untrained Donors				
Day	Rat 1	3	11	13	19	4	10	14[b]	16	18[b]
1	1	60	60	60[a]	60	7	3	1	2	4
2	9	60[a]	60[a]	—	60[a]	13	0	2	1	3
3	60	—	—	—	—	20	0	1	2	1
4	60[a]	—	—	—	—	5	0	1	5	7
5	—	—	—	—	—	8	0	1	3	4
6	—	—	—	—	—	15	0	0	2	2
7	—	—	—	—	—	60	—	—	—	—
8	—	—	—	—	—	60[a]	—	—	—	—

[a] Criterion is 60 responses in less than 30 minutes.

[b] Note that rats 14 and 18 were injected with mix from Trained Donors the first time but with mix from Untrained Donors the second.

Experiment MT-3. The first two experiments suggested strongly that some kind of "transfer" effect could be obtained under the conditions we had employed, but the results of these two studies could easily be attributed to some kind of non-specific, pharmacological agent in the RNA mix. For example, one possible explanation would be that the "trained" RNA contained some kind of "general activator" substance, and that the experimental recipients were therefore more active than the control animals, hence more likely to learn the bar-press task. Also, we had no assurance that, even if we had discovered a genuine "transfer" of acquired behavioral tendencies, RNA was the molecule that mediated the "transfer." Experiment MT-3 was designed to test both these possibilities.

Donor animals were trained as before. Prior to injection, however, the recipients were given two hours of testing. During the first half-hour, each recipient was put on a specially designed activity chamber that had the same internal dimensions as the operant test chamber. Rather than having two bars at one side and a milk dipper at the other, the activity chamber had built into its wooden floor two metallic plates measuring 3.5 × 5 cm. The plates were on opposite sides of the chamber. During all tests in the activity chamber, we measured the total number of times that the animal pressed on either plate, as well as the number of times that it pressed both plates within 20 seconds. Our belief was that these two measures would tell us something about the gross activity levels of the recipients, as well as whether any of them showed a tendency to move rapidly from one side of the chamber to the other (as would be the case if the animals learned during the post-injection tests in the operant chamber).

Following the first 30 minutes in the activity chamber, each recipient was put for an hour into the operant conditioning apparatus for its pre-injection test. Then the animal was given a second 30-minute period in the activity chamber. The animals were then divided into four groups that scored roughly the same number of bar-presses during the pre-injection test. Some 14 hours after the conclusion of the pre-injection tests, the animals were injected with one of four different solutions: Group A received RNA from trained donors, Group B was injected with hydrolyzed RNA from the brains of trained donors, Group C received RNA from the brains of untrained donors, while Group D received hydrolyzed RNA from the brains of untrained donors. Hydrolysis of the RNA was accomplished as it had been during our first series of experiments, described above.

About 14 hours after injection, the entire pre-injection series of tests was given again to each animal; about 23 hours later, this procedure was repeated. Thereafter, the activity measures were dropped and only the daily training session was given. At the end of the fifth day, all of the animals in Group A, but none of the animals in any other group, had reached criterion, so the training was terminated.

The bar-press data for experiment MT-3 are shown in Table VI. Only the animals in Group A, which received "trained" but unhydrolyzed RNA, reached criterion in five days. Since the animals in Group B, which received hydrolyzed "trained" RNA, were if anything somewhat inferior to all other groups, we have tentative evidence that the "transfer factor" in this situation is indeed RNA.

There were no significant differences between the four groups in terms of activity levels before or after injection. The highly significant differences between Group A and the other three groups in bar-press activity following injection do not seem to be the result of any nonspecific, pharmacological activator.

TABLE VI

NUMBER OF BAR-PRESS RESPONSES BEFORE INJECTION AND DURING EACH ONE-HOUR DAILY TEST SESSION FOR NAIVE RATS INJECTED WITH HYDROLYZED OR UNHYDROLYZED RNA FROM TRAINED AND UNTRAINED DONORS (EXPERIMENT MT-3)

	Group A Injected With RNA From Brains of Trained Donors				Group B Injected With Hydrolyzed RNA From Brains of Trained Donors				Group C Injected With RNA From Brains of Untrained Donors				Group D Injected With Hydrolyzed RNA From Brains of Untrained Donors			
Rat	14	9	6	2	13	10	17	16	15	11	3	12	1	7	4	5
Pre-Injection Responses	2	2	4	7	1	2	4	8	1	2	6	6	0	3	4	9
Day																
1	5	8	5	4	0	1	0	4	0	2	3	1	0	1	1	0
2	8	7	3	22	1	2	0	5	0	3	0	3	1	0	1	2
3	60	15	3	60	0	0	0	4	0	1	1	6	1	1	3	16
4	60[a]	60	60	60[a]	0	0	0	8	1	2	2	47	0	2	3	19
5	—	60[a]	60[a]	—	0	1	0	24	0	2	1	60	1	44	4	51

[a] Criterion is 60 responses in less than 30 minutes.

Experiment MT-4. An inspection of the data from Experiment MT-3 suggested to us that, within any one of the four groups, there was a correlation between the animals' pre-injection response rates and the speed with which they reached or approached criterion after injection. In Experiment MT-4 we hoped to maximize the differences between experimental and control recipients by selecting only the very lowest responders on the pre-injection response test.

In order to decrease emotionality in the recipients, they were all given five days of pre-injection (but unreinforced) testing in the chamber. Seven animals were injected with "trained" RNA, and six animals with "untrained" RNA. The two groups were balanced in terms of their responses to the lever on the final pre-injection day of testing. The average pre-injection response rate on the fifth day was less than two responses in both groups.

The subjects were all Long–Evans hooded rats from Blue Spruce Farm Inc., Altamont, New York; the recipients were younger and smaller than usual since the supplier could not provide us with animals of the weight and age we had used in previous experiments.

The yield of RNA was 1 mg/gm of brain tissue, somewhat smaller than usual.

No measurable "transfer" effect was recorded in Experiment MT-4. Only one animal (one that received "trained" RNA) reached criterion within five days. By the end of ten day's training, three experimental animals and three control animals reached criterion, and a fourth experimental animal was close to criterion. Five of these seven animals were the highest responders during the five days of pre-injection testing.

The data from Experiment MT-4 suggest that the "transfer" effect is difficult to demonstrate in animals with very low pre-injection response rates since almost none of the animals reach criterion within five days, no matter with what substance they are injected.

Experiment MT-5. Since Experiment MT-4 failed to yield a "transfer" effect, we believed that we should attempt to repeat our original three experiments using the magazine-training paradigm.

The animals were from Blue Spruce Farm, Inc., but were of the same size and age as were the animals in the first three studies. Only those animals with pre-injection response rates of 4–14 bar-presses were used. We had hoped to use animals with response rates of two to ten bar-presses, but there were not enough animals with these response rates.

The results of this experiment, which are shown in Table VII, are encouraging, but must be taken with a grain of salt since the two groups were not adequately balanced for pre-injection response rates. However, it is pertinent to note that the fastest learners among the experimental animals were not those with the very highest response rates, and that control rats with response rates greater than many experimental rats failed to learn at all.

Experiment MT-6. Donor training and recipient testing in this experiment were the same as in previous experiments, but the recipients received two injections 24 hours apart. Each injection was the equivalent of the RNA extracted from one-half a cerebrum, hence these animals all received about twice as much RNA as did the animals in the earlier experiments. The animals came from Maxfield Animal Supply Co., and were of the same age and weight as those used in the first experiments. The two groups were well-balanced in terms of their pre-injection test scores (2–11 bar-presses).

The results of Experiment MT-6 appear in Table VIII. The two groups differed significantly on the first day of postinjection training ($p < 0.05$) and in when they reached criterion. Four of the five animals in the experimental group reached criterion before any of the five animals in the control group did. Since the "transfer" effect was not significantly stronger than that found in prior studies, we conclude that a double injection of "trained" RNA does not necessarily yield a greater effect than does a single injection as administered in this situation.

TABLE VII

NUMBER OF BAR-PRESS RESPONSES DURING EACH ONE-HOUR DAILY TEST SESSION FOR NAIVE RATS INJECTED WITH RNA FROM TRAINED AND UNTRAINED DONORS (EXPERIMENT MT–5)

	Experimental rats						Control rats				
Rat	1	4	5	6	8	11	2	3	7	9	10
Pre-injection session	5	8	14	14	5	11	4	9	6	5	5
Post-injection session											
1	0	32	2	4	33	42	3	6	2	1	1
2	1	60	5	3	60	60	1	2	1	1	0
3	1	60[a]	60	53	60[a]	60	4	1	0	2	3
4	0	—	60[a]	55	—	60	5	0	2	1	2
5	2	—	—	60[a]	—	60[a]	7	1	1	0	2
6	1	—	—	—	—	—	2	1	1	0	2
7	0	—	—	—	—	—	4	3	1	2	2
8	0	—	—	—	—	—	5	1	1	15	5
9	0	—	—	—	—	—	1	1	1	60	60
10	0	—	—	—	—	—	1	1	2	60	60[a]

[a] Criterion is 60 responses in less than 30 minutes.

TABLE VIII

NUMBER OF BAR-PRESS RESPONSES DURING EACH ONE-HOUR DAILY TEST SESSION FOR NAIVE RATS INJECTED WITH RNA FROM TRAINED AND UNTRAINED DONORS (EXPERIMENT MT–6)

	Injected with RNA From Brains of Trained Donors					Injected with RNA From Brains of Untrained Donors				
Rat	2	5	7	8	9	1	3	4	6	10
Pre-injection	7	10	4	8	4	11	7	6	7	2
Post-injection										
1	23	30	11	22	8	5	7	1	10	6
2	60	60	60	11	2	3	7	3	60	5
3	60[a]	60[a]	60[a]	60	2	13	4	1	60	1
4	—	—	—	60[a]	2	41	3	6	60	0
5	—	—	—	—	0	60	4	21	60[a]	2

[a] Criterion is 60 responses in less than 30 minutes.

Experiment MT-7. One of the variables Ungar (1967) reported as influencing the strength of the "transfer" effect was the amount of training given donor animals. In this experiment, we tested the potency of this variable by giving donor animals varying degrees of training prior to sacrifice. One group of donors was given ten days of magazine training, a second group was given but the usual five days we have employed in all previous experiments. Deprivation, handling, extraction, injection, and training were all as before. Five animals that were given ten days' training and ten animals that were given five days' training were used as donors. Six recipients were injected (subarachnoid space) with RNA from the ten-day-trained donors, whereas 12 animals were injected with RNA from the five-day-trained donors. The groups were balanced in terms of their pre-injection response scores. Both injection and training of recipients were performed "blind."

The curves showing the average number of bar-press responses per session for the recipients are shown in Fig. 2. As can be seen, the animals receiving RNA from the ten-day-trained donors had a significantly higher response rate than did the animals receiving RNA from the five-day-trained donors.

Experiment MT-8. In an attempt to extend the findings of Experiment MT-7, we gave five groups of donor animals different experiences: group 1D donors received five days' magazine training; group 2D received ten days' magazine training; group 3D received 20 days' magazine training; group 4D donors were not given any training, but were housed under normal

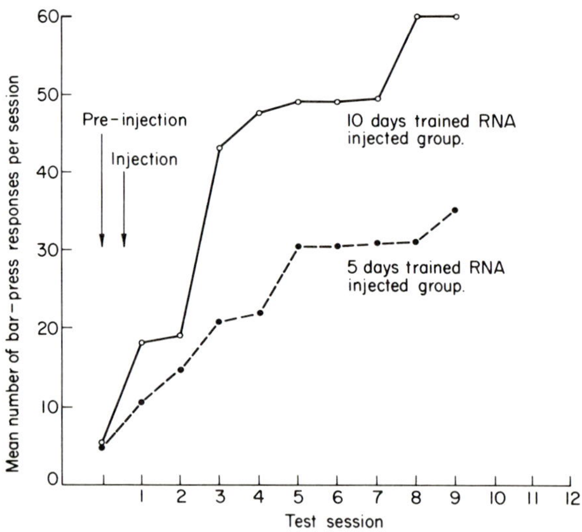

Figure 2 Number of bar-press responses during each one-hour daily test session for naive rats injected with nucleic acid mix (RNA) from 10 days trained and 5 days trained donors (mean of each group).

laboratory conditions; and group 5D donors were kept under conditions of moderate sensory deprivation for a period of 30 days from age 60–90 days, housed in a light-proof box that had constant white noise present to mask external sounds. Deprivation, handling, extraction, injection, and training were all as above. We did make one additional change in procedure, however: a photocell device was added to the cup containing the milk dipper. Each time an animal put its nose (or any other part of its body) into the cup, a relay was broken and a "magazine entry" was recorded automatically. There were six animals in group 1R (that received subarachnoid space injections of RNA extracted from the group 1D donors), five recipients in group 2R, five in group 3R, six in group 4R, and six in group 5R. The groups were balanced in terms of their pre-injection response levels. Injection and testing were performed "blind."

The mean number of bar-press responses per session for each of the five groups is shown in Fig. 3. The group 2R animals were significantly superior to all other groups for the first two days following injection and from groups 3R, 4R, and 5R for the final three days. The group 1R animals were no different from groups 3R, 4R, and 5R on the first two days, but were significantly superior to these groups on days three–five.

It would appear that, as in one case described by Ungar (1967), the relationship between amount of donor training and the strength of the "transfer" effect is U-shaped, with a peak near ten days' training, a result consistent with our findings in Experiment MT-7. The fact that there were no significant bar-press differences between the animals in groups 4R and

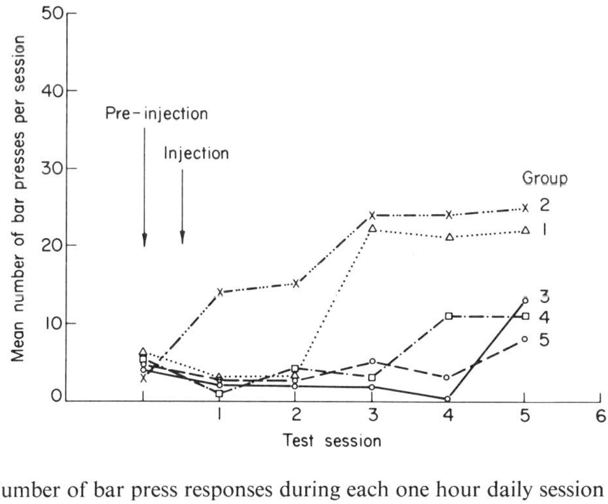

Figure 3 Number of bar press responses during each one hour daily session for naive rats injected with nucleic acid mix (RNA) from 20, 10, and 5 days trained, untrained, and sensory deprived donors.

5R suggest that RNA extracted from the brains of untrained donors does not affect speed of learning a bar-press response in recipients whether or not the donor animals had undergone moderate sensory deprivation prior to sacrifice.

The average number of "magazine entries" recorded by the photocell device on the cup containing the milk dipper are shown in Fig. 4. The group 1 and group 3 animals showed more magazine entries during the pre-injection tests than did the animals in the other three groups (it is difficult, if not impossible, to balance five groups of recipients in terms *both* of their bar-press responses *and* of their magazine entries prior to injection). Since the group 2 animals were inferior to groups 1 and 3 on this measure prior to injection, yet were significantly superior to groups 1 and 3 immediately after injection (in terms of the actual number of bar-press responses made), we may conclude that "magazine entries" are not as meaningful a measure of the "transfer effect" in this situation as are "bar-press responses" following injection. This conclusion is supported by an inspection of the data yielded by individual animals. Almost all rats which eventually learn to press the bar to get milk show a peak number of magazine entries two to three days prior to reaching criterion. As they approach criterion, however, the number of magazine entries drops significantly. Animals that maintain a very high number of magazine entries for several days often do not learn to press the bar at all, but rather spend all their time exploring the food cup area and trying with paw or nose to reach the milk in the reservoir below. The rat that actually learns the problem spends a large proportion of its time shuttling

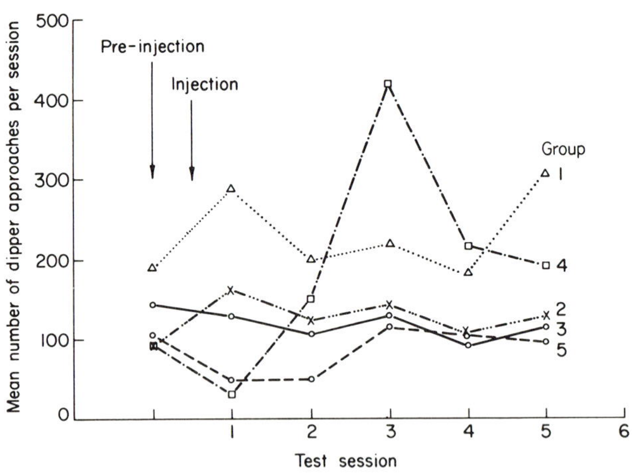

Figure 4 Number of dipper approach responses during each one hour daily test session for naive rats injected with nucleic acid mix (RNA) from 20, 10, 5, days trained, untrained, and sensory deprived donors.

back and forth from the bar to the dipper and in pressing the bar; its magazine entries tend to come only during the period immediately after it has pressed the bar.

Discussion

The "magazine-transfer" paradigm has yielded highly significant "transfer" results in our laboratory for some two years now, and we have performed enough parameter studies that we begin to appreciate what are the critical factors that influence the success of these experiments. These critical factors include the following: (a) the amount of RNA injected should be at least the equivalent of that taken from half a donor brain; (b) the recipients should be well-gentled and well-habituated prior to injection; (c) the recipients should be well-deprived prior to injection; (d) the donors should be given an optimal amount of training prior to sacrifice, and this optimal amount of training must (at the moment at least) be determined empirically in any given training situation; and (e) perhaps most important of all, the pre-injection response tendencies of the recipients must be measured carefully and taken into account if the transfer effect is to be maximized. Additionally, we recommend that the donor animals be sacrificed within minutes after their last training session in a situation such as the one that we have employed, and that the injections be intracranial if RNA is used as the transfer material (McConnell, 1967a).

We have had little success in attempting to use dialyzed materials in the magazine-approach experiments; indeed, we have some tentative evidence that use of dialyzate from the brains of animals trained in an operant conditioning chamber causes what appears to be a "negative" transfer effect. In one pilot study (not reported above), we trained donor animals to press the lever in order to activate the dipper (on a fixed ratio schedule of one reinforcement for each 50 bar-presses); recipients injected with dialyzate from the brains of these trained donors learned significantly more *slowly* than usual, while animals injected with "untrained" dialyzate were not retarded in their speed of learning. Such results are consonant with our findings (in our early attempts to replicate the procedure used by Jacobson and his colleagues described above) that intraperitoneal injections of materials from trained brains could significantly retard learning when food reinforcement was given the animals during training, even though a positive "transfer" effect could be demonstrated in the same recipients during unreinforced trials. It is fascinating to speculate that "approach" tendencies may be mediated by large molecules, such as RNA, while "avoidance" tendencies may be mediated by small molecules, such as would be found in a dialyzate. The fact that most successful "transfer" studies in which dialyzate was used were, in fact, "avoidance" learning paradigms, lends some further credence to this kind of speculation. The "saccharin avoidance" studies to be described below are, in part, an attempt to test this hypothesis.

The magazine-training paradigm we have employed in these experiments does seem to yield striking results, although somewhat more sophisticated equipment is required than has been used in most other laboratories working in this field. We do not as yet know whether or not the "transfer" achieved is stimulus-specific, nor can we guarantee that it is "learning" or "memory" (as opposed to some motivational component) that is being "transferred," but we are relatively sure that the effect is not one that involves a mere change in the recipients' activity level. Altogether, the "magazine-training" paradigm seems worthy of considerable further study.

ATTEMPTS TO TRANSFER AN AVOIDANCE RESPONSE

In the past three years (1964–1967), there have been reports from several laboratories of successful attempts to transfer acquired behavioral tendencies from one rat or mouse to another, by extracting substances from the brains of trained animals and injecting these extracts into untrained recipients. In almost all cases, the training given to the donor animals in these experiments has been of the kind that lasts over many trials and usually over several days. To the best of our knowledge, no one has yet been able to transfer anything like "one-trial" learning, although several investigators have made the attempt. Indeed, in both our laboratory and in Georges Ungar's, evidence has been obtained that suggests that the more training one gives the donor animals, at least up to a certain maximum (as we showed above), the stronger the resulting transfer (Ungar, 1967). In some ways, such a finding is rather surprising, for many of the "one-trial" learning situations employed by experimental psychologists yield rather strong and long-lasting effects, and the "one-trial" situation has proved an effective tool in studies of memory consolidation. One might intuitively think that the stronger the learning and the more easily it comes about, the more likely it is that it could be transferred chemically from one animal to another. (Another charm of a one-trial learning situation, of course, is that it leads to considerable savings of the experimenter's time.) This section of our paper, then, is a progress report of a new series of experiments we have recently begun in which we attempt to transfer an easily-established conditioned aversion to a novel-tasting food in rats.

In the spring of 1967, Samuel H. Revusky visited the Mental Health Research Institute at The University of Michigan and described some research that he and Frank DeVenuto had performed at the U.S. Army

Medical Research Laboratory at Fort Knox, Kentucky. Briefly stated, they trained rats to avoid a saccharin solution by giving x-irradiation to the animals shortly after the animals were given the saccharin solution. After the saccharin aversion had been established, they extracted RNA from the brains of the donor animals using a cold phenol technique, and then injected the material subcutaneously into 40 recipient animals. Another 40 rats received RNA injections from control donors. Both groups of recipients were then tested under 24-hour liquid deprivation to see if any part of the saccharin aversion had transferred chemically. The results of the study by Revusky and DeVenuto were equivocal, but they did find some statistically significant results in the predicted direction (Revusky and DeVenuto, 1967).

As Revusky and DeVenuto point out, the experimental paradigm they employed is interesting in many ways. In the first place, it is a simple matter to train large numbers of donors. Only two or three pairings of the saccharin solution with x-irradiation, or with an injection of a nauseating agent such as apomorphine, are necessary for a very strong and long-lasting aversion to take place. The control donor animals can be given exactly the same experiences as the experimental donors, but if the x-irradiation is paired with tap water rather than with saccharin, the donor animals will actually show an increased consumption of the saccharin solution rather than developing an aversion towards it. We thought that this paradigm had a great deal to commend it, so we set out to replicate the experiment.

The subjects in our first experiment were 18 adult male Sprague–Dawley rats; six of these animals were donors, while 12 were recipients. We began by giving the donor animals ten days of restricted access to water. During these ten days, they were allowed to drink all they wanted during a ten-minute period each day, but were not given water at any other time. After these ten days, the saccharin aversion training began. On days 11, 13, and 15, the donors were given ten minutes' access to a solution of saccharin in drinking water. Fifteen minutes after they had drunk the saccharin solution, the donors were injected intraperitoneally with apomorphine hydrochloride (15 mg/kg) [Garcia *et al.*, 1966]. On days 12 and 14, the donors were given plain tap water and were not injected. As we expected, their response to this training was a dramatic aversion to the saccharin solution. On day 15, the donors drank about one-third as much of the saccharin solution as they did on day 11. Their consumption of plain drinking water on days 12 and 14, however, remained the same as before the injections began—obviously it was the saccharin flavor, and not liquid *per se*, that the rats were avoiding.

Four hours after the apomorphine hydrochloride injection on day 15, the donors were sacrificed by decapitation and their brains immediately frozen. Three of the brains, chosen at random, were used for RNA extraction; the other three donor brains were put through dialysis. The RNA extraction method we employed was the same as that described above. The brains that underwent dialysis were homogenized with five volumes of saline and then dialyzed against 20 volumes of distilled water for 48 hours. The volume of the dialysate was then reduced to one ml per brain by lyophilization. As many of you will recognize, this is the dialysis procedure used by Ungar and his associates (Ungar, 1967).

Our 12 recipient animals were given 14 days of restricted water intake. During this period of time, they were allowed ten minutes of free access to liquid daily, the ten minutes being divided into two contiguous five-minute periods. During the first eight days, the recipients were given ordinary tap water during both of the five-minute periods. On the ninth day, we established a preference baseline for all the recipients. Half of them were given saccharin during the first five-minute period and water during the second; the other half of the animals were given water and saccharin solution in reverse order. In all cases, the recipients consumed more liquid during the first five-minute period than during the second five-minute period. This is, of course, just the sort of behavior one would expect from a very thirsty rat.

For the next five days following the preference test, all recipients were given ordinary water during both five-minute periods of time. On the 14th day, the 12 recipients were split into three groups of four animals each; the groups were balanced for baseline saccharin preference and matched for weight. The four animals in group I received 200 μliter of RNA extract injected into the subarachnoid space of their brains; the animals were injected under pentobarbital anesthesia. The four animals in group II received one ml of dialyzate injected intraperitoneally without anesthesia, while the four animals in group III received one ml of nondialyzable material intraperitoneally, also without anesthesia. About 14 hours after injection, each of the animals was given a repeat of its preference baseline test—that is, the rat received water during one five-minute period and saccharin solution during the other five-minute period. Half the animals received the water first, while the rest of the animals received the saccharin solution first. For the next three testing days, the order of presentation was reversed daily. Thus, half the recipients received water first on test days one and three, while the other half received water first on test days two and four, and the experiment was balanced throughout.

Our analysis of variance did show a triple interaction between extracts and solution between days that was significant at the 0.025 level. Inspection of the behavior of the animals in the various subgroups could lead one to speculate about interesting processes going on here and there, but if there was a transfer effect, it wasn't what one might call outstandingly noticeable.

Perhaps these results were only to be expected. For one thing, we made the classic mistake of trying to "improve" on the paradigm used by Revusky and DeVenuto. These experimenters failed to find a strong transfer effect with RNA injections when their recipients were run under 24-hours water deprivation. However, when they offered one group of animals continual access to both water and saccharin solution, they found a very significant drop in saccharin solution consumption on the second day of testing. It is possible that we would have found significant differences had we also tested our animals under conditions of reduced drive and when they had free access to both water and saccharin solution. We intend to do so immediately.

It is also true that we had fewer donors per recipient than did Revusky and DeVenuto. While this fact might not have mattered as far as the RNA group was concerned (for we injected as much RNA as we have in some two dozen past experiments that yielded positive transfer results), we were certainly using far fewer donors in the dialyzate groups than Ungar and his associates find necessary. And I regret to report that we have some evidence that our RNA extract may not have contained very much intact RNA. Analysis of an RNA extract we obtained on a subsequent experiment showed no intact RNA present at all, just break-down products; we are not entirely sure why there was no intact RNA present. This finding suggests strongly, however, that one should save a portion of the RNA extract from each donor group and analyze it afterwards to make sure how much RNA the extract actually contained.

We are now in the process of repeating the saccharin aversion work with a larger group of animals and making every effort to correct the mistakes that we made in our first experiment. It is our opinion that the evidence obtained so far is sufficiently encouraging to warrant further work on the saccharin-avoidance paradigm. At the moment, however, we cannot recommend it as a paradigm in which it is easy to demonstrate the chemical transfer of acquired behavioral tendencies.

SUMMARY AND CONCLUSIONS

During the past 30 months, we have used more than a thousand rats in a dozen or more experiments to test whether or not acquired behavioral tendencies can be transferred chemically from one animal to another by extracting various substances from the brains of trained animals and injecting them into untrained recipients. In general, our results strongly support the hypothesis that some kind of informational "transfer" can indeed be achieved, and we believe we now know what some of the critical factors are that affect the "transfer" phenomenon. We are not as yet sure what types of "information" can thus be passed from one animal to another, nor how "stimulus-specific" the effect is, but it does appear that the storage and "transfer" of different types of information may well be mediated by different kinds (or at least by different sizes) of molecules. Hopefully, further research in our own and other laboratories will help answer some of the knotty problems as yet unsolved.

We would like to close by emphasizing a point too often forgotten in the controversy that the "transfer" idea has engendered, namely, that our primary interest is in how memories are encoded, stored, and decoded in the brains of the *donor animals*. The transfer technique is primarily useful as a kind of "behavioral litmus paper" to tell us more about the processes going on in the brains of the donors during the learning process. We know by now from the large number of successful reports from laboratories all around the world, that we can get an effect of some kind by injecting untrained recipients with chemicals taken from the brains of trained animals. Even if subsequent experiments should prove conclusively that the effect is not really that of an implantation of "memories" into the brains of the recipients, it is still quite likely that the "transfer" technique itself will be a useful tool in our attempts to learn what goes on in the brains of the donor animals when they learn whatever task we set them.

ACKNOWLEDGMENT

We wish to thank the following people for their assistance in various parts of these experiments: Arthur Aaronson, Richard Block, Patricia Fox, Frank Masiarz, Donald Newman, Barbara Salive, Jessie Shelby, Nancy Slater, Jon Sterngold, William Straugh, Paul Yoder and Kenneth Zuckerman. We also thank Marlys Schutjer for her help in preparing the manuscript.

REFERENCES

Babich, F. R., Jacobson, A. L., Bubash, S. and Jacobson, Ann. Behavioral modifications in naive rats produced by injections of ribonucleic acid extracted from trained rats. *Worm Runner's Digest*, 1965a, **7**, 11–14.

Babich, F. R., Jacobson, A. L. and Bubash, S. Cross-species transfer of learning: effect of ribonucleic acid from hamsters on rat behavior. *Proc. Nat. Acad. Sci.*, 1965b, **54**, 1299–1302.

Babich, F. R., Jacobson, A. L., Bubash, S. and Jacobson, A. Transfer of a response to naive rats by injection of ribonucleic acid extracted from trained rats. *Science*, 1965c, **149**, 656–657.

Fjerdingstad, E. J., Nissen, Th. and Røigaard–Petersen, H. H. Effect of ribonucleic acid (RNA) extracted from the brain of trained animals on learning in rats. *Scand. J. Psychol.*, 1965, **6**, 1–5.

Fried, C. and Horowitz, S. Contraction—a leaRNAble response. *Worm Runner's Digest*, 1964, **6**, 3–6.

Fuchs, A., Harrington, R., Lariviere, R. and Robinson, T. Degree of learning and degree of memory transfer in planarians. *Worm Runner's Digest*, 1966, **8**, 28–31.

Garcia, J., Ervin, F. R. and Koelling, R. A. Learning with prolonged delay of reinforcement. *Psychon. Sci.*, 1966, **5**, 121–122.

Humphries, B. M. and Jacobson, R. The effect of ingestion of conditioned planaria on the response level of naive planaria: II. *Worm Runner's Digest*, 1961, **3**, 165–170.

Jacobson, A. L., Babich, F. R., Bubash, S. and Goren, C. Maze preference in naive rats produced by injection of ribonucleic acid from trained rats. *Psychonom. Sci.*, 1966a, **4**, 3–4.

Jacobson, A. L., Fried, C. and Horowitz, S. D. Planarians and memory: I. Transfer of learning by injection of ribonucleic acid. *Nature*, 1966b, **209**, 599–601.

Jacobson, A. L., Fried, C. and Horowitz, S. D. Planarians and memory: II. The influence of prior extinction on the RNA transfer effect, *Nature*, 1966c, **209**, 601.

Jacobson, A. L., Babich, F. R., Bubash, S. and Jacobson, A. Differential approach tendencies produced by injection of ribonucleic acid from trained rats. *Sci.*, 1965, **150**, 636–637.

John, E. R. Studies on learning and retention in planaria. In Brazier, M.A. (Ed.), *Brain Function*, Vol. II, University of California Press, 1964, 161–182.

Kabat, L. Transfer of training through ingestion of conditioned planarians by unconditioned planarians. *Worm Runner's Digest*, 1964, **6**, 23–27.

Lindberg, O. and Ernster, L. The turnover of radioactive phosphate injected into the subarachnoid space of the brain of the rat. *J. Biochem.*, 1950, **46**, 43–47.

McConnell, J. V. The biochemistry of memory. In Ratner, S. C. and Corning, W. C. (Eds.), *The Chemistry of Learning*, Plenum Press, New York City, 1967a, 310–322.

McConnell, J. V. Factors affecting the "transfer of training" effect in rats. *J. Biol. Psychol.*, 1967b, **9**(1), 40–48.

McConnell, J. V. Comparative physiology: Learning in invertebrates. *Ann. Review of Physiology*, Annual Reviews, Inc., Palo Alto, California, **28**, 1966, 107–136.

McConnell, J. V. Memory transfer through cannibalism in planarians. *J. Neuropsychiat.*, 1962, **3** (supplement), S42–S48.

McConnell, J. V., Jacobson, R. and Humphries, B. M. The effects of ingestion of conditioned planaria on the response level of naive planaria. *Worm Runner's Digest*, 1961, **3**, 41–45.

Pickett, J. B., Jennings, L. B. and Wells, P. H. Influence of RNA and victim training on maze learning by cannibal planarians. *Amer. Zool.*, 1964, **4**, 158; *Worm Runner's Digest*, 1965, **7**, 30–38.

Ragland, Rae S. and Ragland, J. B. Planaria: interspecific transfer of a conditionability factor through cannibalism. *Psychonom. Sci.*, 1965, **3**, 117–118.

Reiniš, S. The formation of conditioned reflexes in rats after the parenteral administration of brain homogenate. *Activitas Nervosa Superior*, 1965, **7**, 167–168.

Reiniš, S. Influence of brain homogenate injection on the speed of the formation of alimentary conditioned reflex in rats. *Worm Runner's Digest*, 1966, **8**, 7–24.

Rieke, J. and Shannon, L. The effects of deoxyribonuclease and ribonuclease on the transfer of learning by cannibalism in planarians. *Worm Runner's Digest*, 1964, **6**, 7–9.

Revusky, S. H. and DeVenuto, F. Attempt to transfer aversion to saccharin solution by injection of RNA from trained to naive rats. *J. Biol. Psychol.*, 1967, **9**(2), 18 22.

Ungar, G. Transfer of learned behavior by brain extracts. *J. Biol. Psychol.*, 1967, **9**(1), 12–27.

Ungar, G. and Oceguera–Navarro, C. Transfer of habituation by material extracted from brain. *Nature*, 1965, **207**, 301–302.

Westerman, R. A. Somatic inheritance of habituation of responses to light in planarians. *Science*, 1963, **140**, 676–677.

Westerman, R. A. A study of the habituation of responses to light in the planarian *Dugesia dorotocephala*. *Worm Runner's Digest*, 1963, **5**, 6–11.

Zelman, A., Kabat, L., Jacobson, R. and McConnell, J. V. Transfer of training through injection of "conditioned" RNA into untrained planarians. *Worm Runner's Digest*, 1963, **5**, 14–21.

BEHAVIORAL TRANSFER VIA INJECTION OF BRAIN HOMOGENATE: ACTIVATION OR SPECIFICITY?*

James A. Dyal

University of Waterloo
Waterloo, Ontario

Arnold M. Golub

Mental Health Research Institute
University of Michigan
Ann Arbor, Michigan

Considerable scientific interest and controversy has been stimulated with respect to the validity of the phenomenon of transfer of behavioral modification from one organism to another through the injection of brain extracts. Since the history and current status of the problem has already been well reviewed by Dr. Byrne and other previous speakers in this symposium, we will concentrate on the report of six experiments which have been completed in our laboratory.

Our first four experiments were addressed to the question: Is there a reliable behavioral effect? The next two experiments attempted to evaluate the degree and type of specificity of the effect. It was our belief that the chemical transfer of information probably takes place by biasing the subject in the direction of certain stimuli, or biasing him to make certain responses. Therefore, we have typically tested for the transfer effect in a nonreinforced

*The research was supported by Texas Christian University Research Foundation: this manuscript was written while the senior author was NSF Science Faculty Fellow at the University of California, Berkeley.

test situation. We felt that this type of test was more likely to reveal subtle biases which might be masked by testing under conditions of primary reinforcement. A similar point of view has been recently expressed by Booth (1967) in his review of vertebrate brain ribonucleic acids and memory retention.

The apparatus used in the first four experiments was a Grayson–Stadler Skinner Box, slightly modified by moving the bar to the side opposite the food magazine, and by passing a photoelectric beam across the entrance to the food cup.* The apparatus was automated in such a way that both number of bar presses (BP) and number of food magazine entries (ME) were accumulated on counters, thereby eliminating the possibility of experimenter bias in judging the responses. In addition, double-blind techniques were used where they were practicable.

The experimental donors received eight to ten days of CRF training followed by three days of extinction, followed in turn by four more days of CRF training. This treatment condition will be referred to as the AER condition (acquisition-extinction-reacquisition). When the animal pressed the bar during CRF he received a 45 mg Noyes food pellet. Simultaneously, there occurred a click of the food delivery mechanism and an instantaneous diminution of the light in the photoelectric circuit across the food cup. During extinction he received only the light-click stimulation.

The control donors were *untrained* in that they never experienced a contingency between bar pressing and the events involved in food delivery. Each control animal was yoked to an experimental animal in such a way that whenever the experimental subject pressed the bar the food delivery mechanism in the control box was pulsed, but a food pellet was not delivered. However, the control animal did receive the light-click stimulation whenever an experimental subject made a response. When a control subject pressed the bar the event was recorded on a counter but the pellet dispenser was not pulsed. Thus, the control group serves to control for the effects of sensory stimulation in the Skinner box. All Ss were trained for 30 min./day. Within one hour following the completion of training on the last training day, the donors were killed with ether. The brain, excluding the olfactory bulbs and the cerebellum, was removed within two minutes of expiration. The tissue was briefly washed with distilled water to remove blood and any bone particles. The brain was then placed into a 25 ml tissue grinder and stored in a refrigerator at 0°F. The brains of the donor rats in each treatment condition were collected in separate tissue grinders. Physiological saline was added in the amount of 1 cc per brain. The brains were then homogenized,

*For additional details of the apparatus and procedures in experiments one–four, see Dyal *et al.* (1967) and Dyal and Golub (1968).

care being taken to avoid excessive heat generation by holding the tissue grinders close to dry ice during the homogenization process. The homogenate was permitted to warm at room temperature for 15 min. prior to injection. Recipient Ss were lightly etherized and injected intraperitoneally with 3.2 cc of the homogenate, and were tested 24 hours after injection. They had been on a 23-hour food deprivation schedule beginning five days before the test, and they were 48–56 hours deprived at the time of the test. The testing was conducted in two successive 30-minute sessions. During the first 30 minutes, the bar was removed from the Skinner box, and the pellet dispenser with the attendant light-click stimulus was pulsed automatically every 60 seconds, but a food pellet was not delivered into the food magazine. During this test period the number of entries into the food magazine was recorded. During the second 30-minute test period the bar was replaced and activated on a CRF schedule, and both bar presses and magazine entries were recorded.

In experiment 1, there were eight donors and four recipients in each treatment condition. The results are presented in Fig. 1. As you can see, the experimental group (AER) made roughly twice as many magazine entries in the first 30-minute test as did the control group. Statistical comparison of the two groups using a Mann–Whitney U-test, resulted in a U value of zero—that is, there was no overlap between the two distributions. The associated p value was 0.01.

In Fig. 1 the group labeled A received homogenate from donors which had been given ten days of CRF training but no extinction or reacquisition. Interestingly enough, these animals did not perform differently from the yoked controls. This datum, together with a review of the literature, suggests to us that the transfer effect is facilitated by the donors having had experience with both reward and nonreward.

The second experiment was an attempt to replicate the results of experiment 1. There were 12 donors and six recipients per group. As can be seen from Fig. 1, the size of the effect as measured by magazine entries in the first 30 minutes was approximately the same as in experiment 1 ($U = 5$; $p = 0.04$).

In our third replication of the experiment, the procedures were essentially the same as in experiments 1 and 2 except that pre-injection operant levels of ME were obtained under the same conditions which were present during the postinjection tests. Also, in experiment 3 the training was done by relatively untrained undergraduate research assistants, and an excessive number of procedural errors and equipment failures occurred. The *n* for each of the recipient groups was 6. The results are presented in Fig. 2. No reliable differences between treatments were obtained on any of the post-injection tests. Furthermore, there were no reliable increases in response rate (ME) from pre- to postinjection in either the experimental (AER) or

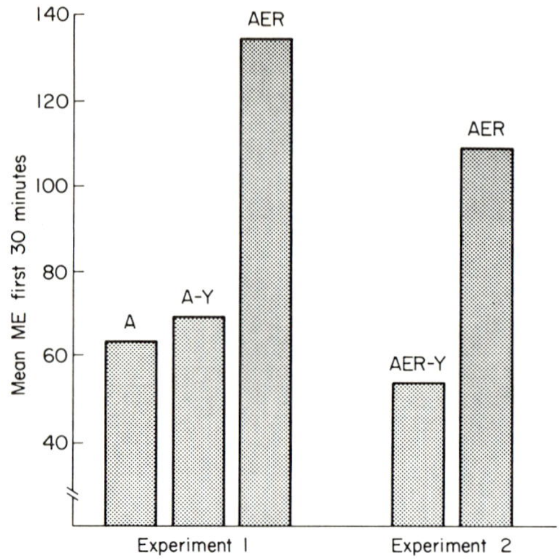

Figure 1 Mean number of ME made during the first 30-minute test period by experimental (AER) and yoked control groups in experiments one and two.

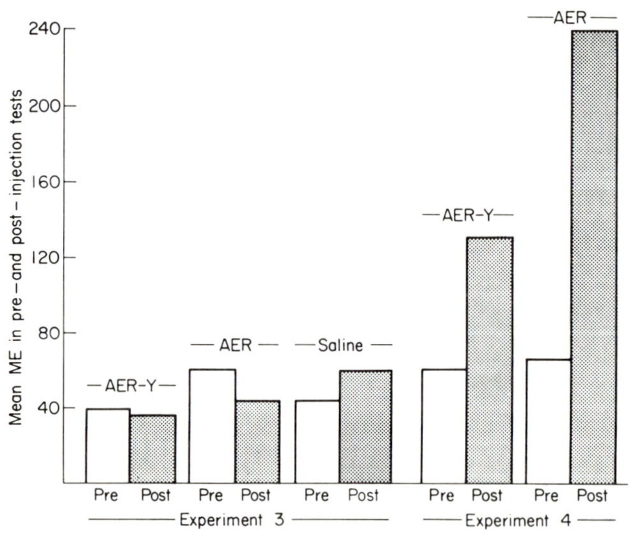

Figure 2 Mean number of MR made during the 30-minute pre-injection baseline test and the first 30-minute postinjection test for experimental (AER) and control groups in experiments three and four.

yoked control (AER-Y) groups. The saline control group on the other hand showed a reliable increase in ME.

At this point we decided that either chance was beginning to catch up with us, or that some extraneous variables had confounded the results. We decided to repeat the experiment with better-trained experimenters. The results of experiment four (Fig. 2) are quite in line with experiments 1 and 2. Both experimental and control groups significantly increased their magazine entries on the postinjection test, but the experimental recipients increased reliably more than the control recipients ($U = 8$; $p = 0.037$).

It should be recalled from the previous description of the test procedures that two postinjection test sessions were conducted. In the second 30-minute session, bar pressing was reinforced on a CRF schedule. Only in the second experiment was there a significant difference between the experimental and control recipients on number of bar presses, and none of the experiments revealed reliable treatment effects as measured by ME during the second test.

Since all four experiments in this series used essentially the same procedure to test the same hypothesis, we ran a test for combined probabilities on the data for the first 30-minute test session (Winer, 1962). The resulting chi square of 25.43 was significant beyond the 0.001 level. Apparently, we can have a good deal of confidence in the statement that rats which have received intraperitoneal injections of brain homogenates from trained rats respond more frequently to stimuli paired with the reinforcement of the donors than do those which receive homogenates from untrained yoked controls. Furthermore, a significant transfer effect has been obtained using our procedure by Golub and McConnell (1968) in the University of Michigan laboratory. We may thus conclude that our single-bar procedure yields statistically reliable and experimentally reproducible transfer effects.

Assuming that there is a real phenomenon to be explained, we may suggest three possible interpretations. The first hypothesis, which is often invoked to "explain" positive transfer effects, is simple sensitization or what we might call the nonassociative put-down. This hypothesis maintains that what is being transferred is an increased sensitivity to the test stimulus. Most important, this sensitivity is not dependent on an association which was formed in the donors between that stimulus and reinforcement, but is simply due to exposure to the stimulus. While this interpretation may have been an appropriate criticism of the original design used by Babich *et al.* (1965), it is not an appropriate explanation of the present results, since the yoked controls received exactly the same stimulus input as the experimentals but without reinforcement.

A second interpretation, one which might legitimately be applied to the results of the first four experiments, is that the obtained effects represent a differential transfer of general activity. It could be argued that because the experimental donors were building up a high rate of bar pressing while the control donors were not, the experimental donors were more active, and that this activity level is being transferred. The present experiments cannot exclude this interpretation. However, it could be argued that if the two recipient groups were differentially aroused, then the experimental group would have been expected to exhibit more bar presses. Taking the four experiments as a whole, this simply was not the case.

The third interpretation holds that some specific learning or memory factor is being transferred. The specificity may be either a response bias or a stimulus bias. This stimulus bias hypothesis is an associative hypothesis since it assumes that the increased sensitivity is dependent upon the test stimulus which has previously been associated with reinforcement of the donors. Although this series of experiments cannot definitely support or refute this interpretation, the results of experiments 3 and 4 suggest that both a general activation and some specific memory factor may contribute to positive results. It will be recalled that in experiment 3 there was no increase in performance over baseline for either group AER or AER-Y; and that furthermore, no significant transfer effect was obtained. In experiment 4 on the other hand, where a differential transfer effect was obtained, *both* groups showed a significant elevation of performance over pre-injection baselines. In order to clearly implicate a specific memory factor, some variety of discrimination task is required. Our next two experiments attempted to test for transfer in situations which might maximize stimulus specificity and response specificity, respectively.

In experiment 5, which attempted to maximize stimulus specificity, we used the black-white discrimination apparatus previously used by Ludvigson and Gay (1966) and shown in Fig. 3. We trained donor animals against their original preference for a total of 128 trials over nine training days. All donors reached a criterion of 18/20 correct responses. The recipients were given 20 nonreinforced preference tests, and were then injected with a brain homogenate from animals which has been trained opposite the recipient preference; *i.e.*, if the recipient had a black preference it was injected with white-trained homogenate. Twenty-four hours after injection the recipients were given ten nonreinforced preference tests.* Twelve of the 18 recipients showed an increase in preference for the side consonant with their injection, three showed no change, and three a shift in the opposite direction. Application of a standard statistical analysis for such data, namely the sign test approximation of the binomial expansion, resulted in a statistic with an associated significant p value of 0.018. However, the sign test does not

*This experiment is more completely reported in Dyal and Golub (1967).

make use of the data for animals which showed no change in preference. Since the experimental hypothesis entails that injection will result in preference change, it was felt that the data for the no-change animals should be included. Since among the animals which showed a change in preference the proportion was 4:1 in the direction specified by the hypothesis, it would seem quite conservative to assign the no-change subjects in the proportion of 2:1 in the direction of the hypothesis. That is, two subjects are taken to support the hypothesis and one to oppose it. Under these conditions the resulting p value is still significant at the 0.015 level. Under the most conservative stance, in which all no-change animals are considered to be negative evidence for the hypothesis, the experimental hypothesis of transfer of stimulus preference must be rejected since the associated p value is 0.13. Considering the controversial nature of the experimental effect, we prefer to take the most conservative position and conclude that transfer was not demonstrated. However, since results supporting stimulus specificity have been obtained in other laboratories (*e.g.*, Jacobson *et al.*, 1966; Bonnet, 1967; Ungar, 1966), we continue to regard the specificity hypothesis as tenable.

In experiment 6 we tried to determine response specificity by training two groups of donors to make morphologically quite different responses. One was trained to bar press, the other was trained to break a photoelectric circuit, by sticking its nose into what would have normally been the food magazine (the food was delivered into a different cup). The logic of the design provides for measurement of both responses during the postinjection test period. If the transfer is response-specific then the recipients which received homogenate from donors trained to bar press should exhibit a greater increase in bar presses over baseline than those which were injected with homogenate from magazine trained donors. Similarly, those recipients injected with magazine homogenate should increase their magazine entries more than the bar-press-injected recipients. The results only partially confirm the transfer of response bias hypothesis. Those recipients which were injected with bar-press homogenate, increased their bar presses over baseline, while those which we injected with magazine entry homogenate decreased their bar pressing (Table I). Evaluation of the pre- vs. postinjection change scores by U-test indicated the difference to be significant at the 0.045 level. On the other hand, the magazine-injected recipients showed no tendency to increase their magazine entries; in fact, as seen in Table II, the postinjection score was somewhat lower than the pre-injection score. The bar-press-injected recipients did not change their ME level on the postinjection preference test.

Following the 30-min. nonreinforced preference tests, half of the recipients which were injected with bar-press homogenate were trained to bar press, and half were trained to stick their nose into the magazine (again, food was

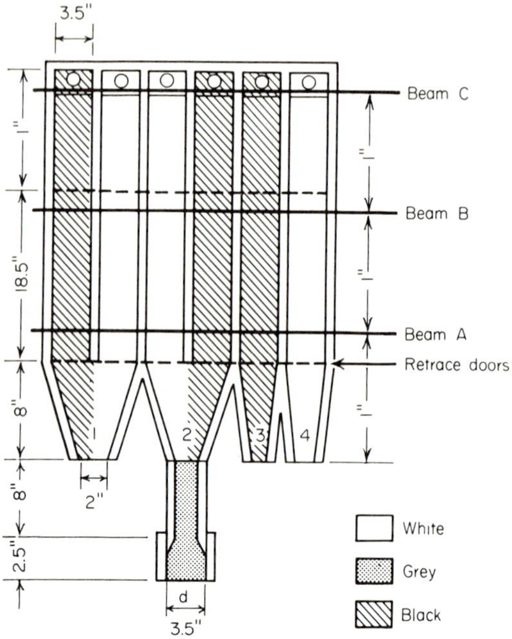

Figure 3 Floor plan of apparatus used in experiment five. Only alleys one and two were used in this study. Retrace doors are indicated by dashed lines and photobeams by solid lines across the apparatus. A start door, not shown, separated the start box from the alley.

TABLE I

Mean Number of BP During 30 Min. Preference Test

Injection Received			
BP		ME	
Pre	Post	Pre	Post
7.45	10.15	8.87	6.19

TABLE II

Mean Number of ME During 30 Min. Preference Test

Injection Received			
BP		ME	
Pre	Post	Pre	Post
57.8	58.5	55.4	46.2

delivered into a different cup). Similarly the magazine-injected recipients were split into two groups; one of which was trained to bar press, and the other to enter the magazine.

Considering first those recipients which were reinforced for bar pressing we found that those subjects which were injected with homogenate from bar-press donors made significantly more bar presses than those which were injected with magazine-trained homogenate. The differences were reliable at the 0.01 level. This result, together with the preference test data, suggests that the transfer is specific to the particular response which has been trained.

This conclusion is somewhat weakened since, as in the case of the preference data, the magazine-injected recipients showed no increase in magazine entries. Furthermore, even the apparent specificity exhibited by the bar-press-injected recipients is questionable, since during the second 30-min. test this group made more ME responses than the magazine-injected recipients. It could be argued that all differences are due to a general activation by the bar-press homogenate, which for some reason did not occur for the magazine homogenate. Although this interpretation cannot be completely refuted, there are some correlational data which argue against the transfer effect being primarily due to activation. During the pre-injection baseline test, the rank-order correlations between BP and ME response measures were +0.68 ($p<0.01$) and +0.59 ($p<0.05$) for the groups which were to be injected with BP homogenate and ME homogenate, respectively. If the effect of the injections is simply an increased activation of all response systems, then it would seem that the injections should have little or no effect on the *correlation* between the two response measures; rather there should simply be an increase in the level of both responses. On the other hand, if the injection does in fact generate a response bias, this should tend to reduce the positive correlation between the two response measures. Consequently, if a subject now has a bias to press the bar, he presumably spends more time around the bar and his responsiveness to the magazine would be reduced proportionately. The data are consistent with this reasoning in that the postinjection correlations between BP and ME during the first 30-min. test are −0.09 and +0.33 for the BP recipients and the ME recipients, respectively. Neither of these correlations is statistically significant.

In conclusion, we would suggest that, with two exceptions, our experiments meet the criteria specified by Byrne in this symposium. The exceptions are: (1) the black-white discrimination experiment did not make use of automatic recording; however, it is quite unlikely that this factor affected the results, since whether the subject enters the black alley or the white alley is an unambiguous response which has been used in hundreds of previous experiments on discrimination in rats; and (2) attempts to replicate our results have been made in only one other laboratory (Golub and

McConnell, 1968). We feel confident of the replicability of the effect, using the procedure of experiments one-four, and will provide detailed procedural protocols for other investigators who may wish to try it.

The tentative conclusions which we draw from our experiments thus far may be summarized as follows:

1. Transfer of behavioral modification can be reliably obtained using our single-bar training method.
2. The behavioral transfer cannot be explained on the basis of simple sensitization.
3. Positive transfer effects probably involve nonspecific differential activation as well as specific stimulus and response biases.

ACKNOWLEDGMENT

We would like to acknowledge the collaboration of Dr. Robert Marrone in the first two experiments.

REFERENCES

Babich, F. R., Jacobson, A. L., Bubash, S. and Jacobson, A. (1965). Transfer of a response to naive rats by injection of ribonucleic acid extracted from trained rats. *Science*, **149**, 656–657.

Bonnett, K. A. Behavioral specificity in chemical interanimal transfer of training. Paper presented at Western Psychological Association Meetings, 1967.

Booth, D. A. (1967). Vertebrate brain ribonucleic acids and memory retention. *Psychol. Bull.*, **68**, 149–177.

Dyal, J. A., Golub, A. M. and Marrone, R. L. (1967). Transfer effects of intraperitoneal injection of brain homogenates. *Nature*, **214**, 720–721.

Dyal, J. A. and Golub, A. M. (1967). An attempt to obtain shifts in brightness preference as a function of injection of brain homogenate. *J. biol. Psychol.*, **9**, 29–33.

Dyal, J. A. and Golub, A. M. (1968). Further positive transfer effects obtained by intraperitoneal injections of brain homogenates. *Psychon. Sci.*, **11**, 13–14.

Golub, A. M. and McConnell, J. V. (1968). Positive transfer effects obtained by intraperitoneal injections of brain homogenates: a replication. *Psychon. Sci.*, **11**, 1–2.

Ludvigson, H. W. and Gay, S. E. (1966). Differential reward conditioning: S-contrast as a function of magnitude of S. *Psychon. Sci.*, **5**, 289–290.

Ungar, G. (1966) Chemical transfer of learning: its stimulus specificity. *Fed. Proc.*, **25**, 207.

Winer, B. J. (1962). Statistical principles in experimental design. New York: McGraw–Hill, pp. 43–45.

INTERANIMAL INFORMATION TRANSFER BY BRAIN EXTRACTS STUDIED IN VARIOUS TESTS

O. L. Wolthuis

Medical Biological Laboratory of the National Defense Research Organization TNO Rijswijk Z.H., The Netherlands

From the experimental results which shall be described it is concluded that the choice of the training method is of crucial importance for the demonstration of interanimal information transfer by means of brain extracts. The results also indicate that transfer phenomena may be observed which are not directly related to the clue of the training method. The choice of relevant pseudotrained controls is thought to be essential.

In principle the methods used to demonstrate the transfer phenomenon are relatively simple: a donor animal is trained to a certain criterion, then decapitated and subsequently an extract is prepared from its brain. This extract is injected into naive recipients. These recipients are placed in the same situation as the donor animal and their behavior is compared with that of recipients which have been injected with a control brain extract. A change in the behavior of the animals injected with the "trained" extract indicates that these recipients have acquired information not present in the animals injected with the control extract. The simplicity of these methods has tempted many investigators to enter this field of research, some obtaining positive and many others negative results.

To increase the chance of detecting a transfer phenomenon, four different test systems were chosen. The test systems were: (a) a one trial, two compartment, light-dark discrimination test, (b) an avoidance test with a light clue in a folded T-maze, (c) an automated drinking test with a light clue and (d) a training method to obtain alternation behavior in a T-maze with a chocolate reward as reinforcement. In all experiments the animals were gentled according to a fixed schedule for three days; except in those experiments in which training of alternation behavior occurred, in those cases gentling lasted five days. No attempts were made to purify the brain extracts, which in all cases were supernatants of centrifuged (19,000–29,000 *g*) brain homogenates.

The one-trial, light-dark discrimination test was used in seven series of experiments, and essentially the same procedure was followed throughout. The apparatus consisted of a lighted and a darkened compartment connected by a door. After the dark preference of the donor animals had been measured during two-minute observation periods for two days, they were locked in the dark compartment and received repeated foot shocks for a total of 15 seconds. The shock intensity was 760 μA (constant current). Control donor animals were treated exactly alike, but received no foot shock. The animals were killed either immediately thereafter or on one of the following days (Table I) and their brains were put on solid carbon dioxide. The brains were then homogenized in ice-cooled saline and centrifuged for two hours at 4°C. The supernatant was injected immediately afterwards. Injection was either intraperitoneally or in the jugular vein. During i.v. injection the animals were anesthetized with hexobarbital (150 mg/kg i.p.).

The natural dark preference in four series of recipients was measured once daily for two minutes before injection of the supernatant. In three other series this was omitted, since any previous experience with the apparatus might interfere with an eventual weak transfer effect. On various days after the injection (Table I) the recipient animals were placed into the apparatus and the time spent in the dark compartment was measured during two minutes. In some experiments this was repeated on several days. None of the recipients received a foot shock.

It was hoped that those recipients which had received "trained" supernatant would spend a smaller portion of their allotted time in the dark compartment than those recipients which had been injected with control supernatant. Although several experimental schedules were used (Table I), a significant difference between the two groups of recipients was not present in any of the seven series.

A second type of experiments made use of a folded T-maze, basically consisting of three parallel alleys connected at one end, as shown schematically in Fig. 1. In the side alleys on either side of the starting compartment, a part of the grid floor could be made "safe," which then was indicated by a light. The choice of the safe place (A or B) was changed at random. Both the experimental donor animals and the recipients were trained to avoid a foot shock by reaching the safe place at the end of the lighted alley within

TABLE I

RESULTS FROM TRANSFER EXPERIMENTS WITH THE ONE-TRIAL LIGHT-DARK DISCRIMINATION TEST. THE FIGURES REPRESENT THE AVERAGE TIME PER GROUP SPENT IN THE DARK COMPARTMENT (SECONDS). IN THE COLUMN UNDER "FOOTSHOCK" THE SYMBOLS + OR − INDICATE WHETHER OR NOT THE ANIMALS HAVE RECEIVED A FOOT SHOCK. THE NOTATION "1 BR/REC." MEANS THAT EACH ANIMAL OF THE GROUP RECEIVED THE EQUIVALENT OF 1 BRAIN PER RECIPIENT. N.S. = NOT SIGNIFICANTLY DIFFERENT.

Donor Animals										Recipients									
Exp. Day −1	0	Foot Shock	+1	+2	+3	+4	+5	+6	N	Injection Route, Dose	Exp. Day Inj. 0	1	2	3	4	5	6	7	N
92	99	+	0.5						10	i.p., 1 br./rec.	81	108	108						10
83	102	−	100						10	i.p., 1 br./rec.	82	88	110						10
99	104	+							6	i.v., 1 br./rec.	—	96							6
82	94	−							6	i.v., 1 br./rec.	—	98							6
104	111	+						12	10	i.p., 1 br./rec.	—	78							10
101	112	−						113	10	i.p., 1 br./rec.	—	78							10
100	116	+			0				5	i.v., 1 br./rec.	—				91	N.S.			5
82	108	−			116				5	i.v., 1 br./rec.	—				110				5
70	87	+						1	20	i.p., 2 br./rec.	63		65		N.S.	85		98	10
69	82	−						78	20	i.p., 2 br./rec.	65		72			106		99	10
67	88	+			7				30	i.v., 3 br./rec.	72				91	N.S.			10
57	76	−			84				30	i.v., 3 br./rec.	66				71				10
69	79	+			4				30	i.p., 3 br./rec.	71				97				8
55	78	−			81				30	i.p., 3 br./rec.	70				98				8

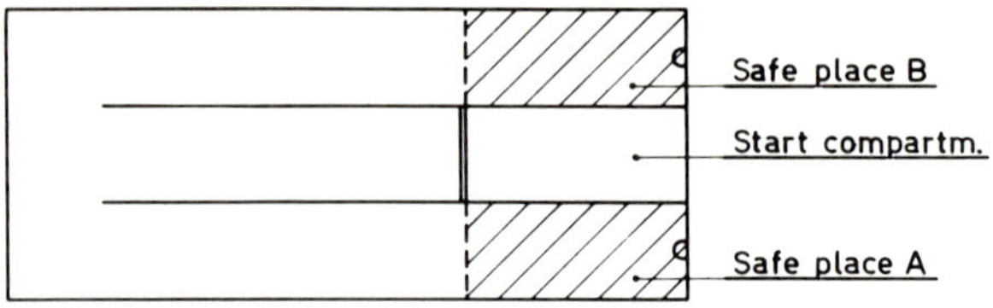

Figure 1 A schematic representation of the folded T-maze used in the avoidance experiments.

12 seconds. An amount of supernatant equivalent to one brain/recipient was injected intraperitoneally.

In the first four series of recipients (totalling 40 experimental and 40 control), the animals which had been injected with "trained" supernatant learned significantly ($p_2 < 0.01$) faster than the controls which had received brain supernatant from nontrained donor animals (Fig. 2A). These results were based on the total number of correct responses of each rat. The ranksum test of Wilcoxon was used to calculate the two-tailed significance of the differences observed.

In the next three series of experiments (totalling 30 experimental and 30 control recipients) the control recipients received brain supernatant from donor animals which had been "pseudotrained", *i.e.*, they were trained in the same maze with both side alleys "safe" and unlighted (ambient lighting only). There was no difference in learning speed between the two groups of recipients, *i.e.*, between those which received "trained" and those which received "pseudotrained" supernatant (Fig. 2B).

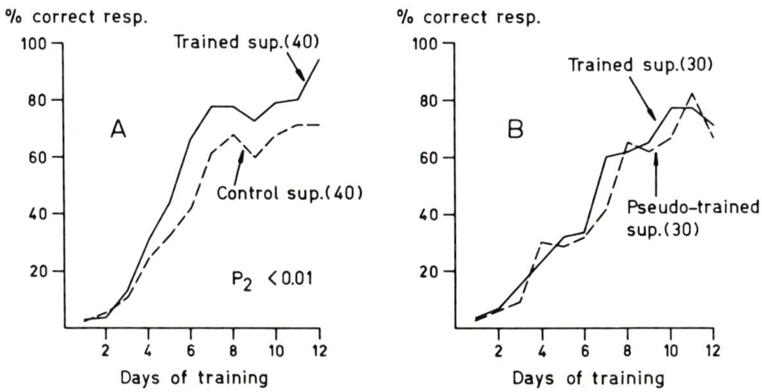

Figure 2a The combined training results of four series of recipients. The control recipients had been injected with a supernatant of nontrained, *i.e.*, naive, donor animals.
Figure 2b The combined training results of three series of recipients. In this case the control recipients received supernatant from "pseudotrained" donor animals (see text).

From these two sets of experiments with the folded T-maze it is concluded that under the conditions mentioned, something is transferred which enhances learning speed without being specifically related to light as a clue. However, this cannot be called aspecific transfer. Such a statement would imply that the differences found in the first four series mentioned above also exist when these recipient animals would be tested in a type of test apparatus with different environmental clues. This experiment has not yet been done. Whatever the case, these results do indicate that the addition of appropriate controls is highly relevant for the evaluation of the results.

With the possibility in mind of purification of greater quantities of brain extracts, a test procedure was devised which is completely automated, has no possibilities for left or right bias, allows various methods such as training, pseudotraining and nonreinforced training, and requires a great number of responses. In fact, the only manipulation needed is to put the animals into the apparatus, clean the cages, supply food every other day, and take out the trained animals after 20 days. The scores are recorded by electronic counting equipment. An additional advantage of the method appears to be that water deprivation of the animals does not take place. After two days, water consumption remains stable during the rest of the training period.

The apparatus consists of two cages in a sound isolated box. Each cage can contain eight to ten animals. Water comes through a standard tube penetrating the wall of each cage. The tubes are connected to peristaltic pumps which can be put into action electronically. When an animal makes an effort to drink, this is counted as a score. After the first touch is scored, the counting system is blocked for five seconds. If the animal is still drinking after five seconds, another score is counted, and so on. Since the calibrated peristaltic pumps deliver a constant volume of water per five seconds, there is a direct relation between the number of those scores which lead to delivery of water and water consumption. The number of scores, and consequently water consumption is not dependent on the licking frequency, or the efficacy with which the animal can suck water out of the tube.

After touching the water tube, the signal triggering the counter blocks the counting system for five seconds, and is also fed into a programmator which determines whether the animal gets water, nothing at all, or a foot shock. The foot shock is delivered by a piece of gridfloor around the water tube. A light (ten W) is mounted in each cage. During each minute, the light is on for 15 seconds. The scores made during the lighted or the dark periods are recorded separately.

After preliminary studies, three experiments on transfer were carried out. Three series of donor animals were trained to drink during the lighted periods only. Upon a drink effort during a "light-on" period, the peristaltic pump immediately started to deliver water for five seconds. During the dark periods a drink effort resulted in a foot shock and no water; the foot shock intensity was increased in steps of 60 μA every two days. After the value of 240 μA had been reached, foot shock intensity was maintained at that level. Evaluation of the training results was made by expressing the number of scores during the lighted periods (correct scores) as a percentage of the total amount of scores. The results of this type of training are shown in Fig. 3. Two cages were used and the results were comparable. In addition, the training curves are reproducible from one experiment to another.

The effects of injection of brain supernatant in a dose equivalent to two brains/recipient were investigated in nonreinforced recipients, injected intraperitoneally with a volume of four ml. Upon each drink effort, these recipients

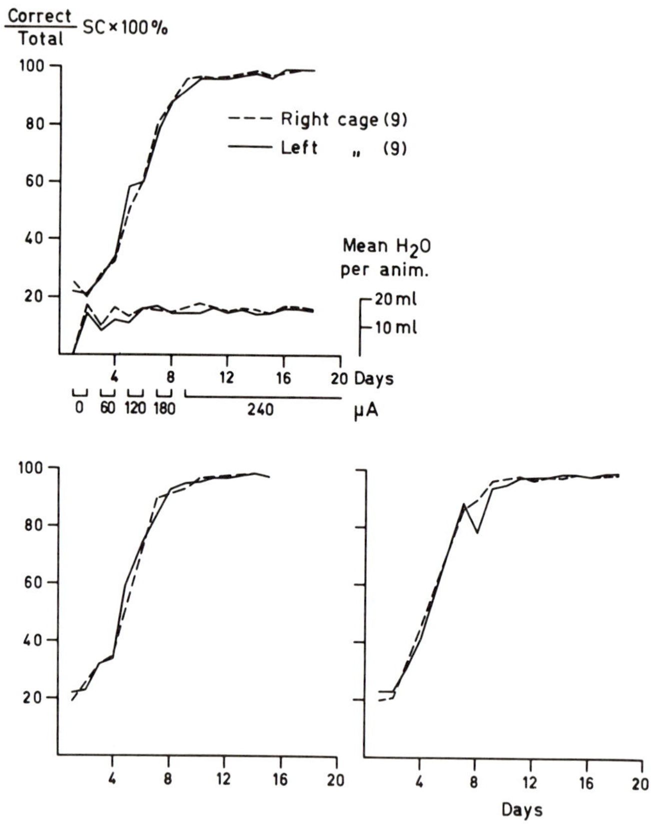

Figure 3 Training results of three series of donor animals in the automated drinking test, each consisting of 18 animals divided over two cages. Water consumption is not shown in the lower two figures. The only aberrant value obtained, on day eight in the third series (bottom right), appeared to be due to a disconnected grid rod. sc = scores.

always received water and never a foot shock. They were subjected to the same light/darkness schedule as the donor animals. The animals in one cage received "trained" supernatant, the animals in the other cage received control supernatant prepared from brains of nontrained donors. The differences between the two groups on each day were calculated by a two by two table. The probabilities of these differences on days two to six were combined according to Rümke and van Eeden (1963).

The results so far look promising (Fig. 4); the tendency to drink during the lighted periods for recipients which received "trained" supernatant was greater than for control recipients. The combined results were significantly different ($p_2 < 0.001$). Water consumption in both groups was not different in any of the series. However promising these results may look, caution is

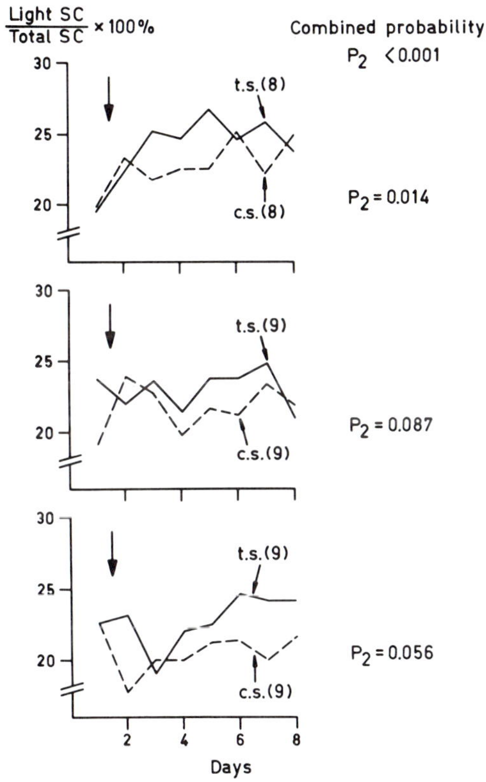

Figure 4 Results obtained with three series of nonreinforced recipients in the drinking test. The large arrows indicate when the animals were injected with supernatant, t.s. = trained supernantant, c.s. = control supernatant prepared from brains of nontrained, *i.e.*, naive, donor animals.

needed for the interpretation, since the controls did not receive a supernatant from pseudotrained donor animals but from nontrained, *i.e.*, naive, animals.

Since all tests described above include some sort of a conditioned reaction, attempts were made to obtain transfer with a completely different type of test, in which animals would not be trained with an external clue (*e.g.*, light) and which would require a more complex response than just a single standard reaction. Food or water deprivation should be avoided, if possible. A test procedure which met these requirements was found in the training of alternation behavior in a T-maze. In this maze, donor rats were trained to find a chocolate reward in a goal box at the end of either the left or the right arm of the T. After the rat had been successful in finding his reward in the left goal box, the next run had to go to the right box to find his reward again, then to the left and so on. Control donor animals were pseudotrained, *i.e.*, in a random sequence they received their reward in the left or the right goal box, or they received no reward at all. These control donor animals did not learn to alternate but they developed a repeating tendency, *i.e.*, they developed a onesided preference for either the left or the right goal box (Fig. 5). There were five runs per training session with an intertrial interval of 15 minutes. Training was continued for 20 days with one session per day.

All recipient animals were trained in the same way as the donor animals, one group received "trained" and the other one "pseudotrained" supernatant, both the equivalent of one brain/recipient. In the first three series of recipients (30 experimentals and 30 controls, see Fig. 6), those animals which had been injected with "trained" supernatant learned the alternating behavior faster than those recipients which had received "pseudotrained" supernatant. Calculated for say 2–16, the difference was in the same direction in all three

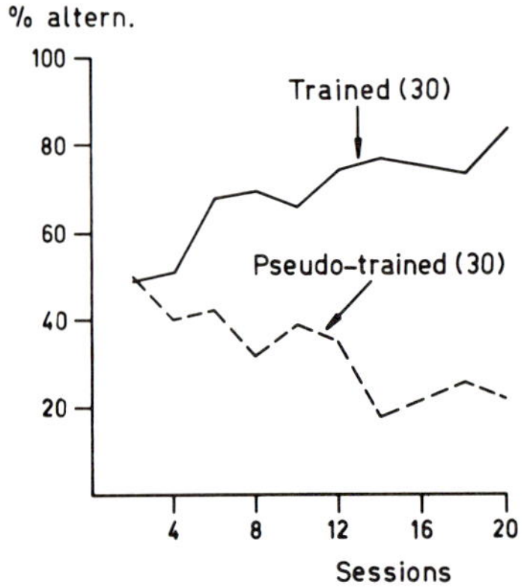

Figure 5 The combined results of three series of donor animals trained for alternation behavior. The controls are pseudotrained and develop a repeating tendency (see text). The results of two sessions are taken together.

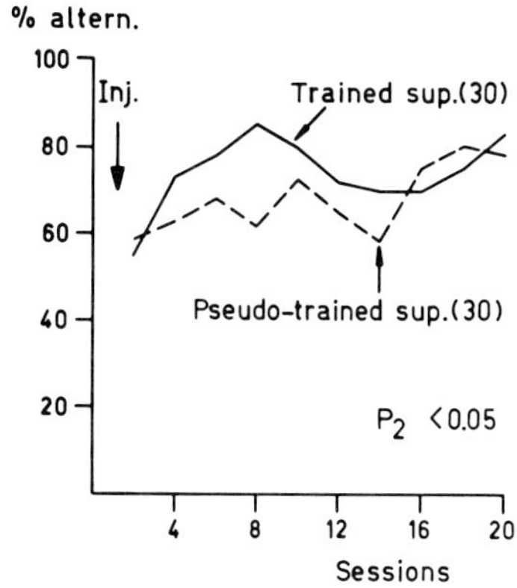

Figure 6 The combined results of training of alternation behavior in three series of recipients. These recipients were either injected with "trained" supernatant or with "pseudo-trained" supernatant obtained from the animals represented in Fig. 5. The results of two sessions are taken together.

series and the combined results showed that the difference was significant ($p_2 < 0.05$). Again the results were evaluated by Wilcoxon's ranksum test and combined according to Kincaid (1963). Subsequently the training method was slightly modified (three instead of five runs per day), which allowed groups of 20 instead of ten rats each. The results of the first series are represented in Fig. 7, again there was a significant ($p_2 < 0.05$) difference in the same direction between the two groups of recipients. In fact a third group of recipients ($n = 20$) had been added, which had received brain supernatant from naive animals. The training curve (not shown in the figure) of these recipients runs between those of the groups represented in Fig. 7. If this can be confirmed in further experiments it might indicate that with the "pseudotrained" supernatant something is transferred which inhibits thc training of alternation behavior. Whether the occurrence of such a factor bears any relation to the repeating tendency which is found in the pseudo-trained donor animals, remains to be elucidated.

Just as in the two compartment test mentioned earlier, which has been extensively studied in our laboratory (v.d. Poel, 1966), other behavioral parameters have also been measured or counted in these T-maze experiments. Running times, choice times, intention movements, "stretched" behavior, and the number of independent alternations have been noted for each animal. Although these data have not yet been completely analyzed, it appears that if there are any differences at all between the various groups, they are very subtle ones. The presence of an olfactory clue in the T-maze is very unlikely. If two groups of rats are trained with the same procedure, and if in one

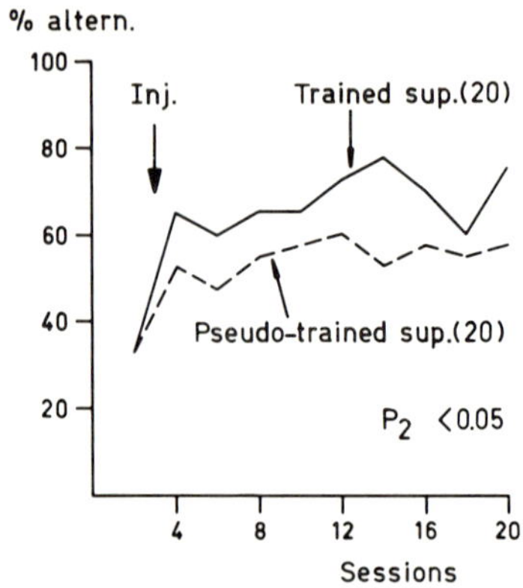

Figure 7 The results of the first series of recipients trained for alternation behavior, according to a more rapid procedure than used for the experiments of Fig. 6. Injection of the supernatant took place after the second session.

group the T-maze is cleaned between each run, the training results are not different. This was investigated in a separate experiment.

In accordance with what has been found by other investigators, these early results, taken together, indicate that a phenomenon consistent with the hypothesis of information transfer was found in the experiments described. Although the last series of experiments, with training of alternation behavior, might upon further investigation offer evidence in favor of task specific transfer, the data obtained so far do not justify any statement on behalf of such a specificity.

ACKNOWLEDGMENTS

These experiments would not have been possible without the criticism and help received from M. Wijnans (statistician), J. F. Anthoni (electronical engineer), A. M. van der Poel (ethologist), W. F. Stevens (biochemist), as well as the excellent technical assistance of M. Remmelts, A. S. M. van Rijsoort, and E. C. Groos, who did most of the training.

REFERENCES

Rümke, Chr. L. and van Eeden, C. In: H. de Jonge, Inleiding tot de medische statistiek, vol. II, p. 773, 1963 (Edited by the Department of Statistics of the Netherlands Institute of Preventive Medicine, Leiden.)

Kincaid, W. M. In: H. de Jonge, Inleiding tot de medische statistiek, vol. II, p. 771, 1963. (Edited by the Department of Statistics of the Netherlands Institute of Preventive Medicine, Leiden.)

v.d. Poel, A. M. Acta Physiol. Pharmacol. Neerl., **14**, 503, 1967.

MEMORY: ANTAGONISTIC TRANSFER EFFECTS

William B. Rucker* and Ward C. Halstead

The University of Chicago
Chicago, Illinois

These investigations began with an NSF grant to Ward Halstead to study induced information transfer in a parabiotic preparation (Fig. 1). Our surgically connected rat pairs shared a common circulation which equilibrated radioactive red cells within two hours. Nevertheless, when we gave black-white discrimination training to four teams of parabiotes with one member blindfolded, and then tested with blindfold exchanged, no information had transferred—either through the blindfold or through the circulation (Halstead, 1963). Thus when we heard that Fjerdingstad *et al.* (1965) had been able to transfer training by intracisternal injections of RNA from trained

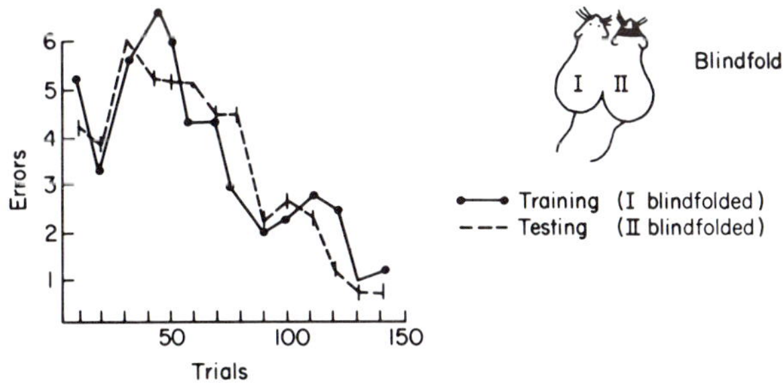

Figure 1 Absence of the transfer of training effect in pairs of rats sharing a common circulation.

*Present Address: Psychology Department, Mankato State College, Mankato, Minn.

donors, we switched to experiments in which extracts from one animal were injected into another.

John McMahon and John Zachary began pilot studies using techniques similar to those employed by the Danes. Their findings with RNA injections suggested that the transfer effect appeared early in recipient training. When ultracentrifuge fractions were employed, it appeared that the nuclear-mitochondrial fraction enhanced learning, and that the postmicrosomal material retarded learning. This was our first hint of negative transfer, defined as a change in the recipient's behavior in the direction opposite to donor training.

More efficient training techniques were required for larger-scaled experiments. Fig. 2 shows the training apparatus used for subsequent experiments. The subject, a young adult, male, Wistar–Furth/MAI rat, began each trial from a platform in the start-box from which a ramp led downward into the running area to allow the rat's natural downward focus to encompass the three stimulus doors. A response was counted if and when the rat got his ears between the two columns framing a stimulus door.

Animals were rewarded for entering any of the three doors during 18 trials of pretraining. During pretraining, the unmarked stimulus doors, initially left open, were raised toward the closed position in three steps to teach the rat how to nudge them over. Two 45 mg Noyes' pellets were available on top of a screen covering a trough containing about 20 unattainable pellets. Behind the trough in each of the three goal compartments was a light used only for discrimination training.

During a discrimination trial, two latched doors would be either lighted or darkened simultaneously, and the odd door unlatched and rewarded.

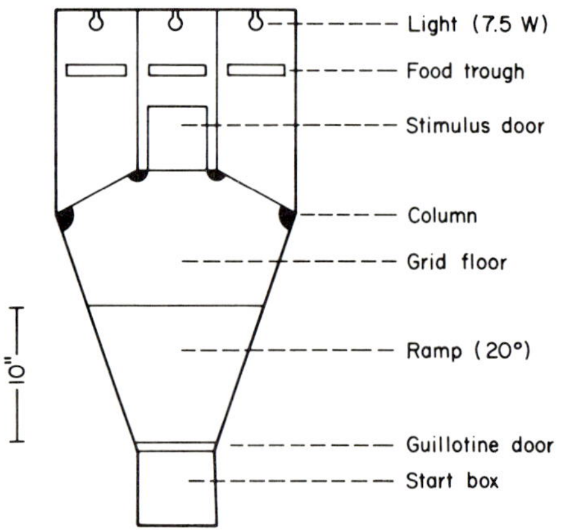

Figure 2 Three-choice discrimination apparatus.

During an approximately 20-second intertrial interval, the position of the odd door could be changed according to a pre-arranged sequence. This sequence was statistically random (Brownlee, 1965) but was actually balanced by restrictions on randomness suggested for the two-choice situation by Gellerman (1933). Rats trained with our sequence of rewarded alternatives, but without light cues (insoluble problems), never performed better than chance. However, almost all rats trained on a soluble discrimination reached by the 54th trial a criterion of seven correct in a running block of nine trials. There is only a five percent risk that a subject would perform so well by chance alone (Bogartz, 1965). That the subjects were learning to respond to the visual cue, rather than to the oddity cue, is evident from the great difficulty that criterion-trained rats had when the visual discrimination was reversed.

We used this apparatus to test the effects of yeast RNA injections on discrimination learning. We hoped to find experimental conditions for later transfer experiments which would allow us to use adjunct injections of yeast RNA as a means of balancing and cancelling those effects possibly carried by brain RNA which were general and not specifically related to the training of the donors. For this purpose, we made intraperitoneal injections of yeast RNA at acute dose levels (up to 1.0 g/kg) comparable to the sum of chronic doses reported to enhance learning in the rat (Cook, 1964; Goren, 1965).

Table I outlines the experimental variables for our yeast RNA experiments. Some of the injections were given at a high pH because alkalosis may weaken the blood-brain barrier (Gilbert, 1965). Gilbert reported that the blood-brain barrier could be disturbed in selected sensory areas by repetitive and specific sensory stimulation. Hence, in the first four yeast RNA experiments, recipients were subjected to various durations (9 to 22 hours) of a light flickering for a half second each second. We were unable to find any set of conditions under which yeast RNA affected discrimination learning.

When we returned to transfer work, we felt that the difficulty that others were experiencing in their attempts to replicate the effect could be attributed to several factors (Carlton, 1966; Booth, 1967). First of all, the transfer agent should arrive at the right place in adequate amounts and in adequate form. Secondly, recipients should be tested under conditions which minimize behavioral variance and which do not evoke strong responses that might overrule transferred instructions. Thirdly, there was additional evidence to support the possibility of negative transfer which might under some circumstances cancel concurrent positive effects (Fjerdingstad *et al.*, 1966). These considerations led to four transfer experiments, T-1 through T-4, which have been numbered in their logical, rather than in their actual, order.

In the first of these experiments, T-1, we began by giving donors one of

TABLE I

EXPERIMENTAL DESIGNS FOR YEAST RNA STUDIES

Experiment	N per Cell	Cells	Independent Variables Tested	Constant Conditions
Y-1	7	3	Dose	High pH, short injection-training interval, short flicker duration
Y-2	10	3	Dose	Low pH, long injection-training interval, short flicker duration
Y-3	7	2	Injection to training interval	High pH, long flicker duration
Y-4	4	6	Dose, duration of flicker, pH	Long injection-training interval
Y-5	3	4	Crossover dose	High pH, short injection-training intervals, no flicker
Y-6	11	2	Dose	No flicker, short injection training interval

No error measure was significant for any experiment. Subjects in Y-1, -2, -3, and -4 were trained with light rewarded. Subjects in Y-6 were trained with dark rewarded. Light and dark cues were reversed after dose in experiment Y-5.

two kinds of training. Nine donors were trained with the lighted door as the positive stimulus. They took an average of 35 trials to reach criterion. Nine other donors were trained for 54 trials on the same sequence of rewarded doors, but since all doors were lighted on each trial, the problem was insoluble. Practice on this insoluble task resulted in an average of 18.9 rewarded trials for these donors, as compared to 19.8, the mean number of reinforced trials for the donors given discrimination training.

The brains were harvested onto powdered dry ice, pooled according to donor training, and then stored in the freezer until RNA was extracted by a method detailed in Appendix I. This method involves a single phenol extraction, conducted at high pH (7.6) and at high temperature (60°C). Four pretrained rats were assigned to each of three recipient groups. The subjects, according to group designation, were injected intraperitoneally either with RNA (approximately 15 mg/kg) from one of the two donor sources or with an equal volume of buffer alone.

For T-1 and for the other experiments of the T-group, the experimenter, who tested recipients 12 hours after injection, did not know the correlation between subjects and treatments. All recipients in T-1 were trained with light positive. The treatment groups did not differ in terms of the errors committed before reaching criterion. They did differ, however, in terms of the errors committed in the first nine trials ($p < 0.025$). Finer analysis, by Scheffé's method of multiple contrasts (Brownlee, 1965), indicated that recipients of RNA from *either* source made significantly fewer errors in the first nine trials than did recipients injected with buffer ($p < 0.10$).

For the remaining experiments of the T-group, we introduced two new techniques: adjunct injections of protamine and a parallel extraction for RNA. Recipients in these experiments were injected with 50 mg of protamine per kg, one-half hour before treatment with either RNA or buffer. Protamine may act, at least in cell suspensions, both to protect RNA from RNase and to aid RNA to enter cells (Amos and Kearns, 1963).

Wishing to increase the yield and quality of our RNA extracts, we designed a parallel extraction technique, detailed in Appendix II, and based in part on work by Brown and Littna (1964). The first step is a phenol extraction conducted at low temperature (1–4°C) with a low pH (5.5) buffer. The supernatant, removed by centrifugation, should contain biologically active RNA of high molecular weight. Secondly, the phenol phase and interphase are extracted at a high temperature (60°C) with a high pH (8.2) buffer. RNA extracted under these conditions is reported to be biologically active, but of low molecular weight (Trakatellis *et al.*, 1964). The RNA yield from the pooled supernatants was five to ten percent greater than that achieved by a single extraction conducted at a high temperature with a high pH buffer.

Since the yeast RNA experiments had failed to uncover any nonspecific effects of RNA on training errors, it seemed possible that the extracts from *both* donor sources in T-1 had carried transfer agents. In particular, the donors who practiced on the insoluble problem had always received their reward in a lighted chamber; and light may have developed as a secondary reinforcement for them. The donors for the next experiment, T-2, were consequently trained for 54 trials *either* on the same insoluble problem *or* on a new insoluble problem. Only one door was lighted on each trial of the new insoluble sequence, and it coincided with reward in a pseudo-random manner. After protamine injections, eight recipients were injected with RNA (14.3 mg/kg) from donors trained on the first insoluble sequence, eight with RNA (22.7 mg/kg) from donors trained on the second sequence, and eight with buffer. When trained to light positive, however, the three groups differed *neither* in terms of errors to criterion *nor* in terms of errors in the first nine trials. We shall return to this result shortly.

All of the donors for T-3 had been trained to criterion with dark positive in other experiments (Y-6 and T-4). Their brains had been harvested into isopentane for storage at liquid nitrogen temperatures and then pooled for extraction by the parallel method. The yield from this extraction allowed us to use two dose levels, 9.2 and 27.5 mg/kg, and to inject 15 recipients at each dose level. Furthermore, ten additional recipients were injected with buffer following the usual protamine injection, and ten more recipients received only buffer injections.

Twelve hours after injection, the fifty recipients were given six trials for which neither stimulus was rewarded. The responses to light and dark stimuli were used to construct a Light Preference Index which is proportional to the latency of responses to the dark stimuli, and inversely proportional to the latency of responses to the light stimuli. The multiple contrasts for this measure were *not* significant. Differences in the raw latency data were highly significant ($p<0.005$), but this could be attributed reliably ($p<0.05$) only to the generally more rapid responses of the RNA recipients.

Fortunately, there seems to be an O'Henry ending. The last experiment, T-4, suggests an interpretation for the results of the other T-experiments. Although a few of the donors for this experiment had received yeast RNA injections before training, this had not affected their performance, as reported earlier. For each donor group, 12 rats were pretrained and then trained with light positive for either 0, 9, 54, 108, or 162 trials. Brains from these donors were harvested over dry ice and stored in the freezer. In addition, 12 brains were collected after 54 trials, and another 12 brains after 162 trials, for storage in the isopentane-liquid nitrogen system. For all donor groups, brains were pooled for extraction by the parallel technique. Frozen aliquots of the extract were set aside in order to inject recipients on four separate occasions, according to a randomized-block design. In all, there were eight recipients in each group and a total of eight groups: in each of seven groups, all rats were injected with 9.4 mg/kg of RNA from a particular donor source, and in one group the rats were injected with buffer.

As in T-3, the recipients in T-4 were tested without reward for their responses to the light and dark stimuli. Fig. 3 charts the highly significant ($p<0.005$) change in the Light Preference Index of these recipients, as a function of the amount of donor training. We set up three multiple contrasts, shown at the bottom of the figure, to define and to test a model of transfer of three effects, indicated as A, B, and C. These correspond to the three inflections of the graph. The model assumes that all extracts carry effect A. Only those extracts prepared from donors given at least some discrimination training are assumed to carry effect B. The model further assumes that overtraining is necessary to develop effect C. Only the extract prepared from donors given 162 trials of practice should carry all three effects. Since the three contrasts were significant ($p<0.05$), the three effect model is tenable, at least for brains stored in the freezer. Note that storage in isopentane-liquid nitrogen eliminated effect B from the extracts prepared from donors trained for 54 trials.

Within four hours after preference testing, these same recipients were

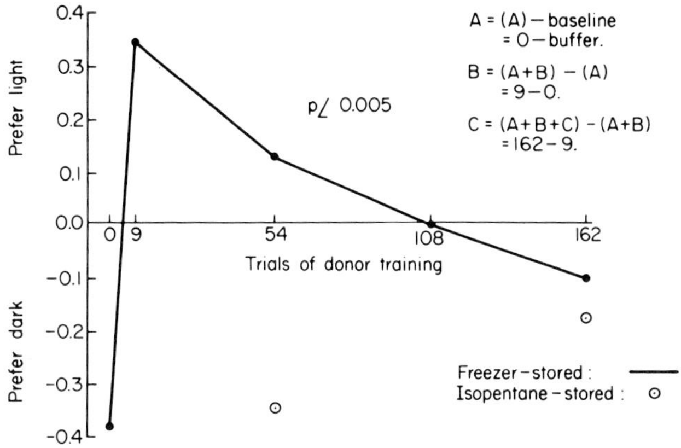

Figure 3 Light Preference Index for recipients in T–4 as a function of the amount of donor training.

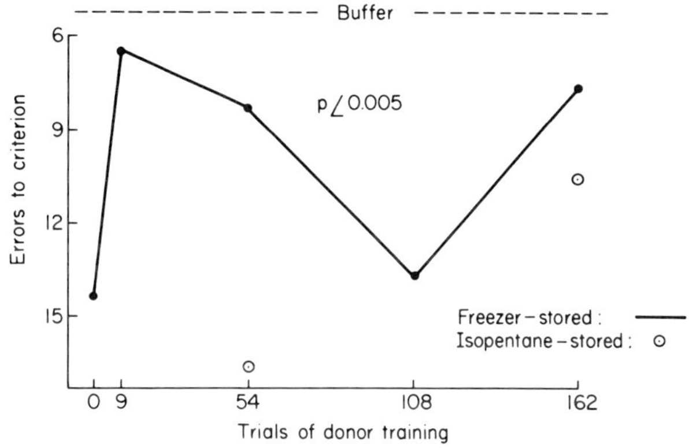

Figure 4 Training errors for recipients in T–4 as a function of the amount of donor training.

trained to criterion with the *darkened* door rewarded—the task opposite to donor training. Fig. 4 shows the mean errors to criterion made by these recipients as a function of the amount of training given to their respective donors. This measure is highly significant ($p < 0.005$). Recipients of buffer made the fewest errors; therefore, the rather poor performance of the recipients of RNA from pretrained donors becomes a negative baseline for transfer effects. This negative baseline corresponds, we think, to the A effect measured before training. The two humps in the error curve show that recipients of RNA from *either* undertrained *or* overtrained donors perform better on the task. All of the inflection points are significant ($p < 0.10$ or better) by multiple contrasts, and the two humps seem to correspond to the B and C effects identified in the preference data from these recipients. Consistent with this notion is the observation that the B effect is again missing from recipients of RNA from brains stored in isopentane.

Fig. 5, which superimposes Figs. 3 and 4, shows that the assumption of three linear and additive factors is rather too parsimonious. First of all, the B factor is in the direction of donor training when recipients were tested without reward, but acts in the direction *opposite* to donor training when recipients were tested with reward. Secondly, the B and C factors, which individually seem to aid recipients during rewarded testing, seem to interact antagonistically at intermediate levels of donor training. This may account for the poor performance of the recipients of RNA from donors given 108 training trials. We suggest, then, that the B and C factors are qualitatively different, and that either factor may be employed in an entirely different manner by recipients tested under different conditions of reward. Such transfer effects may eventually tell us something about how macromolecular information is processed *by* the recipient.

The results of T-4 seem to be consistent with the first three experiments in the T-group. Referring to Table II, it seems possible that the massive failure of T-3 may have resulted from a loss of the B factor from isopentane-stored brains. T-2 may have failed as a result of the antagonistic interaction

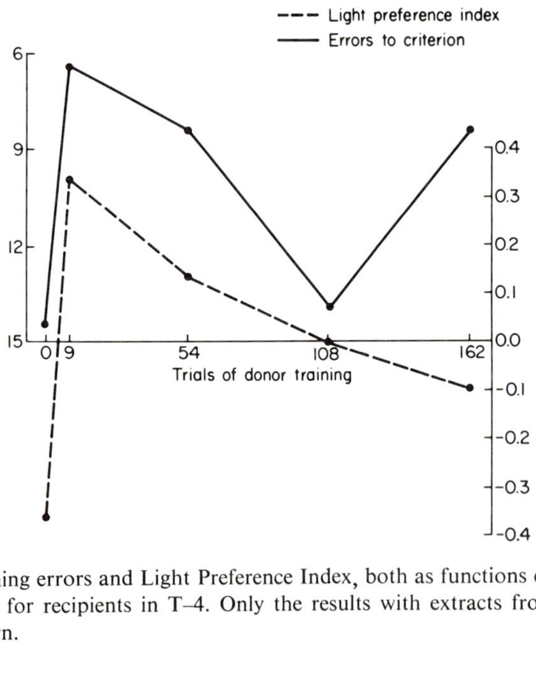

Figure 5 Training errors and Light Preference Index, both as functions of the amount of donor training, for recipients in T–4. Only the results with extracts from freezer-stored brains are shown.

TABLE II

CHARACTERISTICS OF TWO AGENTS HYPOTHESIZED TO TRANSFER TRAINING

Experiment	Donor Training	Trials	Transfer Agents Developed	What Transfers to Recipient	Agent Lost in Preparing Extract
T-1	Light Rewarded	35	$B > C$	$B \gg C$	*C lost* through single extraction
	Insoluble (I)	54	$B \geq C$	$B > C$	
T-2	Insoluble (I)	54	$B \geq C$	$B \geq C$ (Net transfer: Zero)	
	Insoluble (II)	54	— —	— —	
T-3	Dark Rewarded	45	$B > C$	$B = C$ (Net: Zero)	*B lost* through isopentane storage
T-4	Light Rewarded	0	— —	— —	
	Light Rewarded	9	$B \gg C$	$B \gg C$	
	Light Rewarded	54	$B > C$	$B > C$	
	Light Rewarded	108	$B = C$	$B = C$ (Net: Zero)	
	Light Rewarded	162	$B < C$	$B < C$	
	Light Rewarded	54	$B > C$	$B = C$ (Net: Zero)	*B lost* through isopentane storage
	Light Rewarded	162	$B < C$	$B \ll C$	

MODEL ASSUMES:

1. Transfer agent B develops earlier in donor training than does transfer agent C.
2. A single extraction is sufficient to partition agent B, but this method allows transfer agent C to remain in either the phenol phase or interphase.
3. Agent B is inactivated by storage in isopentane-liquid nitrogen.
4. The net transfer effect of an extract is the sum of the individual effects of agents B and C (as determined by the kind and amount of donor training) *minus* the antagonistic interaction between these two agents.

between the B and C factors. If so, the *success* of T-1 should be attributed to a difference in the extraction technique—namely that the C factor may not have been extracted by the single high temperature and high pH method. If different transfer agents may be differentially extracted, this would allow us to avoid antagonistic interactions—except when they are the object of the investigation.

A discussion of these findings in relation to what the rat may be expected to learn early and late in training and in relation to other antagonistic transfer effects has been included in a theoretical note earlier in these proceedings (Halstead and Rucker, 1969).

ACKNOWLEDGMENT

We gratefully acknowledge the substantial and expert contributions of Mr. Robert Shapiro and Miss Linda Howell in the behavioral part of this work. We are obliged to Dr. Charles Wyttenbach and Dr. Richard Haselkorn for their advice and generous loans of equipment and space for the biochemical part of this study.

APPENDIX I

Single RNA Extraction Method
(Modified from Haselkorn, 1962)

1. Ten rat brains are ground frozen in dry ice and the powder placed in a glass centrifuge tube.

2. Just prior to thawing, add a mixture of 30 ml water, 2 ml of 25% sodium dodecyl sulfate (SDS), five ml of 1.0 *M* tris-maleic acid (TM, Mathieson, pH 7.6), and two ml of 1% sodium bentonite (resuspended from the precipitate formed at 10,000 RPM in a Beckman 30 Rotor). Stir briefly and add 40 ml of water-saturated, redistilled phenol. Stir well for two minutes and then stir gently in a 60°C water bath for 60 minutes.

3. Centrifuge for ten minutes at 2,500 RPM and discard interphase and phenol layers.

4. To the aqueous phase add two ml of the bentonite solution and one volume of saturated phenol. Extract for 30 minutes at 60°C.

5. Repeat step 3.

6. To the aqueous phase add one volume of saturated phenol and extract for 15 minutes at 60°C.

7. Repeat step 3.

8. Extract the aqueous phase three times with 24:1 (v/v) chloroform-isoamyl alcohol. Add one ml of 3 *M* sodium acetate buffer at pH 5.0 per 20 ml of aqueous phase. In the cold, add two volumes of cold absolute ethanol and let stand for 15 minutes.

9. Centrifuge in the cold for 15 minutes at 6,000 RPM. The pellet is RNA. (The supernatant may be checked for further precipitation.)

10. The RNA is washed twice in absolute ethanol, dried, and dissolved in a minimal volume of TM (0.01 *M*, pH 7.5) and dialyzed in RNase-free dialysis tubing for 24 hours against two changes of 2 L buffer. (U.V. Spectrophotometric analysis of similarly prepared rat brain dialysates, performed by John Zachary, indicate an average yield of 1.33 mg of RNA per rat brain. The RNA may be reprecipitated for storage.)

APPENDIX II

Parallel Extraction Method
(Based in part on Brown and Littna, 1964)

1. For ten brains: Remove the brains from the freezer or liquid nitrogen (invert the beaker to melt and drain off isopentane through a funnel). Weigh and then grind the frozen tissue in dry ice and place the powder in a 250 ml. flask.

2. Just prior to thawing, add 30 ml of cold low pH buffer (ten ml 1.0 *M* TM, pH 5.5; five ml one percent sodium bentonite (prepared as in I), ten ml 25% SDS; 75 ml twice distilled H_2O) and stirred gently in an ice bath for 15 minutes before adding 30 ml of water-saturated, redistilled phenol. Stir gently for 25 minutes, maintaining the temperature between 1–4°C.

3. Centrifuge for 15 minutes in the cold at 10,000 RPM.

4. Remove aqueous phase to cold flask by suction with pipette. (This is made easier with rubber tubing attached to the mouth end of the pipette.)

5. To the remaining material, including the interphase, add 15 ml of warm high pH buffer (same as low pH buffer except that the pH of the TM is 8.2). Stir gently for 45 minutes at 60°C.

6. Centrifuge in the cold at 10,000 RPM for 15 minutes.

7. Remove the aqueous phase to a cold flask.

8. Wash each aqueous phase three times with one to two volumes of 24:1 (v/v) chloroform-isoamyl alcohol (l-octanol is an adequate substitute). Add two volumes of cold absolute ethanol to the last separation and store overnight in the freezer.

9. Centrifuge for 20 minutes at 20,000 RPM, pour off the supernatant and allow the precipitate, RNA, to drain almost dry. The precipitate is removed to magnetically stirred 0.1 *M* TM (pH 8.2) and stirred for several hours without foaming in an ice bath. The extractions for each sample were combined prior to the RNA determinations.

10. An aliquot of the combined solutions was taken for successive dilutions to be analyzed colorimetrically for RNA by the method of Mejbaum (Discussed by Dische, 1955) and the remainder portioned out for freezing until just before use.

REFERENCES

Amos, H. and Kearns, K. E. Influence of bacterial ribonucleic acid on animal cells in culture. II. Protamine enhancement of RNA uptake. *Exp. Cell. Res*, 1963, **32**, 14–25.

Bogartz, R. S. The criterion method: Some analyses and remarks. *Psych. Bull.*, 1965, **64**, 1–14.

Booth, D. A. Vertebrate brain ribonucleic acids and memory retention. *Psych. Bull.*, 1967, **68**, 149–177.

Brown, D. D. and Littna, E. RNA synthesis during the development of *Xenopus laevis*, the South African Clawed Toad. *J. molec. Biol.*, 1964, **8**, 669–687.

Brownlee, K. A. *Statistical Theory and Methodology in Science and Engineering* (2nd Ed.). New York: John Wiley and Sons, Inc., 1965.

Carlton, P. L. (Ed.), Memory transfer and RNA. Privately circulated manuscript. Rutgers, the State University, New Brunswick, New Jersey, 1966.

Cook, L. Effects of drugs on operant conditioning. In H. Steinberg (Ed.) *Animal Behavior and Drug Action*. Boston; Little, Brown, and Company, 1964.

Dische, Z. Color reactions of nucleic acid components. In E. Chargaff and J. N. Davidson (Eds.), *The Nucleic Acids*, Vol. I. New York: Academic Press, 1955, p. 285–305.

Fjerdingstad, E. J., Nissen, T. and Røigaard-Petersen, H. H. Effects of ribonucleic acid (RNA) extracted from the brain of trained animals on learning in rats. *Scand. J. Psych.*, 1965, **6**, 1–6.

Fjerdingstad, E. J., Nissen, T. and Røigaard-Petersen, H. H. Facilitation of learning in rats by intra-cisternal injections of RNA from the brain of trained rats. In O. Walaas (Ed.), *Molecular Basis of Some Aspects of Mental Activity*, Vol. I. New York: Academic Press, 1966, p. 165–179.

Gellerman, L. W. Chance orders of alternating stimuli in visual discrimination experiments. *J. genet. Psych.*, 1933, **42**, 206–208.

Gilbert, G. J. Focal breakdown of the blood-brain barrier by specific sensory stimulation. *Trans. Am. Neurolog. Assoc.*, 1965, **90**, 246–248.

Goren, C. Ribonucleic acid: Influence on the maze-learning ability of rats. *Worm Runner's Digest*, 1965, **7**, 28–31.

Halstead, W. C. Paper (unpublished) presented at Symposium on Nucleic Acids and Behavior: 35th Annual Meeting of the Midwestern Psychological Association, Chicago, Illinois, 1963.

Halstead, W. C. and Rucker, W. B. The molecular biology of memory. In W. Byrne (Ed.) *Molecular Approaches to Learning and Memory*. New York: Academic Press, 1969, 1–14.

Haselkorn, R. Studies on infections RNA from turnip yellow mosaic virus. *J. molec. Biol.*, 1962, **4**, 357–367.

Trakatellis, A. C., Axelrod, A. E., Montjar, M. and Lamy, F. Induction of immune tolerance with ribosomes and ribonucleic acid extracts in new-born mice. *Nature*, 1964, **202**, 154–157.

PURINE METABOLITES IN MEMORY CONSOLIDATION*

Walter B. Essman

Queens College of the City University of New York

Several experimental approaches have provided indications for a functional role(s) for the ribonucleic acid (RNA) molecule in cognitive processes. This role, however, has not always been clearly a causal relationship between RNA synthesis, concentration, turnover, or binding in the brain and cognitive functions. It has rather been derived from evidence, which on one hand suggests a facilitatory capacity for parenterally injected RNA (yeast-derived) to enhance "learning" (Cook *et al.*, 1963; Goren, 1965; Brown, 1966), and on the other hand, has been relied upon to account for "transfer" of acquired behavior when parenterally-injected brain "RNA extracts" from a trained donor animal were given to a naive recipient (Babich *et al.*, 1965, 1965b; Jacobson *et al.*, 1965; Byrne and Samuel, 1966). There has been evidence to strongly obviate the possibility that peripherally administered RNA can migrate in any measurable quantity to the central nervous system (Roll *et al.*, 1949; Sved, 1965; Eist and Seal, 1965). It appears still less likely that any intact RNA molecules that were successful in escaping hydrolysis *en route* to the brain, could still succeed in facilitating or mediating learning and/or memory by migrating to specific loci within the brain. The most probable disposition of parenterally-introduced RNA molecules is their hydrolyzation by liver RNA, thus making free bases of the molecule available for subsequent metabolism. In the central nervous system, RNA has been attended to in one respect, in that changes in the base ratios of the nucleotide in neurons have been observed under conditions of physiological (Hydén, 1962), chemical

*This research was supported, in part, by grants from the City University of New York and Biomedical Sciences Support Grant NIH No. 5–SO5–FR–07064–02.

(Egyházi and Hydén, 1961; Hydén, 1963) and behavioral (Hydén and Egyázi, 1962) stimulation. It is of interest that these alterations have consistently involved the purine bases, adenine and guanine. Since the purines occur as free bases centrally or peripherally, only as a result of degredative processes which lead to their terminal metabolism, it appeared reasonable that the purine metabolites should be given some attention, both in terms of the substrates from which they derive and in the independent functions that they may serve.

Purines, once cleaved from pentose phosphate, are oxidized to uric acid; specifically, adenine, upon deamination by adenase, forms hypoxanthine which is oxidized by xanthine oxidase to xanthine, and then to uric acid. Uric acid is similarly formed from xanthine, which results from the deamination of guanine by guanase. The terminal metabolite of purine metabolism in man is uric acid; however, in most other mammals this is further degraded to allantoin by the presence of uricase.

Xanthine and uric acid are both endogenous central nervous system stimulants (Orowan, 1955; Ritchie, 1965), being chemically related to the xanthine alkaloids; caffeine, theobromine, and theophylline. The latter have probably been more extensively studied as exogenous central nervous system stimulants and appear to exert a wide range of behavioral effects. It is therefore surprising that xanthine and uric acid have been largely overlooked in this respect. There have been several suggestions that uric acid serves to maintain a high level of cognitive and intellectual functioning (Orowan, 1955), is related to intelligence test performance (Stetten and Hearon, 1959), and appears well correlated with achievement (Brooks and Mueller, 1966). On the other hand, serum uric acid levels were not found to be appreciably lower in senile and senescent patients (Kral *et al.*, 1967) and are elevated in cases of Down's syndrome (Coburn *et al.*, 1967). There is also evidence to suggest that certain forms of hyperuricemia are attended by severe intellectual impairment. Such evidence, in general, is rather inconclusive, and perhaps at best may again implicate performance *per se* rather than specific cognitive functions.

There are several specific findings upon which the present study was based: (1) the relationship between parenterally-injected RNA and serum uric acid levels; (2) the relationship between increased uric acid concentration and elevation in brain RNA; and (3) the facilitation of maze learning by uric acid and its impairment by uricase. These findings, which are presented below, suggested that uric acid, not RNA or xanthine, had specific effects upon learning and memory, and that these effects were possibly mediated through the central action of uric acid.

Since it might be expected that parenterally-injected RNA, once hydrolyzed would subsequently lead to an increase in the concentration of the endogenous metabolites of its free bases, this was initially investigated. Previous data (Essman and Lehrer, 1967) have suggested that maze performance of mice was facilitated by four hours following the i.p. administration of RNA extracted from mouse brain. These findings could possibly be attributed to increased concentration of endogenous C.N.S. stimulants deriving from the purine bases of the hydrolyzed nucleotide. In the present study, RNA was prepared by phenol extraction of whole mouse brains, was purified and made relatively free of protein and DNA, and was injected i.p. into male CF-1 strain mice weighing approximately 27 gm. Each mouse was given approximately 0.5 mg. RNA dissolved in 0.9% saline. Another group of mice was given an equivalent dose and volume of purified yeast RNA, and a third group of animals was given an equivalent volume of 0.9% saline. Four hours following injection, blood samples were drawn and a micro-assay for plasma uric acid was performed. The concentration of uric acid was determined by measuring the increase in fluorescence of allantoin formed, against a standard, from the oxidation of the extracted product by uricase. The results, summarized in Table I, indicate that RNA generally leads to increased plasma uric acid levels by four hours following parenteral administration, although the brain RNA led to appreciably greater concentration increases than did the yeast RNA. Since the base ratios of these RNA molecules differ appreciably

TABLE I

PLASMA URIC ACID LEVEL FOLLOWING RNA OR SALINE INJECTION

Treatment	Plasma Uric Acid (μg/ml)	P
Brain RNA (100 mg/kg)	11.80	$<.001$
Yeast RNA (100 mg/kg)	3.28	$<.05$
0.9% Saline	3.10	—

from one another, one might also expect to find differences in the concentration of base metabolites.

On the basis of the above finding and prior findings of a facilitative maze performance effect produced with brain RNA injection, another study concerning the maze acquisition effects of uric acid was conducted. Male CF-1 strain mice were injected, i.p., with either 0.75 mg/kg of uric acid, 1.65 mg/kg of uricase, or an equivalent volume of 0.9% saline. One hour following injection, all animals were given four acquisition trials in a simple water maze (Essman and Jarvik, 1961), followed by four habit-reversing trials. Latency to escape from the water (interval between entry and successful traversal onto an escape ramp) and maze errors (entry into incorrect corner of maze opposite to that on which escape ramp was provided) were measured for each of the acquisition and reversal trials. These data are summarized in Figs. 1 and 2; analysis of these data indicated significant positive savings (facilitation) conferred by uric acid treatment, and significant negative savings (inhibition) brought about when endogenous uric acid was oxidized by uricase.

The possibility that a similar facilitatory effect upon maze acquisition might be brought about by the administration of xanthine was investigated

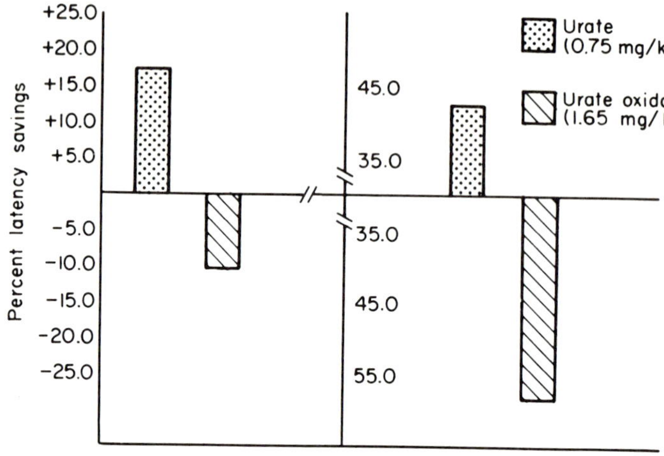

Figure 1 Percentage latency savings for maze acquisition and habit reversal in mice treated with uric acid or uricase.

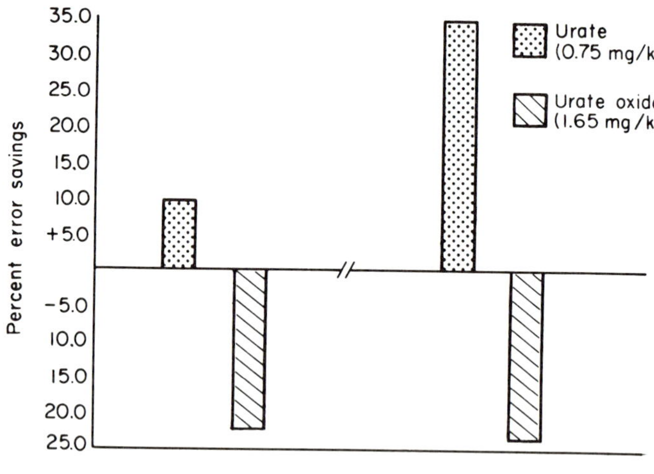

Figure 2 Percentage error savings for maze acquisition and habit reversal in mice treated with uric acid or uricase.

in another study. Mice were injected i.p. with 2.0 mg/kg or 5.0 mg/kg of xanthine, or an equivalent volume of 0.9% saline. Acquisition and habit reversal training was given, as described in the previous study. It was found that xanthine administered at either dose, one hour prior to training, failed to provide any significant savings in either escape latency ($p > 0.70$) or in errors ($p > 0.10$).

The effect of either uric acid or xanthine in possibly accounting for maze performance effects was also considered with respect to locomotor activity. If this were altered by either or both of these compounds, then maze performance could possibly be partially accounted for in terms of motor effect. Groups of mice were given uric acid, urate, xanthine, or saline in the doses used in the above studies. A cumulative record of locomotor activity measured in a closed field following i.p. injection did not indicate any alteration in locomotor activity over a 90-minute period. Sequential analysis of the activity records for animals under all treatment conditions, for each of six 15-minute activity records, indicated that none of the drug-treatment conditions led to any alteration in locomotor activity within or between the segmented time intervals.

Another study dealt with the question of whether the drug effect observed in previous studies might be specific for either motor learning or cognitive learning in this maze situation. Previous data (Essman, 1967c) have shown that savings in water-maze escape-response acquisition was a function of the amount of motor training given prior to actual escape trials. Utilizing the same water-maze situation described above, mice were given motor training without escape 24 hours prior to escape training trials. One hour following i.p. injection of either uric acid, uricase, xanthine, or saline at the doses previously indicated, four escape-training trials were given with escape latency and maze errors noted for each trial. The results, summarized in Table II, indicate that significant error savings were conferred by treatment with uric acid, whereas uricase clearly produced significant negative error

TABLE II

PERCENTAGE INCIDENCE OF TOTAL MAZE ERROR SAVINGS (SALINE-TREATED GROUP/SALINE ERRORS × 100)

Treatment	N	Per Cent Total Error Savings	P
Uric Acid (0.75 mg/kg)	20	+7.0	<.05
Uricase (1.65 mg/kg)	20	−33.3	<.01
Xanthine (2 mg/kg)	20	−30.0	<.05
Xanthine (5 mg/kg)	20	−3.0	>.20

savings, or impairment of maze acquisition. The effect of xanthine in this situation appeared to be more disruptive, rather than being facilitatory. Therefore, in the present paradigm, by maximizing the degree of motor learning involved in the maze acquisition behavior prior to the drug treatment, the observed drug effect—enhancement of maze acquisition by uric acid—could be more directly linked to its role as a stimulant in cognitive rather than motor behavior. The bi-directional effects of an apparently increased C.N.S. stimulation by exogenous uric acid, and a likely decrease in endogenously mediated stimulation by partial removal of uric acid by uricase were observed behaviorally. It appeared possible, therefore, that alterations in stimulation level, provided pharmacologically, could affect the synthesis of neural RNA. This suggestion has been previously made, and has found support in the case of several specific compounds (Hydén, 1963). It was the purpose of the following experiment to determine the extent to which the concentration of whole brain RNA could be affected by several of the compounds utilized in the above studies.

Groups of mice were given i.p. injections of either 0.75 mg/kg of uric acid, 1.65 mg/kg of uricase, or an equivalent volume of 0.9% saline. One hour following injection, RNA concentration was estimated for whole brain, excluding cerebellum and olfactory bulbs, by pentose analysis (Schneider, 1957). The resulting data, summarized in Fig. 3, suggest trends in concentration differences approximating statistical significance ($p < 0.10$) for uric acid-induced brain RNA elevation and uricase-induced brain RNA reduction. Since previous data (Essman, 1965, 1966a) have indicated that alterations in neural RNA may contribute to the rate at which neural memory consolidation may occur, the present data suggested that some preliminary attention be given to the question of the memory consolidation process.

Groups of CF-1 male mice were treated with either uric acid (0.75 mg/kg),

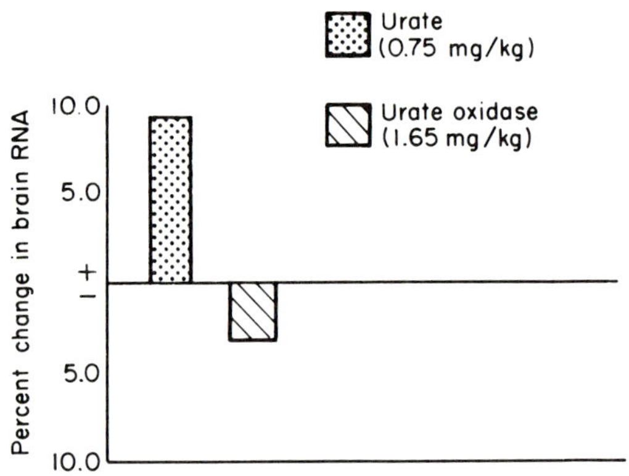

Figure 3 Percentage in brain RNA one hour following injection of uric acid or uricase in mice.

uricase (1.65 mg/kg), xanthine (2.0 mg/kg and 5.0 mg/kg), yeast RNA (50 mg/kg and 100 mg/kg), or an equivalent volume of 0.9% saline; administered i.p. one hour prior to being given a single training trial in an apparatus designed to provide for conditioning of a passive avoidance response (Essman and Alpern, 1964). Two groups of mice under each of the drug-treatment conditions were given a conditioning shock (3mA), while two control groups under all drug-treatment conditions were given no conditioning shock ($\bar{S}$). One shocked group and one unshocked group, under all drug-treatment conditions, was given a single electroconvulsive shock (ECS) applied transcorneally (10mA, 200msc., 700V), and one of each group under all drug-treatment conditions was given sham-ECS ($\overline{ECS}$). ECS or $\overline{ECS}$ was given ten seconds following the training trial. Twenty-four hours following training, a single testing trial was given to determine the incidence of conditioned response retention or ECS-induced retrograde amnesia. Whereas the unconditioned response on the training trial occurred within ten seconds an elevation in this latency to 30 seconds or more on the testing trial was taken as the criterion for conditioned response retention. Similarly, ECS-induced retrograde amnesia could be defined by testing trial latencies which approximated those of animals that has never received a training trial foot shock. The results of this study are summarized in Table III. The percentage incidence of conditioned response retention is shown for all drug-treatment conditions, with the training and posttraining conditions. The amnesic effect

TABLE III

PERCENTAGE INCIDENCE OF CONDITIONED RESPONSE RETENTION FOR SALINE- AND DRUG-TREATED GROUPS

Treatment (1 hour Pretraining)	Condition[a]			
	S–ECS	$\bar{S}$–ECS	S–$\overline{ECS}$	$\bar{S}$–$\overline{ECS}$
Saline (0.9%)	10	0	100	0
Uric Acid (0.75 mg/kg)[b]	60	10	100	0
Uricase (1.65 mg/kg)	5	0	100	0
Xanthine (2.0 mg/kg)	0	0	100	0
Xanthine (5.0 mg/kg)	5	0	100	0
Yeast RNA (50 mg/kg)	20	10	100	0
Yeast RNA (100 mg/kg)	10	10	100	0

[a]S— 3 mA training foot shock;
$\bar{S}$— no training shock;
ECS — electroconvulsive shock, 10mA, 700V, 200msc., 10 seconds following training;
$\overline{ECS}$ — sham ECS, no current passed.

[b] $\chi^2 = 16.84$; $p < .001$.

of posttraining ECS was reduced in about 60% of the mice tested; this effect was highly significant. Uricase, xanthine, and yeast RNA at the dosage indicated, had no apparent effect upon the amnesic effect of ECS. The data further indicate that none of the drugs affected the performance of the unconditioned response, as observed in animals given no training shock, regardless of whether ECS was given. The high incidence of conditioned response retention among mice given training shock, but no posttraining ECS, was also unaffected by any of the drug conditions. Finally, it may also be noted that the full clonic-tonic convulsion produced by ECS was unaltered under any of the drug-treatment conditions.

The process of memory consolidation was specifically considered in a subsequent study. The above data indicated that the memory trace, usually vulnerable to the disruptive effect of postlearning electroconvulsive shock, was only minimally impaired in uric acid-treated mice. This finding indicated that either uric acid increased the rate at which the memory trace was consolidated, or attenuated the central effects of electroconvulsive shock (ECS) that account for disruption of memory consolidation and appearance of retrograde amnesia for a previously acquired response.

In several previous studies (Essman, 1965, 1966a, 1966b, 1967a, 1967b), a reduction in ECS-induced retrograde amnesia by pharmacological modification of C.N.S. metabolism has indicated that the ECS effects upon brain RNA and serotonin were altered. The effect of ECS was to reduce the regional concentration of RNA in the brain and to elevate brain serotonin and modify its metabolism. This relationship was most apparent in the diencephalon, and was significantly correlated ($\rho = -0.90$). As a related concomitant of ECS-induced retrograde amnesia and possible correlating factor of the memory consolidation process, brain RNA and serotonin were investigated with regard to how they were affected by uric acid treatment, how they were modified by ECS that led to retrograde amnesia, and how they changed under conditions wherein uric acid treatment effectively reduced the incidence of retrograde amnesia produced by ECS.

Groups of male CF-1 strain mice, weighing approximately 27 gm, were randomly assigned to one of 14 possible groups, each comprised of 20 mice. Each set of two groups was injected with either 0.5, 1.0, 2.0, 5.0, 10.0, or 20.0 mg/kg of uric acid dissolved in 0.9% saline, and the remaining two groups were given a control injection of 0.9% saline. All injections were given intraperitoneally one hour before the administration of a single training trial in the apparatus previously referred to (Essman and Alpern, 1964). This apparatus specifically consisted of a small clear lucite vestibule, adjoining a larger darkened chamber, with a 2-inch diameter hole separating the two. A mouse, when placed in the small transparent vestibule surrounded by overhead fluorescent lighting, shows a tendency to rapidly exit from this compartment and enter the darkened chamber through the hole. Administration of a 3.0mA foot shock, provided through a grid floor in the large chamber for three seconds, has been shown to be sufficient to establish upon retesting 24 hours later, a passive avoidance response; *i.e.*, the animal remains in the clear vestibule and avoids the chamber in which it has been previously shocked. The response latency of mice on the training trial has usually been less than ten seconds, whereas on the testing trial conditioned avoidance can be defined by long delays in approaching the chamber in which the animal has been shocked. A delay exceeding 30 seconds has been taken as a criterion for conditioned response acquisition, and retention; and the absence of any response by 60 seconds has provided for the cutoff point at which animals were manually removed from the apparatus. In the present procedure, one group at each of the uric acid doses, and one of the saline-treated groups, were given a single transcorneal electroconvulsive shock, (ECS) ten seconds following their entry into the larger chamber (10mA, 700V, 200msc.) and the administration of the conditioning foot shock. The remaining groups were given a sham (ECS) $\overline{\text{ECS}}$ at the same posttraining interval; this consisted of application of the transcorneal electrodes without passage of any current.

All of the ECS-treated animals showed a full clonic-tonic convulsion which appeared to be unaffected in any way by drug treatment. Twenty-four hours following the conditioning trial all groups of animals were tested for their retention of the conditioned avoidance response. The testing trial consisted of re-introducing the animal to the outer vestibule and timing the interval between placement in the outer chamber and possible entry into the larger chamber.

The results of this experiment are summarized in Table IV in which the median response latency differences and percentage incidence of conditioned response retention are shown. The median response latency difference was based upon a comparison of the difference in testing trial latency and training trial latency for each group. As might be expected, all groups given sham-ECS showed a large latency difference, indicating appreciable elevation in testing trial latency, suggestive of the predicted manifestation of conditioned avoidance behavior on the testing trial. This observation is further supported by the consistent incidence of conditioned response retention of 100% of all the animals on each of the groups given sham-ECS. As for the ECS-treated animals, the median response latency difference is very small for the saline-treated animals, whereas the only comparable small difference occurred for the group of animals treated with 1.0 mg/kg of uric acid. For all other uric acid treatment conditions the latency differences were elevated. These data are somewhat clarified by the fact that the percentage incidence of conditioned response retention among saline-treated animals was quite low, suggesting that 80% of the animals in this group showed a retrograde amnesia induced

TABLE IV

MEDIAN RESPONSE LATENCY DIFFERENCE (TESTING-TRAINING TRIAL LATENCY) AND PERCENTAGE INCIDENCE OF CONDITIONED RESPONSE RETENTION IN URIC ACID-TREATED MICE WITH AND WITHOUT POSTTRAINING ELECTROCONVULSIVE SHOCK

Treatment	$\overline{ECS}$		ECS	
	Median Response Latency Difference	Percentage Incidence of CR Retention	Median Response Latency Difference	Percentage Incidence of CR Retention[a]
Saline—0.9%	54.0	100	5.0	20[b]
Uric Acid				
0.5 mg/kg	53.0	100	20.0	73
1.0 mg/kg	52.0	100	0.0	50
2.0 mg/kg	51.0	100	16.5	47
5.0 mg/kg	51.0	100	23.0	70
10.0 mg/kg	52.0	100	17.0	68
20.0 mg/kg	53.0	100	23.0	60

[a] $\chi^2_{df=6} = 16.16$; $p. < 02$.

[b] $\chi^2_{df=1} = 12.11$; $p < .001$.

by posttraining ECS. Among the uric acid-treated animals the incidence of conditioned response retention was significantly higher, although there were no significant differences between the uric acid-treated groups as a function of dosage.

These data indicate that the amnesic effect of posttraining ECS is appreciably antagonized in animals that have received prior treatment with uric acid over a considerable dosage range. This statistically significant reduction in the incidence of retrograde amnesia suggests that either the rate at which the memory trace is consolidated has been increased by the central effects of peripherally-administered uric acid (*i.e.*, the interval following training during which the memory trace is vulnerable to the disruptive effects of electroshock has been shortened), or that the central effects of cerebral electroshock, which account for the emergence of retrograde amnesia, have been modified by uric acid treatment. The present data further clarify the point that uric acid treatment over the dose range utilized in the present study has no effect upon task acquisition and retention.

The phenomenon of electroshock-induced retrograde amnesia is one for which a relationship between brain RNA concentration changes and alterations in serotonin concentration and metabolism have been indicated. In the latter respect, it is of interest to note that a previous report in this symposium

(Appel) has indicated that the incorporation of amino acid precursors into synaptic proteins was appreciably inhibited by serotonin. In this respect, one might speculate as to the specific role that presumed serotonogenic synapses might have in memory consolidation, since it is apparent that retrograde amnesia produced by ECS is accompanied by brain serotonin concentration increases that occur during the period within which the memory trace is presumably consolidated.

The following study was undertaken in order to investigate the relationship between ECS and changes in brain RNA and serotonin metabolism that occur subsequent to ECS.

Fourteen groups of 20 mice each were injected i.p. with either the doses of uric acid indicated in the previous experiment, or with an equivalent volume of 0.9% saline. One hour following injection, groups at each dose level of drug or saline were given a single transcorneal 10mA ECS, whereas the remaining control groups were given sham-ECS ($\overline{\text{ECS}}$) without passage of current. All animals were killed within 20 minutes following ECS or $\overline{\text{ECS}}$ treatment by cervical section; the whole brain, excluding olfactory bulbs and cerebellum, was removed, weighed, and rapidly frozen. One hemisphere was prepared for estimation of RNA concentration, utilizing a pentose analysis method (Schneider, 1957), and the other hemisphere was prepared for solvent extraction of serotonin (5-HT) and 5-hydroxyindoleacetic acid (5-HIAA), utilizing fluorometric techniques (Bogdanski *et al.*, 1956; Udenfriend *et al.*, 1958).

For six animals in each of the 14 treatment groups outlined above, identical procedures were carried out. Immediately following the administration of ECS or $\overline{\text{ECS}}$ and i.p. injection of 20 mg/kg of tranylcypromine was given in order to effect inhibition of monoamine oxidase activity. This provided a means by which a steady state kinetics estimate of brain serotonin could be made (Tozer *et al.*, 1966) by obtaining brain tissue from these animals at 10, 30, or 60 minutes following injection, and deriving rate constants for the increase in serotonin concentration and decrease in 5-HIAA concentration over time. This technique provided for estimation of serotonin turnover rate and turnover time under each of the experimental conditions. Previous experiments (Essman, 1968,) have indicated that ECS, *per se*, did not alter the activity of monoamine oxidase in the brain under drug or saline treatment conditions and, furthermore, that the distribution of tranylcypromine over time following its administration was not altered by ECS.

The mean concentrations of whole brain RNA obtained for saline- and uric acid-treated mice, given ECS or $\overline{\text{ECS}}$ one hour following drug treatment, are summarized in Table V. It is apparent from these data that basal whole brain RNA concentration in uric acid-treated mice did not differ significantly from control levels obtained in the saline-treated animals. ECS, however,

TABLE V

MEAN ($\pm\sigma$) RNA CONCENTRATION ESTIMATED FROM WHOLE MOUSE BRAIN FOLLOWING SALINE- OR URIC ACID TREATMENT WITH (ECS) OR WITHOUT ($\overline{ECS}$) SUBSEQUENT ELECTROCONVULSIVE SHOCK

Treatment	$\overline{ECS}$	ECS	P_Δ (df = 18)
Saline (0.9%)	0.730 (0.100)	0.430 (0.080)	<.001
Uric Acid			
0.5 mg/kg	0.703 (0.118)	0.673 (0.095)	>.40
1.0 mg/kg	0.754 (0.666)	0.654 (0.078)	<.01
2.0 mg/kg	0.706 (0.055)	0.756 (0.087)	>.10
5.0 mg/kg	0.897 (0.203)	0.813 (0.090)	>.10
10.0 mg/kg	0.773 (0.035)	1.004 (0.053)	<.001
20.0 mg/kg	0.722 (0.045)	0.902 (0.068)	<.001

led to a significant reduction ($p < 0.001$) in whole brain RNA in the saline-treated mice. This reduction of approximately 41% was in no way paralleled by the ECS effect observed in uric acid-treated animals. In the latter instance, ECS generally did not lead to a lowered concentration of whole brain RNA, with the possible exception of the animals treated with 1.0 mg/kg. In this latter respect, it is of interest to note that in the behavioral study mice treated with 1.0 mg/kg of uric acid also showed no change in median latency difference from the training to the testing trial, and the incidence of conditioned response retention was somewhat lower than the average incidence at other uric acid doses. At higher doses of uric acid (10 and 20 mg/kg) ECS led to a significant increase in brain RNA concentration. This finding could suggest a possible interaction between uric acid at the higher doses and electroshock, such that under these conditions a maximal degree of C.N.S. stimulation could be provided, and RNA synthesis under these conditions might be expected to increase.

Fig. 4 shows the percentage change in brain RNA following ECS. A trend could be suggested in such changes as a function of dosage, when the variability of the RNA change at each dose level is taken into account.

The data for brain serotonin and 5-hydroxyindoleacetic acid concentration, and for serotonin metabolism at each of the drug dosages administered and for the saline-treated animals under ECS and $\overline{ECS}$ conditions are summarized in Table VI. With electroconvulsive shock, brain serotonin concentration in saline-treated control animals was elevated by approximately 12%. Under these same conditions there was an increase in serotonin turnover rate of approximately 28% and a decrease in serotonin turnover time of approximately 11%. These findings, in general, indicate that ECS,

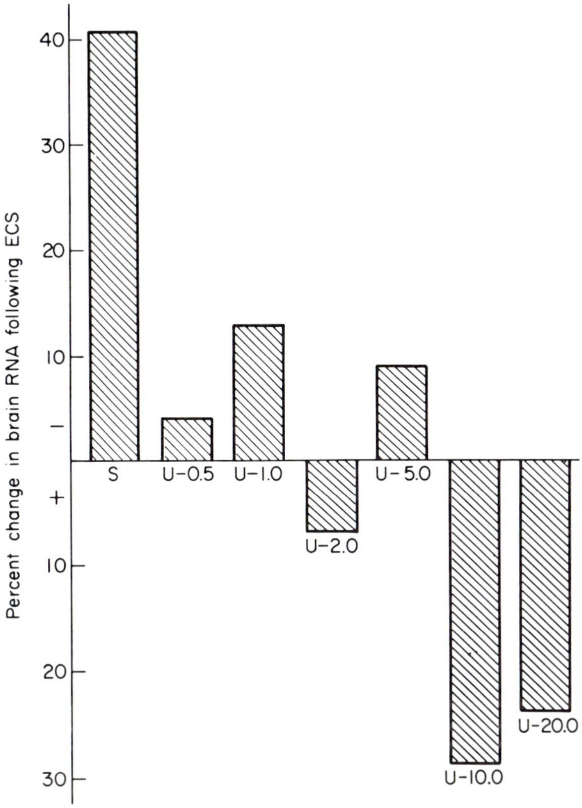

Figure 4 Percentage change in whole brain RNA concentration following electoconvulsive shock (ECS) administration to Saline (S)- and uric acid (U)-treated mice.

under control treatment conditions, leads to an accelerated synthesis and degradation of brain serotonin. Under conditions in which animals were treated with uric acid, it may be seen that prior to the administration of ECS brain serotonin concentration was elevated from between 26 to 141 % above control levels. Serotonin metabolism, under these drug treatment conditions, led to a decrease in turnover rate at doses above 2.0 mg/kg and an increase in turnover time. These data, in general, suggest that uric acid treatment leads to an appreciable reduction in the synthesis of serotonin and 5-hydroxyindoleacetic acid. This alteration in serotonin metabolism and concentration, which was manifest at the time animals were trained for the acquisition of the avoidance response, clearly had no effect upon the acquisition of the task, nor did it in any way modify the retention of the response tested 24 hours later. With the administration of ECS to mice, injected one hour prior to

TABLE VI

MEAN (±σ) CONCENTRATION OF 5-HYDROXYTRYPTAMINE AND 5-HYDROXYINDOLEACETIC ACID AND 5-HT TURNOVER IN BRAIN FOR URIC ACID-TREATED MICE WITH AND WITHOUT ELECTROCONVULSIVE SHOCK

Treatment	$\overline{\text{ECS}}$		5-HT Turnover		ECS		5-HT Turnover	
	5-HT (μg/g)	5-HIAA (μg/g)	Rate (μg/g/hr)	Time (Min)	5-HT (μg/g)	5-HIAA (μg/g)	Rate (μg/g/hr)	Time (Min)
Saline—0.9%	0.68 (0.03)	0.32 (0.15)	0.35	114	0.76 (0.04)	0.41 (0.16)	0.45	101
Uric Acid								
0.5 mg/kg	1.63 (0.33)	0.39 (0.14)	0.31	315	1.35 (0.39)	1.35 (0.04)	0.22	366
1.0 mg/kg	1.64 (0.33)	0.46 (0.16)	0.47	208	1.53 (0.26)	0.28 (0.03)	0.21	438
2.0 mg/kg	1.40 (0.19)	0.30 (0.11)	0.20	420	1.20 (0.34)	0.34 (0.09)	0.20	360
5.0 mg/kg	0.86 (0.14)	0.49 (0.19)	0.21	245	1.53 (0.24)	0.30 (0.02)	0.23	396
10.0 mg/kg	0.98 (0.25)	0.28 (0.03)	0.13	450	1.44 (0.41)	0.32 (0.05)	0.22	390
20.0 mg/kg	1.24 (0.13)	0.28 (0.03)	0.17	362	1.11 (0.42)	0.32 (0.05)	0.17	390

this treatment with uric acid, it appears that no significant alteration in brain serotonin concentration occurred. Generally, it was observed, that serotonin turnover rate was decreased by approximately 16% and serotonin turnover time was increased by approximately 10%. It therefore seems apparent that the administration of ECS to animals previously treated with uric acid leads to a further decrease in the synthesis and degradation of serotonin.

When the relationship between the observed reduction in brain RNA concentration and the elevation in brain serotonin was determined as a function of uric acid dosage, these changes were significantly correlated ($\text{rho} = -0.69$; $p < 0.02$). This significant relationship between ECS-induced changes in brain RNA and serotinin concentration appears to still be preserved from the appreciably higher degree of relationship between these changes observed under nondrug treatment conditions; the reduced, although still significant, relationship indicates that these changes must be dose-dependent in several respects.

The data obtained in the present study may be summarized by concluding the following:

1. Memory consolidation, as defined within the context of an interval of temporal vulnerability to the amnesic effect of posttraining ECS, is facilitated or accelerated by parenterally-introduced uric acid.

2. Concomitant with these changes in memory consolidation, produced by an endogenous central nervous system stimulant—a product of purine metabolism—are marked modifications in the metabolism of two molecules implicated in memory consolidation: RNA, which is usually decreased in concentration in the brain by ECS when the latter produces retrograde amnesia, is unaltered with low doses of uric acid, and is increased, suggestive of pronounced central nervous system stimulation, at higher doses of uric acid. This observation suggests a synergistic stimulatory effect. The concept of molecularly-coded information must therefore also consider those metabolic processes with which the molecule itself is involved, and those molecular changes which it becomes functionally capable of initiating.

Serotonin, as previously indicated, can play a role in glial nucleic acid metabolism and regulate the incorporation of amino acids into proteins at the synapse. It is also modified metabolically by uric acid, and the effect generally is one of reduced synthesis and degradation of this indole. With ECS, which in control animals results in retrograde amnesia, the usual elevation of serotonin in the brain and the increase in its synthesis and degradation in the brain do not occur in uric acid-treated animals; in this instance there was generally a further decrease of synthesis and degradation of brain serotonin.

The present data suggest the dependency of the memory consolidation process, and its disruption by agents or events like electroconvulsive shock, upon the metabolism of molecules, the functional character of which may itself be prescribed by both molecular and synaptic events.

ACKNOWLEDGMENT

The author wishes to acknowledge the valuable technical assistance of M. I. Golod and M. L. Steinberg.

REFERENCES

Appel, S. Macromolecular synthesis in synaptosomes. In: W. L. Byrne (Ed.) "Molecular Approaches to Memory and Learning." N.Y.: Academic Press, 1969.

Babich, F. R., Jacobson, A. L., Bubash, S. and Jacobson, A. Transfer of a response to naive rats by injection of ribonucleic acid extracted from trained rats. *Science*, 1965a, **149**, 656–657.

Babich, F. R., Jacobson, A. L. and Bubash, S. Cross-species transfer of learning: effect of ribonucleic acid from hamsters on rat behavior. *Proc. U.S. Nat. Acad. Sci.*, 1965b, **54**, 1299–1302.

Bogdanski, D. F., Pletscher, A., Brodie, B. B. and Udenfriend, S. Identification and assay of serotonin in brain. *J. Pharmacol. Exp. Ther.*, 1956, **117**, 82–88.

Brooks, G. W. and Mueller, E. Serum urate concentrations among University professors. *J. Amer. Med. Assn.*, 1966, **195**, 415–418.

Brown, H. Effect of ribonucleic acid (RNA) on the rate of lever pressing in rats. *Psychol. Rec.*, 1966, **16**, 173–176.

Byrne, W. L. and Samuel, D. Behavioral modification by injection of brain extract prepared from a trained donor. *Science*, 1966, **154**, 418 (Abstract).

Coburn, S. P., Seidenberg, M. and Mertz, E. T. Clearance of uric acid, urea, and creatine in Down's syndrome. *J. appl. Physiol.*, 1967, **23**, 579–580.

Cook, L., Davidson, A. B., Davis, D. J., Green, H. and Fellows, E. J. Ribonucleic acid: effect on conditioned behavior in rats. *Science*, 1963, **141**, 268–269.

Egyházi, E. and Hydén, H. Experimentally induced changes in the base composition of the ribonucleic acids of isolated nerve cells and their oligodendroglia cells. *J. biophys. biochem. Cytol.*, 1961, **10**, 403–410.

Eist, H. and Seal, U. S. The permeability of the blood-brain barrier and blood-CSF barrier to C^{14} tagged ribonucleic acid. *Amer. J. Psychiat.*, 1965, **122**, 584–586.

Essman, W. B. Facilitation of memory consolidation by chemically induced acceleration of RNA synthesis. *Proc. XXIII Internat. Congr. Physiol. Sci., Tokyo*, 1965, 470.

Essman, W. B. Effect of tricyanoaminopropene on the amnesic effect of electroconvulsive shock. *Psychopharmacologia*, 1966a, **9**, 426–433.

Essman, W. B. Modification of neural RNA with amnesic agents and events. *Proc. XVIII Int. Congr. Psychol., Moscow*, 1966b, **20**, 65–67.

Essman, W. B. Changes in memory consolidation with alterations in neural RNA. *Proc. V Int. Congr. C.I.N.P.*, In: *Exerpta Medica*, 1967a, *ICS*, **129**, 185–190.

Essman, W. B. The temporal limits of memory consolidation as a function of alterations in brain serotonin. *Proc. VII Int. Congr. Biochem., Tokyo*, 1967b, J-204.

Essman, W. B. Water maze savings by mice as a function of motor training. *Psychol.* Rep., 1967c, **21**, 8–10.

Essman, W. B. Changes in ECS-induced retrograde amnesia with DBMC: behavioral and biochemical correlates of brain serotonin antagonism. *Physiol. and Behav.*, 1968, **3**, 527–531.

Essman, W. B. and Alpern, H. Single trial learning: methodology and results with mice. *Psychol. Rept.*, 1964, **14**, 731–740.

Essman, W. B. and Jarvik, M. E. A water-escape test for mice. *Psychol. Rep.*, 1961, **8**, 58.

Essman, W. B. and Lehrer, G. M. Facilitation of maze performance by "RNA extracts" from maze-trained mice. *Fed. Proc.*, 1967, **26**, 263.

Goren, C. Ribonucleic acid: influence on the maze-learning ability of rats. *Worm Runner's Digest*, 1965, **7**, 28–31.

Hydén, H. A molecular basis of neuron-glia interaction. In: Schmitt, F. O. (Ed.) *Macromolecular Specificity and Biological Memory*. Cambridge, Mass.: The M.I.T. Press, 1962, Pp. 55–69.

Hydén, H. Biochemical and functional interplay between neuron and glia. In: Wortis, J. (Ed.) *Recent Advances in Biological Psychiatry*, 1963, **6**, Pp. 31–54.

Hydén, H. and Egyházi, E. Nuclear RNA changes at nerve cells during a learning experiment in rats. *Proc. Nat. Acad. Sci.*, 1962, **48**, 1366–1373.

Jacobson, A. L., Babich, F. R., Bubash, S. and Jacobson, A. Differential approach tendencies produced by injection of RNA from trained rats. *Science*, 1965, **150**, 636–637.

Kral, V. A., Solyom, L. and Enesco, H. E. Effect of short-term oral RNA therapy on the serum uric acid level, and memory function in senile versus senescent subjects. *J. Amer. Geriat. Soc.*, 1967, **15**, 364–372.

Orowan, E. The origin of man. *Nature*, 1955, **175**, 683–684.

Ritchie, J. M. The Xanthines. In: Goodman, L. S., and Gelman, A. *The Pharmacological Basis of Therapeutics*. N.Y.: The Macmillan Co., 1965, Pp. 354–366.

Roll, P. M., Brown, G. B., DiCarlo, F. J. and Schultz, A. S. The metabolism of yeast nucleic acid in the rat. *J. biol. Chem.*, 1949, **180**, 333–340.

Schneider, W. C. Determination of nucleic acids in tissues by pentose analysis. In: Colowick, S. P. and Kaplan, N. O. (Eds.) *Methods in Enzymology*, Vol. III, N. Y.: Academic Press, 1957, Pp. 680–684.

Stetton, D. and Hearon, J. Z. Intellectual level measured by Army Classification Battery and serum uric concentration. *Science*, 1959, **129**, 1737.

Sved, S. The metabolism of exogenous ribonucleic acid injected into mice. *Canad. J. Biochem.*, 1965, **43**, 949–958.

Tozer, T. N., Neff, N. H. and Brodie, B. B. Application of steady state kinetics to the synthesis rate and turnover time of serotonin in brain of normal and reserpine-treated rats. *J. Pharmacol. Exp. Ther.*, 1966, **153**, 177–182.

Udenfriend, S., Weissbach, H. and Brodie, B. B. Assay of serotonin and related metabolites, enzymes and drugs. In: *Methods of Biochemical Analysis*, 1958, **6**, Ed.: D. Glick, New York. Interscience Publishers, Inc., 95–130.

MEASUREMENT OF THE EFFECTS OF BRAIN EXTRACT ON INTERORGANISM INFORMATION TRANSFER*

Kenneth P. Weiss

Department of Psychology
Kings College
Wilkes-Barre, Pennsylvania

The difficulties reported by various laboratories in attempting to replicate interanimal information transfer experiments are not unreasonable when evaluated with respect to the intrinsic problems associated with the experimental, environmental, and methodologic variables between laboratories. The importance of these differences in experimental conditions between laboratories is accentuated (a) by the current lack of a consensus with respect to which chemicals transferred via the brain extract are responsible for the observed effects, and (b) by our present inability to demonstrate the degree of specificity obtainable in the transferred behavior.

An additional area of concern is the recipient's experiential and learning history. In the experimental situation where the recipient is traditionally measured, it is improbable that *only* the specific behavior to which the donor animal has been trained will be elicited distinguishable and measurable from the possible plethora of random and similarly transferred information. It *is* probable that the response to be measured may be antagonistic to or in competiton with behavior inherent in the recipient's normal, pre-injection repertoire. Similarly, a behavior which anthropomorphically should be

*This research is supported in part by an ESSO Foundation Grant and Nasson College.

prepotent in either the donor or recipient's repertoire may, in fact, be a recessive component of the animal's experience and measurable behavior. Thus, the rather limited classification of responses to which donors are trained and on which recipients are measured must be analyzed and chosen carefully. Whenever possible, responses should be chosen which, logically, with training, will represent a major portion of the animal's experience or survival exigencies.

The procedure outlined, in the experiment reported here, is presented in an attempt to demonstrate how these and several other factors affecting the reliability and replicability of the phenomena may be controlled and evaluated.

The experiment consisted of first training ten male, 70-day-old (second generation), Sprague–Dawley, albino rats (Donors) to avoid approaching or drinking from a glass water- delivery tube, when an indicator light attached above the delivery tube was lit. The training was accomplished while the animals remained in their home cages. These cages were altered by mounting a small indicator light 7 cm above the glass water delivery tube, which was placed in the front center of each standard cage. An electrode, which came in direct contact with the water in each animal's water bottle, was inserted along the side of the delivery tube through the rubber stopper. An automatic timer outside, and sound insulated from, the animal room was adjusted daily in a random sequence to insure that the indicator lights would alternately be on and off, establishing a mean interval of three minutes for each condition. The range of possible intervals was from 30 seconds to six minutes. The timer also energized the (shock) electrode in the water bottle whenever the indicator light was in the on condition (the maximum possible shock was limited to 50 mA). Thus, if the animals attempted to approach and drink from the delivery tube when the cue light was on, they would receive a shock as a result of their completing a circuit between the water in the delivery tube and the wire cage floor. Following 48 hours of continuous exposure to these conditions, all animals learned to criterion (100% correct responses) to drink water only when the cue light was off. To insure that this training became a major and significant portion of the animal's total environment, they were exposed continuously to these conditions while in their home cages, for an additional 50 days.

Following the 50-day training period, the animals were anesthetized with Nembutal and subsequently expired during the process of excising the entire brain. The brains from the ten donor animals were immediately frozen in solid carbon dioxide.

Twenty-four hours later a pooled brain extract was prepared by the cold phenol extraction procedure described by Babich *et al.* (1965). (Complete evaporation of the ethanol was not accomplished and the isotonic saline solution was contaminated with a small trace of ethanol.) The extract was stored in solid carbon dioxide for twenty-four hours and then allowed to come up to room temperature, at which time approximately one brain equivalent (1 ml) was injected intraperitoneally to each of ten naive, male, 120-day-old, Sprague–Dawley, albino rats.

These recipient animals were then placed in the cages previously occupied by the donor animals. The cue lights over the delivery tubes were now programmed to be alternately on and off for equal intervals of three minutes. The electrode, which had previously been used to condition the avoidance (shock) response, was now connected to a ten-channel, high impedance, sensing device which recorded, on electromagnetic counters, the tendency for each animal to approach the water delivery tube. Each animal had its approaches, under the light on and light off condition, recorded on separate counters. An approach was operationally defined and counted as the first time the animal completed a circuit to ground, with its tongue, as it attempted to drink during any one interval. No shock was given to these animals, under either the light on or off condition, nor had they been water or food deprived. The cumulative tendency for each animal to approach the water delivery tube was continuously recorded for 48 hours. During this period there were 960 trial periods for each animal. 480 alternately under each condition. In order to insure that the animals would respond as often as possible, special glass delivery tubes were drawn with very small openings in the ends (approximately one-sixteenth inch diameter).

Ten days after the completion of this phase of the experimentation, an additional (saline) group of animals was measured for their performance under the two possible drinking conditions. This group of 130-day-old naive animals was treated exactly the same as the recipient group, *i.e.*, they were measured for 48 hours after injection. These animals were injected with 1 ml of sterile, 0.9% saline solution. The results indicating the preferences of the three groups to approach the water delivery tube under the two conditions are summarized in Table I.

From Table I it is evident that only the preference to drink water during the light-off condition differs significantly ($P_2 < 0.01$) from all other conditions. Table I also indicates considerable variability in the means of all groups, although not significant at the 0.01 level.

Table II is a summary of the data; the scores for each of the ten animals represent actual approaches under the alternating conditions. The sensing device was programmed to lock out after the animal's initial contact with the water in the delivery tube. The apparatus was reset (unlocked) after termination of the trial. This table presents also the means of the scores for each condition and group. The rather wide variation in the total number of approaches between animals is partially due to the normal variability the animals exhibit resulting from the amount they consume during any one trial.

Figures 1, 2, and 3 present the preferences for approaching and drinking under the two conditions demonstrated by the animals in the three groups. The bar graphs represent the percentage of each animal's total measured approaches under each condition.

TABLE I

Means	(1) Recipient Light On 64.7	(2) Donor Light On 76.8	(3) Saline Light On 88.4	(4) Saline Light Off 88.5	(5) Donor Light Off 92.5	(6) Recipient Light Off 115.0	Shortest Significant Ranges
(1) 64.7	—	12.1	23.7	23.8	27.8	50.3	$R_2 = 36.1$
(2) 76.8	—	—	11.6	11.7	15.7	38.2	$R_3 = 36.5$
(3) 88.4	—	—	—	0.1	4.1	26.6	$R_4 = 37.7$
(4) 88.5	—	—	—	—	4.0	26.5	$R_5 = 38.3$
(5) 92.5	—	—	—	—	—	22.5	$R_6 = 38.8$
	Recipient On	Donor On	Saline On	Saline Off	Donor Off	Recipient Off	

Treatment means not underscored by the same line are significantly different ($P_2 < .01$). Duncan's Multiple Range Test (McGuigan, 1960).

TABLE II

NUMBER OF APPROACHES

Animal Number	Donor Group 48-Hour Pretraining (No Shock) Fixed Interval 3 min. Light On	Light Off	Saline Group 48 Hours (Post Saline Injection) Fixed Interval 3 min. Light On	Light Off	Recipient Group 48 Hours (Post Injection) Fixed Interval 3 min. Light On	Light Off
1.	81	109	91	83	110	114
2.	41	32	71	79	80	171
3.	76	164	138	124	32	97
4.	54	78	73	94	50	66
5.	46	63	92	86	81	133
6.	103	109	113	94	88	126
7.	89	100	58	70	35	89
8.	68	90	80	93	102	91
9.	126	103	62	82	26	118
10.	84	77	106	80	43	145
Mean	76.8	92.5	88.4	88.5	64.7	115.0

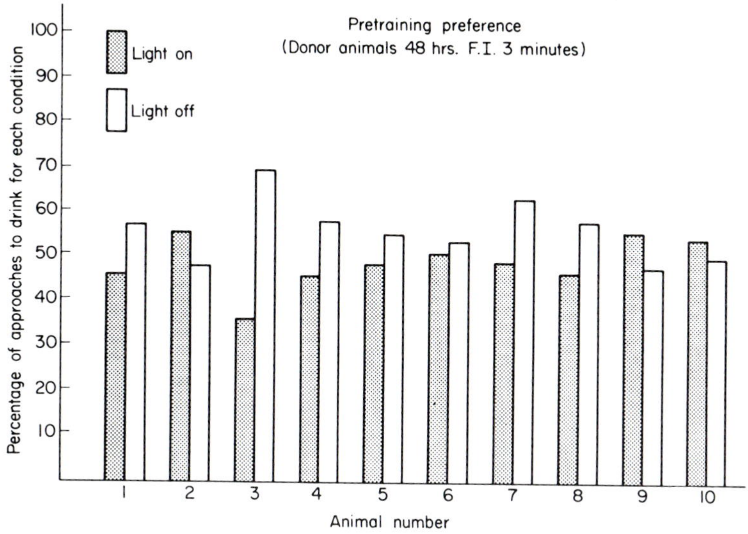

Figure 1

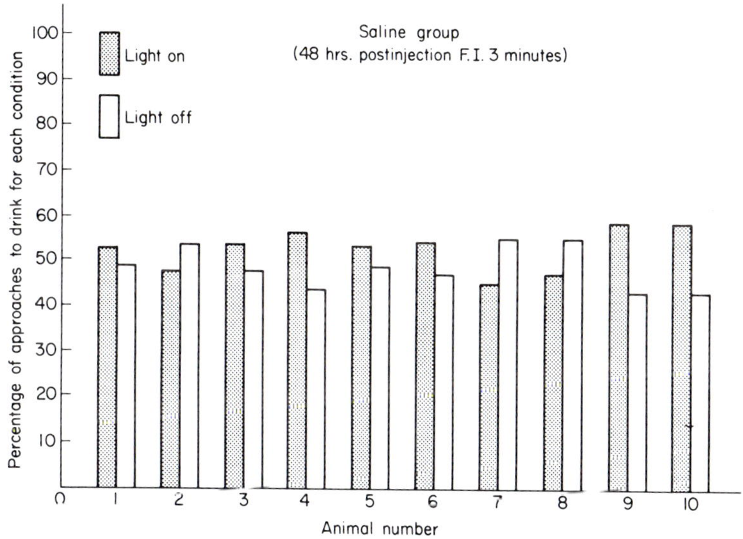

Figure 2

From the data it can be seen that the recipients exhibit an apparent preference to approach the delivery tube, during the light off condition. This preference is significant ($P_2 < 0.01$) and in the direction predicted. Thus, the information which the donors had learned, through training, was apparently transferred.

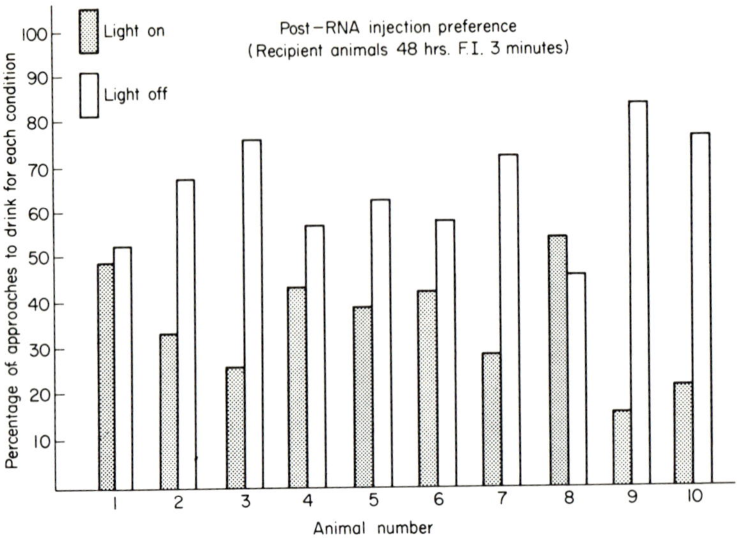

Figure 3

DISCUSSION

Procedural difficulties evident during early unsuccessful experiments were eliminated in the design of this experiment. These early pilot studies did indicate, however, that naive animals do not demonstrate a significant preference for drinking in either the light-on or light-off condition. The light-off condition was, therefore, not counterbalanced in this experiment.* The rationale employed in choosing the light off as the drinking condition was determined with respect to the possibility of a power failure during the rather lengthy training conditions. Thus, in the event of loss of power the cue light and shock would be inoperable simultaneously, and would represent, for the animal, a relatively long light-off (drink) interval. This precaution insured that the donor animals would receive no extinction trials and that the training would be continuous and uninterrupted. During the 50 days of donor training this problem occurred three times totaling approximately twenty hours of power failure. The total duration of the power failures were excluded from the computation in the determination of the mean three-minute on and off intervals. It is interesting to note that the records of the animal's average

* A counter-balanced (light on to drink) condition was conducted, with results similar to those reported, subsequent to the preparation of this report. This group was compared to a control group receiving an injection of brain extract from animals trained on an irrelevant task.

consumption do not appear to be affected during these periods. The mean water consumption appears to be relatively stable and uninfluenced by the different scheduled conditions.

General Assumptions and Experimental Criteria

As was indicated earlier, the design of this experiment was predicated on the analysis of several extraneous variables which appear to be operating and causing difficulties throughout much of the research conducted in this area. This design is an attempt to either mitigate or eliminate many of these difficulties.

There are several assumptions which may be appropriately posited in an attempt to establish a systematic criteria for the design, interpretation, and evaluation of these experiments.

The present difficulties in reliably demonstrating the transfer phenomena and the current lack of a complete and adequate explanation of the nature of the underlying neurological and chemical mechanisms involved, dictates:

Firstly, that an experimental paradigm having any heuristic value must lend itself to reliable, efficient, and expedient replication. Such a paridigm should also be sufficiently uncomplicated and robust to facilitate its utility in experimentally elucidating upon and answering the perplexing questions which surround the research in the area. The variables employed should be amenable to experimental manipulation with little likelihood of confounding data and results.

Secondly, recognition of the necessity for objective and automated training and measurement procedures to increase efficiency, reliability and tend to eliminate variability, is an essential prerequisite for a general paradigm.

Thirdly, the utilization of these objective automated procedures implies the procedural capability for training and measuring large numbers of animals simultaneously.

Fourthly, the last general criteria which appears many times not to have been considered appropriately, concerns the need for easily controllable and counter-balanced experimental conditions and variables.

DONOR ANIMALS

In addition to the latter, there are certain criteria which relate most directly to the donor animals.

1. The trained behavior should be as simple and specific as possible, with consideration given to the goals of the particular experiment.

2. Unless the strength or efficiency of the learned behavior is part of the independent variable, a behavior which is well learned to some objective

criterion should be more effective as a transferrable response than one less well or completely learned.

3. It is axiomatic that the more "important" the training is to the animal and the larger the portion of the animal's total experience and environment this training represents, then the more probable will be the successful measurement of the transfered response.

RECIPIENT ANIMALS

The criteria which seem most appropriate for the recipient animal are somewhat more difficult to meet, but are probably more critical than any of the others to the success of a particular experiment.

1. The need for continuous, simultaneous, postinjection measurement of all animals is indicated by the relative paucity of information available regarding (a) the latency between the injection and measurement of the first transferred response. (b) the effects of a lapse of time on the efficiency and strength of a transferred behavior and (c) the total length of time the animal is effected by transferred information (the shape of the extinction curve).

2. The major difficulty in many of the experiments reported, appears to be related to the experimenter's inability to measure partial and subtle transfer effects and transferred predispositions or tendencies to respond. One explanation for the difficulties encountered in attempting to replicate research in this area is probably related to a choice of dependent variables which require measurement of rather gross and discrete responses from the recipient animals. These responses may be operative in the form of a subtle predisposition or partially transferred information, but are not observable because of their incomplete, weak or attenuated nature. Thus, any behavior which may be measured in terms of a continuous or nondiscrete response will tend to be more sensitive to these subtle tendencies or predispositions which may be transferred but are not measurable.

3. Wherever possible, it would seem efficacious to choose a paradigm which yields data amenable to statistical quantification and appropriate powerful statistical procedures.

4. Several experimental paridigms, such as those demonstrated by Luttges *et al.* (1966), may be loaded against the demonstration of transfer phenomena by employing designs in which recipient animals must be over-motivated, excited, or sensitized by the experimental situation. Animals in situations and under conditions such as these may have a tendency to respond in the desired direction, but their ability to demonstrate this information transfer may very likely be over-shadowed by the stronger, more directed (or motivated) active behavior that is a result of this excitement. Thus, a very hungry (highly motivated) animal or one that is escaping from shock or immersion

in water is not likely to demonstrate much more than reflexive or autonomic behavior. Best (1963) has discussed, in detail, the problems associated with the detrimental effects of generalized excitement on performance.

SUMMARY AND CONCLUSIONS

A cold phenol, pooled-brain extract was prepared from ten donor rats trained using a simplified, automatic, long-term training procedure, and injected intraperitoneally into ten naive recipient animals. The recipients demonstrated a significant preference for drinking during a stimulus light-off, water-drinking condition, to which the donors had previously been trained. Their approaches to drink during the light on and off condition were measured separately on automatic sensing devices. The recipient animals preference was compared to (a) the donors pretraining preference, (b) a saline-injected groups preference and (c) their own relative preference for each condition.

The range of contradictory results and the consistent difficulty evident in the attempts at replication is indicitive of the current lack of understanding of the underlying mechanisms which are responsible for the transfer phenomenon. Presently, there is a sufficient quantity of research which has been conducted in an attempt to elucidate further on the nature of chemical information transfer, which demonstrates clearly that there is, in fact, something being transferred between organisms.

The ability exhibited by several researchers to replicate their own experimentation and various aspects of other research, suggests that the difficulty is not entirely attributable to the lack of a consensus in the identification of, or inability to isolate the responsible chemical mediator. The evidence suggests, however, that these contradictory results and difficulties in replication are due to the confounding between identification of the nature of the underlying mechanisms and the rather significant experimental deviations between laboratories and paradigms.

The procedure outlined in this paper represents an attempt to suggest a possible set of criteria which should facilitate replication and standardization between laboratories.

It is crucial that a few, easily replicable, demonstrations be developed to enable researchers to systematically manipulate the variables in a proven paradigm, in an attempt to isolate and identify the various components of the phenomenon which, at present, remain an enigma.

The two most important questions emerging from the experimental evidence concern (1) the analysis and identification of the chemistry involved in information transfer and (2) the determination of the degree of informational specificity which can be transferred between organisms.

There is presently a reasonably wide range of behavioral situations in which transfer has been demonstrated. It appears possible to attribute the phenomenon to some kind of generalized state induced in the recipient animal through the injected extract. It is becoming more difficult, however, to attribute the phenomenon to a generalized state in light of the diversity of behavioral situations in which the phenomena can be demonstrated.

One attempt to determine if the phenomenon involves a specific transfer of information might be to design an experiment in which the behavioral response to be transferred represents a scheduled, or form of sequential or alternation behavior. It would appear that if a response based on a fixed interval or fixed ratio schedule or a sequential behavior can be transferred between organisms that an explanation in terms of a generalized tendency would be inappropriate. We are presently involved in such an experiment.

ACKNOWLEDGMENTS

The author is indebted to Miss Nancy Nettleton, Miss Susan Jacobson, and Mr. David Bardsley, for their continued invaluable assistance without whose contribution this project could not have been completed.

The author is indebted also to Dr. Herbst from the University of New Hampshire and Mrs. Phyllis Sinman, at the Worcester Institute of Experimental Biology for assistance in the biochemical preparations.

REFERENCES

Babich, F. R., Jacobson, A. L., Bubash, S. and Jacobson, A. Transfer of a response to naive rats by injection of ribonucleic acid extracted from trained rats. *Science*, 1965, **149**, 656–657.

Best, J. B. Proopsychology. *Scientific American*, 1963, **208**, 2, 55–62.

Byrne, W. L. Memory Transfer. *Science*, 1966, **153**, 658–659.

Luttges, M., Johnson, T., Buck, C., Holland, V. and McGaugh, V. An examination of "Transfer of Learning" by nucleic acid. *Science*, 1966, **151**, 834–837.

McGuigan, F. J. *Experimental Psychology: A Methodological Approach.* New Jersey: Prentice-Hall, Inc. 1960.

SUMMARY: SYMPOSIUM ON MEMORY TRANSFER AMERICAN ASSOCIATION FOR THE ADVANCEMENT OF SCIENCE
New York, December, 1967

E. Roy John

Brain Research Laboratories
Department of Psychiatry
New York Medical College
New York, New York

This symposium provided an overview of the present status of research on memory consolidation and transfer. The first three papers, those of Jarvik, Barondes, and Agranoff, were primarily concerned with attempts to interfere with consolidation. After summarizing the evidence for the existence of a consolidation phase, Jarvik considered the problem of whether post-ECS amnesia was due to erasure of the memory trace or due to a block of performance. Spontaneous recovery shows that storage sometimes persists, although ECS blocks memory for a one-trial CR for at least six weeks. Geller, collaborating with Jarvik, provided conclusive evidence that ECS does not erase the memory trace, and showed that retention was perfect one hour after ECS, with a gradual decay of performance to complete amnesia by 24 hours. This evidence indicates the existence of a temporary holding phase, capable of mediating retrieval for periods up to one day, but insufficient for permanent storage. Similar conclusions come from the work of McGaugh and his group, who have shown that strychnine can block the usual amnesia if injected during a period of time following an otherwise effective ECS. A number of other reports are in accordance with these findings, and support the stages of memory storage shown in Fig. 1.

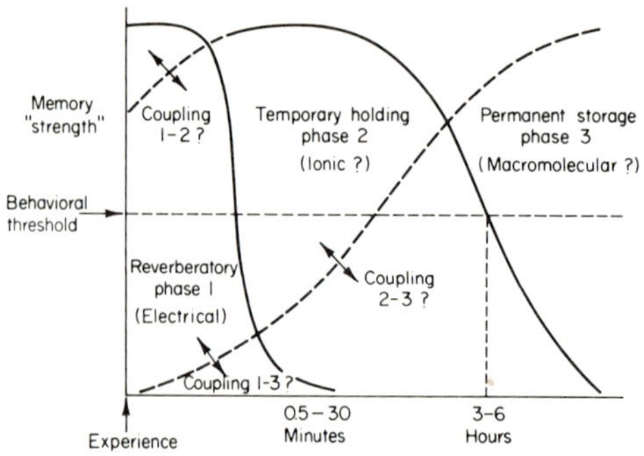

Figure 1

Apparently, ECS interferes with the mechanism which eventually mediates permanent storage, but a different mechanism is responsible for holding the memory during the period before long-term storage is accomplished. This holding mechanism does not seem to involve RNA or protein synthesis. Both Jarvik and Barondes report that conditioned responses continue to occur for more than three hours after one-trial learning, even if RNA or protein synthesis was previously blocked with drugs. The fact that retention then gradually decays, as shown by the reports of Barondes and Agranoff, indicates that protein synthesis is needed for long-term storage to occur.

It is interesting that detention of animals in the region where punishment was delivered can prolong the labile period during which the long-term storage mechanism can be disrupted. This was reported both for the effects of ECS on mice (Rebitelli and Jarvik) and puromycin on goldfish (Agranoff). The mechanism underlying this phenomenon remains obscure.

In a series of careful studies, Barondes has succeeded in clarifying some of the reasons for differences in the effects of various agents which interfere with protein synthesis. The most striking contradiction was that cycloheximide or acetoxycycloheximide caused no deficits in retention, although protein synthesis was impaired as effectively as by a dose of puromycin which blocked retention. Further, cycloheximide plus puromycin substantially improved retention. Also anomalous was the observation that saline injections could reverse the puromycin-induced blockade of behavior weeks later. Further analysis showed that puromycin caused abnormal hippocampal electrical activity, while the other agents caused no obvious electrical abnormality. Parametric studies showed that the initial failure to block retention with cycloheximide or acetoxycycloheximide was probably due to the fact that the animals had been overtrained. Use of a lower criterion for training resulted in more marked interference with retention after blockade of protein synthesis. These studies show that learning is not dependent upon protein synthesis, but that retention involves a protein which probably requires about 45 minutes to complete synthesis. The short-term holding mechanism decays in about four to 24 hours, depending upon experimental variables. The reported effects of overtraining on susceptibility to inhibition of protein synthesis suggest that some of the variability in attempts to obtain behavioral transfer by injection of brain extracts may reflect failure to control this aspect of the experiment.

The work reported by Agranoff is substantially in agreement with the above results, although there is some discrepancy in the duration of susceptibility and decay. An interesting feature of this paper was the observation that KCl injections interfere with consolidation and also block protein synthesis. The observation that repeated injections of acetoxycycloheximide slowed the decay of the short-term memory was also noteworthy. Agranoff proposes that short-term memory lasts until long-term storage is accomplished and is actually converted to long-term memory. In contrast, other workers believe that the two processes are substantially independent.

In an investigation focused on the possible role of RNA rather than protein, Essman observed that injection of uric acid blocked the amnesic effects of ECS on a one-trial step-through shock avoidance task. He reported further that ECS caused a marked reduction in brain RNA, which was blocked by uric acid pretreatment. The implication that adequate RNA levels are necessary for retention is in accordance with the finding by Agranoff that blockade of RNA synthesis with actinomycin D interferes with retention, but is in conflict with the absence of effects observed by Jarvik and Barondes after actinomycin. Variations in task, dose, and species make these contradictions hard to evaluate.

The possible roles of RNA in memory storage were further discussed by Zemp. Using a yoked-control technique, and differentially labeled uridine, a significant increase in diencephalic RNA was observed with training. The observed differences might have been caused by a variety of differences between the trained animal and the yoked control, including stress, attention, amount of jumping, or instrumental learning. A complex series of control experiments provided by Adair suggested that a large portion of the observed increase in RNA synthesis might be attributed to the escape movements rather than to learning per se, since classically conditioned animals were essentially comparable to yoked controls. Since the yoked control receives systematically associated stimuli, the procedure must effectively establish a classical CR in the yoked animals. Thus, although differences in RNA synthesis were observed between the groups, they seemed to relate to the difference between instrumental and classical conditioning rather than to the difference between learning and nonlearning situations.

The papers thus far discussed represent the portion of the symposium concerned with evidence that macromolecular synthesis plays a role in the mediation of long-term memory. Although most of the current lines of investigation related to the role of protein were represented among the participants, no systematic review of studies related to the possible role of RNA in memory was provided. Such studies have been summarized elsewhere and will not be discussed here (E. R. John, *Mechanisms of Memory*, Academic Press, New York, 1967). Although inconsistencies and contradictory results exist, and although the species and situations studied still lack the desirable breadth, the evidence indicates that the synthesis of RNA or of protein probably plays an essential role in the establishment of permanent memory. The evidence from experiments of this sort is insufficient for any decision as to whether macromolecules play an unspecific role (such as the facilitation of transmission, synaptic changes, and the establishment of selectively responsive circuits) or whether their role is somehow more specific, involving the actual coding of information.

The bulk of the remainder of this symposium was concerned with the attempt to transfer learned behavior by injection of macromolecular extracts from the brains of trained animals. Demonstration of critical macromolecular synthesis in animals during learning is an essential prerequisite for such studies to be undertaken with any possibility of success. The first part of this symposium showed that this prerequisite has been fairly well satisfied. Successful transfer of specific learned responses by injection of brain extracts would establish that the role of macromolecules in permanent memory probably involved the specific coding of information in some fashion. This general topic was introduced by Corning, who provided a review of the planarian conditioning controversy, including evidence from his own and other studies that transfer of training could occur by cannibalism or by injection of RNA extracted from trained worms. Initial studies of this sort provided the impetus for attempts to accomplish transfer by injection of materials extracted from brains of trained higher animals.

A review of the transfer studies to date was provided by Byrne. He discussed replication of reportedly successful transfer experiments agreed upon at a previous work session. Of six procedures submitted, two were tested in a number of laboratories. One of these methods was successfully replicated in two laboratories with significant results, while the second was successfully replicated in four out of the five laboratories which undertook this task. The latter method, initially developed by Gay–Raphelson, utilizes a one-trial avoidance procedure. The brains of trained animals were extracted for RNA, which was then injected I.P. into a naive experimental group. Animals in the experimental group showed significantly shorter response latencies (mean = 32 sec.) than animals in the control group (mean = 107 sec.).

Concerning interesting new work, Byrne reported on studies in which radioactively-labelled extracts from various organs were injected. These extracts were relatively localized in the corresponding organs of the recipient animals, suggesting that cellular components could somehow "recognize" the organ from which they came. This study was carried out on material from kidney or liver, but not from brain. In spite of the impressive specificity thus demonstrated for various organs, the blood-brain barrier provides reason to doubt that this phenomenon also exists with respect to the movement of peripherally-injected substances to the brain. No evidence for such migration has yet been provided.

A variety of positive findings were reported by Fjerdingstad, using an essentially protein-free RNA extract, injected intracisternally or intraperitoneally. In general, the overall effect of "trained" RNA injection was to reduce the total errors to criterion. Under some of the circumstances described, an initially inhibitory effect was observed, which gradually reversed to a positive effect. The procedure of Gay–Raphelson was successfully replicated by this laboratory at the $p = 0.02$ level. Perhaps most interesting among the results reported by Fjerdingstad was successful transfer of a left-right discrimination by injection of RNA extracted from trained animals. This effect was significant at about the $p = 0.002$ level.

Among the positive results described by Ungar were successful transfer of morphine tolerance, habituation, and differential habituation. Attempts to transfer a left-right discrimination produced inconsistent results, including some positive cases. Greater consistency was achieved in a repetitively replicated attempt to transfer choice of the lighted alley of a Y-maze in order to avoid shock. The Gay–Raphelson experiment was replicated, with strong positive effects which lasted 48 hours. Ungar utilizes an extract which seems to contain a peptide not destroyed by RNAase. He believes that RNA may play a transient role in regulating the relevant protein synthesis, but cannot be the molecule responsible for stable storage. Left-right discrimination was also used by Rosenblatt and his coworkers in a carefully parametric series of studies. Baseline preferences were established for experimental subjects before injection, serial testing was carried out, and dose response curves were constructed. The results showed that the observed effects ranged from facilitatory to inhibitory through the set of postinjection intervals studied, and varied with the dose. These findings indicate that the ability to find positive effects will depend upon dose and time after injection. Variations in the time course of effects within a mixed population could easily obscure any consistent trend. Were these considerations generally applicable to transfer experiments, they would highlight the need for parametric study and make obvious the desirability of designs in which each animal serves as his own control.

Positive results in transfer of a variety of discriminative response tasks were reported by Dyal. Some of these experiments produced extremely significant effects, while providing controls which ruled out the possibility that the effects reflected some unspecific sensitization. Some of these procedures were successfully replicated by McConnell. In the initial experiments, the differences might possibly be due to the lack of comparability between experimental and yoked control groups with respect to amount of activation. Attempts to transfer a black-white discrimination showed a positive but not significant trend. This worker reached the conclusion that transfer effects can be reliably obtained, that they are not due to sensitization, and that the effects may involve factors related to the general level of activity, the specific conditioned stimulus, and the particular conditioned response which are selected.

Wolthuis also reported the results of a number of different transfer studies. Attempts to change light-dark preference were repeatedly unsuccessful. Attempts at transfer of a maze task for shock avoidance produced some positive results. A task involving discrimination between cues for water and for shock avoidance yielded extremely significant transfer effects, and an alternation task yielded somewhat less significant results. Wolthuis concluded that transfer effects can be demonstrated with some but not all tasks, and that the transfer may be relatively specific.

Similar conclusions can be drawn from the set of results reported by Halstead and his collaborators. In a series of experiments involving different tasks and extraction procedures, positive transfer was obtained in some experiments. The most striking of these was a change in light preference as a function of donor training, significant at the $p = 0.005$ level. These workers identified antagonistic effects and suggested them as a possible explanation for their failures to obtain transfer in some experiments.

A series of six replications of a memory reinstatement procedure was described by Krech and his collaborators. Although positive results were sometimes obtained, these were highly unspecific. In spite of their extremely variable results, these workers concluded that homogenate of whole trained brain may contain a compound which somehow acts as a memory booster.

Using a light-dark preference procedure, positive results were reported by Røigaard–Petersen and his colleagues. These results were characterized by inhibition of the preference established in the donor. Using a light-dark discrimination involving drinking and shock punishment for drinking, Weiss reported positive transfer of dark preference, significant at the $p = 0.01$ level.

In addition to the experiments thus far discussed, which involved direct interference with macromolecular synthesis or attempts at transfer of training by injection of extracts from trained animals, a number of other papers were presented. Bennett described a series of changes observed to occur in the brains of animals reared in an "enriched" environment. Appel discussed ways in which protein synthesis might contribute to neuronal structure or function, and described elegant methods for the study of protein synthesis in a synaptosome preparation. These methods may provide a way to identify the cellular level at which significant alterations occur in behavioral experiments. Cohen analyzed the factors which might determine connections between two neural units in a developing system. He then described quantitative methods for the study of connectivity in the insect thoracic ganglion. Applications of these methods in the conditioning paradigms currently used with isolated thoracic ganglion should provide valuable insights into the anatomical and chemical features of the mediating mechanisms.

The symposium was of value largely because it brought together in one extended session the majority of those investigators working in the field of macromolecular involvement in memory storage, and those workers active in attempts to establish whether learned responses can be transferred by injection of brain extracts from trained animals. Most of the results reported had previously been presented elsewhere. Few attempts were made to construct theoretical models which might provide a framework for further analytical efforts. A profusion of phenomenological detail, contradictions, and agreements was generated.

Yet, in spite of these shortcomings, some general conclusions seem warranted. Certainly, negative findings in transfer experiments exist in abundance, most of them unreported at this symposium. However, many of the careful studies reported here provide evidence that inhibitory as well as excitatory effects can be obtained, that variable dose-response effects take place, and that results vary with time after injection when testing occurs, as well as with the task. Certain agreements begin to emerge between results reported by several groups of researchers. Perhaps we should be more impressed by the fact that so many laboratories report positive results in this so implausible and improbable domain of endeavor, rather than by the failures. It is true that success in a transfer experiment seems intuitively improbable to most neural scientists, but it is also true that we have no significant level of comprehension of how learning occurs in the brain.

On balance, the work thus far completed supports the idea that macromolecular synthesis plays an important role in long-term memory. It seems probable that some substance extractable from brain is capable of facilitating learning or performance in a naive recipient, in an unspecific fashion. It is possible that the action of such extracted substances might even be specific to a particular learned response.

Instead of the disreputability all too frequently assigned to studies in these areas by workers in more established and better understood fields, it might be more appropriate to acknowledge that there are extremely difficult problems involved in an adequate assessment of these phenomena. Knowledge would be better served by the attempt to develop reliable and replicable techniques to analyze both the behavioral and biochemical aspects of these problems, and to criticize inadequate studies so that they will be adequately done, rather than by continuing the somewhat widespread tendency to ridicule and disparage those who are interested in investigating the phenomenon further. We need not appeal to orthodoxy to accomplish progress in this field. The facts will do it for us.

AUTHOR INDEX

Numbers in italics refer to the pages on which the complete references are listed.

SUBJECT INDEX

K

L

M

N

O

P

R

U

V

W

X

Y